MOLECULAR MEDICINE

For Churchill Livingstone

Publisher: Timothy Horne
Project Editor: Dilys Jones
Copy Editor: Jane Ward
Production Controller: Nancy Arnott
Sales Promotion Executive: Marion Pollock

MOLECULAR MEDICINE

An Introductory Text for Students

R. J. Trent

PhD BSc (Med) MB BS (Syd) DPhil (Oxon) FRACP FRCPA

Professor of Molecular Genetics, Department of Medicine, University of Sydney;
Head, Department of Molecular Genetics, Royal Prince Alfred Hospital, Camperdown,
New South Wales, Australia

Churchill Livingstone

EDINBURGH LONDON MADRID MELBOURNE NEW YORK AND TOKYO 1993

CHURCHILL LIVINGSTONE
Medical Division of Longman Group UK Limited

Distributed in the United States of America by Churchill Livingstone Inc., 650 Avenue of the Americas, New York, N.Y. 10011, and by associated companies, branches and representatives throughout the world.

First published 1993

ISBN 0-443-04635-2

British Library Cataloguing in Publication Data
A catalogue record for this book is available from the British Library

Library of Congress Cataloging in Publication Data
Trent, R. J.
Molecular medicine : an introductory text for students / R.J. Trent.
p. cm.
Includes index.
ISBN 0-443-04635-2
1. Medical genetics. 2. Molecular biology. 3. Pathology, Molecular. I. Title.
[DNLM: 1. Genetics, Medical. 2. Molecular Biology. QZ 50 T795m 1993]
RB155.T73 1993
616.042--dc20
DNLM/DLC
for Library of Congress 93-9992

The publisher's policy is to use **paper manufactured from sustainable forests**

Printed in Great Britain by The Bath Press, Avon

CONTENTS

PREFACE

Those who work in the field of medicine will need increasingly to communicate in terms of DNA. To do so effectively will require both core knowledge and an understanding of the potential applications of DNA technology. The aim of *Molecular Medicine* is to provide an overview of the impact that recombinant DNA technology is having and will continue to have on the practice of medicine. Speculative statements have been made in respect of future developments. Some of these may prove to be incorrect in detail but hopefully will be appropriate in substance. A major difficulty with a rapidly advancing field such as molecular medicine is that a book becomes outdated even before it is published, e.g. the section on molecular genetics of Huntington disease fell into this category with the announcement that the 10-year saga to clone the gene had finally met with success. Changes in the text were made at the proof stage to include this vital news. Nevertheless, whether genes are cloned or not becomes irrelevant in the context that diseases have been used as models to illustrate the applications of recombinant DNA technology in medicine.

The different chapters attempt to describe molecular medicine within disciplines. This is artificial as the topic (which I have taken to mean DNA as it applies to medicine) is a basic component of clinical and laboratory medicine as well as medical research. Chapter 2 deals with technical details since these are frequently difficult to find in journal articles. The broad nature of *Molecular Medicine* has meant that many important subjects, e.g. HLA, have received brief mention.

I would like to thank a number of my colleagues who took time to read various chapters and make helpful suggestions. They include: Drs J. Lanser, A. Smith, C. MacLeod, T. Woodage, Professors R. Kefford, D. Sillence, Y. Cossart and Sr R. Dunne. Professor J. Buchanan from Auckland initiated my interest in the educational aspects of molecular medicine. My fellow workers at the MRC Molecular Haematology Unit in Oxford first exposed me to this exciting aspect of medicine.

My sister, Dr Lynette Trent, tried to keep me to the basics. Miss Carol Yeung, my secretary, always maintained her enthusiasm and provided untiring help. Finally, I should acknowledge Ms Dilys Jones, the senior project editor for Churchill Livingstone, for her helpful advice which was much appreciated.

Camperdown, 1993 R. J. T.

1

HISTORICAL DEVELOPMENTS

A HISTORY OF MOLECULAR MEDICINE

THE EARLY DAYS TO THE MID-1980s

DNA is characterised

In the late 1860s, Miescher isolated nucleoprotein. It was established to be an acidic material and was called 'nuclein'. From this came *nucleic acid*. In the 1940s, Avery and colleagues were the first to show that the genetic information in the bacterium *Pneumococcus* was found within its DNA. X-ray crystallographic studies by Franklin, Wilkins and work by Watson and Crick in the 1950s, led to the identification of the *double-stranded* structure of DNA. Subsequently, it was shown that the complementary strands which made up the DNA helix separated during replication. In the mid 1950s, a new enzyme was discovered by Kornberg. It was called *DNA polymerase* and enabled small segments of double-stranded DNA to be synthesised. Discoveries during the 1960s included: the finding of *mRNA* (messenger RNA) which provided the link between the nucleus and the site of protein synthesis in the cytoplasm and identification of autonomously replicating, extrachromosomal DNA elements called *plasmids*. These were shown to carry genes such as those coding for antibiotic resistance in bacteria. Plasmids would later be used extensively by the genetic engineers. A landmark in this decade was the definition of the full *genetic code* which showed that individual amino acids were encoded in DNA as nucleotide triplets (see Ch. 2 for details). In 1961, Lyon proposed that one of the two X chromosomes in female mammals was normally inactivated. The process of *X-inactivation* enabled males and females to have equivalent DNA content despite differing numbers of X chromosomes.

Technological developments

The dogma that DNA → RNA → protein could be held no longer when Temin and Baltimore discovered in 1970 that *reverse transcriptase*, an enzyme found in some RNA viruses (called retroviruses), allowed RNA to be copied into DNA. This enzyme would later allow the genetic engineer to produce DNA copies (known as complementary DNA or cDNA) from RNA templates. Reverse transcriptase also explained how some viruses could integrate their own genetic information into the host's genome.

Enzymes called *restriction endonucleases* were isolated from bacteria by Smith and colleagues during the late 1960s and early 1970s. Restriction endonucleases were shown to digest DNA at specific sites determined by the underlying nucleotide base sequences. A method now existed to produce DNA fragments of known sizes. At about this time an enzyme called *DNA ligase* was described. It allowed DNA fragments to be joined together. The first *recombinant DNA* molecules comprising segments that had been 'stitched together' were now produced. Berg was later awarded the Nobel Prize for his contribution to the construction of recombinant DNA molecules. This Nobel Prize was one of many that were directly relevant to molecular medicine (Table 1.1). It was subsequently shown by Cohen and Boyer that DNA could be inserted into plasmids which could then be reintroduced into bacteria. Replication of the bacteria containing the foreign DNA enabled relatively large amounts of a single fragment to be produced, i.e. DNA could be

Table 1.1 Molecular medicine and some Nobel Prize winners

Year	Recipients	Subject
1959	S Ochoa, A Kornberg	In vitro synthesis of nucleic acids
1962	J D Watson, F Crick, M H F Wilkins	Molecular structure of nucleic acids
1975	D Baltimore, R Dulbecco, H M Temin	Reverse transcriptase
1978	W Arber, D Nathans, H D Smith	Restriction endonucleases
1980	P Berg, W Gilbert, F Sanger	Creation of first recombinant DNA molecule and DNA sequencing
1989	J M Bishop, H E Varmus	Oncogenes
1989	S Altman, T R Cech	RNA and 'ribozymes'

cloned. The first eukaryotic gene to be cloned was the rabbit β globin gene in 1976.

The development of DNA *probes* followed from a 1960 observation that the two strands of the DNA double helix were able to be separated and could then be made to reanneal to each other. Probes comprised small segments of DNA which were labelled with a radioactive marker such as ^{32}P. DNA probes were able to identify specific regions in DNA through their annealing (or *hybridisation*) to complementary nucleotide sequences. The specificity of the hybridisation reaction relied on the fact that base pairing was predictable, i.e. the nucleotide base adenine (usually abbreviated to A) would always anneal to the base thymine (T) and guanine (G) would anneal to cytosine (C). Thus, because of nucleotide base pairing, a single-stranded DNA probe would hybridise in solution to a predetermined segment of single-stranded DNA. Solution hybridisation gave way in 1975 to hybridisation on solid support membranes when DNA digested with restriction endonucleases could be transferred to such membranes by a process known as *Southern blotting*. The ability of radiolabelled DNA probes to identify specific restriction endonuclease fragments enabled DNA maps to be constructed (see Ch. 2 for details).

During the mid 1970s, public and scientific communities expressed concern about the potential dangers of genetic engineering. In 1975, a conference was convened at Asilomar in California to discuss these issues. Subsequently, regulatory and funding bodies issued *guidelines* for the conduct of recombinant DNA work. These guidelines dealt with the type of experiments allowable and the necessity to use both vectors (e.g. plasmids) and hosts (e.g. bacteria such as *Escherichia coli* which carried the vectors) that were safe and could be contained within laboratories certified to undertake recombinant DNA work. Guidelines began to be relaxed during the late 1970s and early 1980s when it became apparent that recombinant DNA technology was safe if carried out responsibly. However, State and private funding bodies insisted that a form of monitoring be maintained which has continued to this day.

The latter half of the 1970s saw the development of the biotechnology industry based on the ability to manipulate DNA. The first genetic engineering company called Genentech was formed in 1976. The structure of the gene became better defined with the availability in 1975 of methods used to sequence individual nucleotide bases in DNA. The significance of this achievement led to the award of the Nobel Prize to Sanger and Gilbert. In 1977, an unexpected observation revealed that eukaryotic genes were discontinuous, i.e. coding regions called *exons* were split by intervening segments of DNA called *introns*.

Medical applications

Variations in DNA sequence between normal individuals were described in the mid-1970s although their full potentials were not realised until the early 1980s. These variations were called DNA *polymorphisms*. The first human genetic disorders to be diagnosed by identifying the mutant gene in the fetus whilst still in utero were α thalassaemia (1976) and sickle cell anaemia (1978) by Kan and colleagues.

In the 1980s, many different *transgenic* mice were produced by microinjecting foreign DNA into the pronucleus of fertilised oocytes. Injected DNA became integrated into the mouse's own genome and the transgenic animal formed in this way expressed both the endogenous mouse genes and the foreign gene. A 'supermouse' was made when a rat growth hormone gene was microinjected into a mouse pronucleus. Human insulin became the second genetically engineered drug to be produced following the success of recombinant human growth hormone (see Ch. 7).

RECENT ADVANCES

Functional and positional cloning

Until the mid-1980s, conventional approaches to understanding genetic disease relied on the identification of an abnormal product such as a protein. An additional step became available when recombinant DNA techniques made it possible to utilise information from the characterised protein to clone the relevant gene. From the cloned gene more information could then be obtained about the underlying genetic disorder. This DNA strategy was called *functional cloning*.

In the late 1980s, an alternative approach to study genetic disease became possible through *positional cloning*. This method enabled direct isolation of genes on the basis of their chromosomal location and certain characteristics which identified a segment of DNA as 'gene-like'. Identification of the mutant gene as well as knowledge of the genetic disorder could then be inferred from the DNA sequence. Hence, this strategy was initially called 'reverse genetics'. Subsequently, the name was changed to the more appropriate 'positional cloning'. The first successful application of positional cloning occurred in 1987 with the isolation of the Duchenne muscular dystrophy gene on the X chromosome. Further successes soon followed:

- Duchenne muscular dystrophy
- Chronic granulomatous disease
- Retinoblastoma
- Wilms' tumour
- Cystic fibrosis
- Neurofibromatosis type 1
- Testis determining factor
- Fragile X mental retardation
- Familial adenomatous polyposis
- Myotonic dystrophy
- Aniridia

A variation of positional cloning enabled genes causing genetic disorders to be identified on the basis that they were potential 'candidates' for the underlying disorders (Table 1.2) (see Chs 2, 3 for further discussion).

Table 1.2 Genes which have been cloned and characterised by the candidate gene approach.
A candidate or 'likely' gene is used to narrow the field when searching for DNA markers to be isolated on the basis of their chromosomal location by positional cloning

Genetic disorder	Candidate gene
Retinitis pigmentosa	Rhodopsin
Familial hypertrophic cardiomyopathy	β-myosin heavy chain
Malignant hyperthermia	Ryanodine receptor
Li–Fraumeni syndrome	p53
Marfan syndrome	Fibrillin
Alzheimer disease (early onset)	β Amyloid precursor protein
X-linked spinal and bulbar muscular atrophy (Kennedy syndrome)	Androgen receptor

Physical mapping

The exciting potential of the positional cloning strategy increased the impetus to find better methods for measuring *actual distances* along the genome. This is called physical mapping. Three techniques were developed during the mid to late 1980s which enabled longer segments of DNA to be identified and mapped in relation to each other. The three were: **p**ulsed **f**ield **g**el **e**lectrophoresis (PFGE), **y**east **a**rtificial **c**hromosomes (YACs) and **f**luorescent **i**n **s**itu **h**ybridisation (FISH).

Pulsed field gel electrophoresis is a variation of DNA mapping. It enables megabases rather than kilobases of DNA (1 megabase or 1 Mb is equivalent to $1x10^6$ base pairs; 1 kilobase or 1 kb is equivalent to $1x10^3$ base pairs) to be measured. Similarly, DNA fragments cloned with the yeast artificial chromosome vectors were ten or more times larger than was possible with conventional cloning vectors. Fluorescent in situ hybridisation began to provide a bridge between the cytogenetic analysis for gross chromosomal changes and the discrete defects detectable by DNA mapping. These techniques quickly became established as important components in positional cloning strategies and will be described in greater detail in Chapters 2 and 3.

Polymerase chain reaction

Work by Mullis and colleagues in the US-based Cetus Corporation made it possible in 1985 to target segments of DNA with oligonucleotide primers and then amplify these segments with the polymerase chain reaction. *PCR*, an abbreviation for the polymerase chain reaction, soon became a 'routine' procedure in the molecular laboratory. In a short period of time, this technology has had a profound and immediate effect in both diagnostic and research laboratories. Reports of DNA patterns which were obtained from single cells by 'PCR' started to appear. Even the dead were not allowed to rest as it soon became possible to identify by 'PCR', DNA patterns from ancient Egyptian mummies, old bones and preserved material of human origin. The availability of automation meant that DNA amplification had unlimited potential for the detection of genetic disorders as well as DNA from infectious agents. A patent was obtained to cover the use of this technology and illustrated the growing importance of commercialisation in recombinant DNA technology (see Ch. 2 for a more extensive technical description of the polymerase chain reaction). Applications of the polymerase chain reaction are now to be found in most areas of molecular medicine.

RNA

The discovery that RNA could function as an enzyme and thus as a regulator of gene expression diverted some of the attention from DNA. The division within the cell between the informational components (DNA, RNA) and the catalytic components (proteins acting as enzymes) was broken when it was shown in 1981 that RNA had enzyme-like activity (called *ribozyme*) in the protozoan *Tetrahymena thermophila*. It was demonstrated by Altman and Cech that some naturally occurring RNA sequences were able to cleave RNA targets at a specific triplet comprising guanine–uracil–cytosine (G–U–C where U or uracil is the RNA's equivalent of thymine in DNA) following hybridisation between target RNA and a ribozyme. This opened up the possibility that not all catalytic steps in cellular activity were protein-dependent (see Ch. 7).

Additional manipulation of cell function became possible with the finding of *antisense RNA*. Only one of the two DNA strands is normally the template for transcription to produce mRNA. This strand is known as the antisense strand and from this sense mRNA is produced. However, the sense strand of DNA in viruses and bacteria was also shown to have the potential to be transcribed. This produced antisense RNA which had the capacity to regulate or inhibit gene activity. Antisense RNA molecules were shown to bind specifically with the sense mRNA. The net effect was inhibition of protein production at the ribosomal level. The part, if any, played by antisense RNA in controlling gene expression of more complex cells still remains to be defined. However, the possible effects of naturally produced antisense RNA or synthetically derived antisense RNA segments were soon being tested to attempt control of cancer growth or the inhibition of viral replication (see also Chs 2, 7).

Oncogenes/tumour suppressor genes

In 1910, Roux was the first to implicate viruses in the aetiology of cancer when he showed that a filterable agent (virus) was capable of inducing cancers in chickens. In the early 1980s, a DNA sequence from a bladder cancer cell line was cloned and shown to have the capacity to induce cancerous transformation in other cells. The cause of the neoplastic change in both the above cases was subsequently demonstrated to be one of many dominantly acting cancer genes which were called *oncogenes*. These have assumed increasing importance in our understanding of how cancers arise and progress. For their work on oncogenes during the mid 1970s, Bishop and Varmus were awarded the Nobel Prize in 1989. More recently, the identification of cellular sequences which normally repress cellular growth led to the discovery of *tumour suppressor genes*. Loss or mutation of DNA belonging to the latter through genetic and/or acquired events could be associated with unregulated cellular proliferation.

During the 1980s, molecular changes detectable in various cancers provided evidence consistent with earlier epidemiological data from which Knudson proposed a *two hit model* to explain the evolution of a number of cancers. Here an initial predisposing DNA change can be inherited through the germline and will only result in tumour formation when a second hit or somatic mutation occurs to inactivate the normal allele inherited from the other parent. Research programmes no longer had to concentrate solely on the small animal or tissue culture model of tumourigenesis but could investigate individual DNA sequences which were likely to be playing a role in the pathogenesis of cancer (see also Ch. 6).

Therapeutic implications

In 1987, the first *recombinant DNA vaccine* against the hepatitis B virus was produced in a yeast host. Concern about manipulation of the human genome through a *gene therapy* approach was followed by a moratorium and extensive public debate until guidelines for the conduct of this type of recombinant DNA work became firmly established. Permission for the first human gene therapy trial to proceed was obtained in 1990. A child with the fatal adenosine deaminase (ADA) deficiency disorder was given somatic gene therapy using her own lymphocytes which had been genetically engineered by retroviral insertion of a normal adenosine deaminase gene. Lymphocytes with the genetically engineered gene were then reinfused into the individual. Initial response was gratifying and long-term follow-up will determine the effectiveness of this form of treatment (discussed further in Ch. 7).

During the late 1980s and early 1990s, novel approaches to gene therapy which avoided the use of retroviruses began to be discussed as potential therapeutic agents. Methodologies which enabled genes to be targeted to their normal genetic locus through a process called *homologous recombination* were also described. This was attractive since once developed it would enable therapeutic manipulations to be conducted with greater accuracy (see Ch. 7).

The cloning of cattle by nuclear transplantation in 1987 highlighted ethical and social issues which could arise from irresponsible use of recombinant DNA technology in the human. Legislative prohibitions relating to certain types of embryo experimentation and forms of human gene therapy, which would involve the germline, were enacted in many countries (see Ch. 9).

DNA polymorphisms

The first example of a human *DNA polymorphism* (a neutral variation in DNA sequence) was called a **r**estriction **f**ragment **l**ength **p**olymorphism (RFLP). Subsequently, it was shown that there were other DNA polymorphisms which had greater variability. One of these was known as a VNTR (**v**ariable **n**umber of **t**andem **r**epeats). DNA polymorphisms soon became essential factors in most studies of genetic disorders (see Chs 2, 3).

The inherent variability in DNA polymorphisms led to the concept of DNA *fingerprinting* in 1985 when Jeffreys described how more complex DNA polymorphisms (*minisatellites*) were able to produce unique DNA profiles of individuals. DNA testing for minisatellites has subsequently had an increasingly greater role to play in forensic practice. The courts of law became involved in DNA technology when the potential for identification of individuals on the basis of the DNA patterns produced by minisatellites was realised. In 1987, DNA 'fingerprints' were allowed as evidence in the first court case (see Chs 2, 8). *Microsatellites*, another type of DNA sequence variation, were shown to be dispersed throughout the human genome. These DNA markers have now assumed considerable significance when it comes to mapping the human genome (see Chs 8, 10).

Technological developments

The development and then automation of DNA amplification has already been described. *Oligonucleotides*, small single-stranded segments of DNA, required for DNA amplification and DNA sequencing became increasingly easier to synthesise from the mid 1980s. Another important technological milestone occurred in 1986 when Hood and colleagues described a method to label DNA primers with fluorescent dyes and then sequence DNA by automated means. In a short period of time *automated DNA sequencing* has become a rapid and accurate technique. Costs are presently the major drawback to its universal utilisation.

The Human Genome Project, a multidisciplinary, multinational project coordinated by the US Department of Energy and National Institutes of Health was given the go-ahead in 1988. The ultimate aim of the Project is to clone and sequence the entire human genome and some model organisms by the year 2005. Information resulting from this project will have scientific and social consequences as yet unrealised (see Chs 9, 10 for further discussion of the Human Genome Project).

THALASSAEMIA–A MODEL FOR MOLECULAR MEDICINE

INTRODUCTION

Clinical features

Many of the developments which have occurred in molecular medicine can be illustrated by reference to a clinical disorder known as thalassaemia. Haemoglobin, the pigment in red blood cells, is made up of iron and a protein called globin. Globin comprises four polypeptide chains: two α chains and two β chains. A genetic defect which impairs globin synthesis produces thalassaemia. The clinical picture in the thalassaemia syndromes is very diverse and ranges from an asymptomatic disorder which is detected fortuitously, to a life-long blood transfusion-dependent anaemia or an intractable anaemia which is fatal in utero or soon after birth.

The word thalassaemia comes from the Greek Θαλασσα which means 'the sea'. The name arose since it was initially considered that thalassaemia was a disease which affected those who lived near the Mediterranean sea. The first accurate clinical description of thalassaemia was given by Cooley in 1925. Cooley's anaemia, as it was then called, was shown to be genetic in origin during the late 1930s and early 1940s when relatives of severely affected individuals were observed to have similar changes in their red blood cells.

Protein studies

Extensive protein analyses were undertaken to characterise the individual globins in the 1950s and 1960s. These studies showed that there were a number of different globins and so multiple genetic loci were likely. During the 1960s, the biochemical defect in the thalassaemias was identified to be an imbalance in the number of α and β globin chains. Failure to produce α globin gave α thalassaemia, which is fatal in its severe form. Failure to produce β globin (β thalassaemia) was associated with a life-long blood transfusion-dependent anaemia. Carriers of the thalassaemia defect were usually asymptomatic. However, a notable feature of the thalassaemias was the considerable variation in their phenotypes (i.e. very varied clinical and laboratory pictures) (see also Ch. 3 for further discussion of the thalassaemias).

DNA STUDIES

Gene characterisation

With the discovery of reverse transcriptase, it was possible to take immature red blood cells from patients with homozygous β thalassaemia (i.e. two identical forms of the gene) and from these nucleated cells identify α globin gene specific mRNA which could then be converted to complementary DNA (cDNA) (Fig. 1.1). In this way DNA probes specific for the α globin genes became available. Using these probes and solution hybridisation techniques it was possible to show that there were distinct abnormalities involving mRNA in the α thalassaemias. DNA probes specific for the β globin genes were isolated next. This was assisted by the cloning of the rabbit β globin gene since this had considerable homology (similarity) to its human equivalent. A number of different abnormalities in mRNA were also found in the β thalassaemias. Differences or heterogeneities observed clinically were now beginning to be seen at the molecular level (Box 1.1).

With the availability of restriction endo-

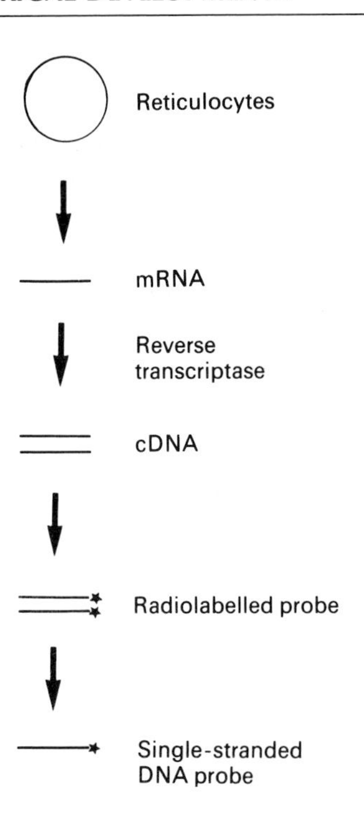

Fig. 1.1 Isolation of an α globin gene-specific DNA probe from mRNA.
Reticulocytes are immature red blood cells that are transcriptionally very active. Therefore, there will be a lot of mRNA present. If the reticulocytes are derived from an individual with homozygous β thalassaemia there will be no β globin gene-specific mRNA. Thus, all globin gene-specific mRNA will be from the α genes. cDNA = copy or complementary DNA; * = radiolabelled DNA.

Box 1.1 Genetic abnormalities in thalassaemia

In the α thalassaemias, it was possible to utilise the α globin gene probes to quantitate gene dosage. For example, a form of α thalassaemia called haemoglobin H (HbH) disease was shown to have 25% of the normal gene-specific activity. This was subsequently confirmed when gene mapping demonstrated that in HbH disease there is only 1 of the 4 normal α globin genes present. In the β thalassaemias, a number of different defects at the mRNA level was found. DNA mapping later showed that the β globin genes were intact and so the mRNA defects had resulted from point mutations in the gene rather than deletions. Thus, the thalassaemias showed that a molecular classification of genetic disorders was possible on the basis of (1) deletional or (2) non-deletional mutations in DNA.

Box 1.2 DNA mapping with restriction endonucleases

Restriction endonucleases are enzymes isolated from bacteria. The unique features of the restriction enzymes are that they recognise specific nucleotide sequences in double-stranded DNA and cleave both strands of the duplex. For example, the restriction endonuclease *Eco*RI (derived from the bacterium *Esch. coli* strain *R*YI) will recognise the palindrome G▾AATTC and digest at the site marked ▾ (a palindrome describes a sequence of DNA that is the same when one strand is read from left to right or the second strand is read right to left – Table 2.2 illustrates the double-stranded recognition site for *Eco*RI and the palindrome). In the cell of origin each restriction endonuclease is part of a restriction-modification system consisting of the restriction endonuclease and a matched modification enzyme which recognises and modifies (usually by methylation) the same nucleotide sequence in DNA recognised by the restriction endonuclease. Modification protects cellular DNA from cleavage whilst foreign (unmodified) DNA is cleaved. The restriction endonuclease system is widespread in bacteria and the names of the enzymes are derived from the source bacteria, e.g. *Hpa*II, *Haemophilus parainfluenzae*; *Bam*HI, *Bacillus amyloliquifaciens H*; *Not*I – *Nocardia ottidis* etc. The usefulness of restriction endonucleases in the analyses of DNA lies in their specificity for nucleotide sequences which are usually four to six base pairs in length. Restriction endonucleases are analogous to specific proteolytic enzymes and are extensively used to digest DNA into known fragments (a further description of these enzmes is given in Ch. 2).

nuclease enzymes (Box 1.2) it became possible to construct DNA maps for the globin genes. DNA maps for the α thalassaemias were quite abnormal, indicating that the underlying gene defects involved loss, i.e. *deletions*, of DNA. On the other hand, the maps in the β thalassaemias were normal. Therefore, the molecular (DNA) abnormalities in the β thalassaemias were either *point mutations* (changes involving a single nucleotide base) or *very small deletions*. Once DNA probes became available the next step was to

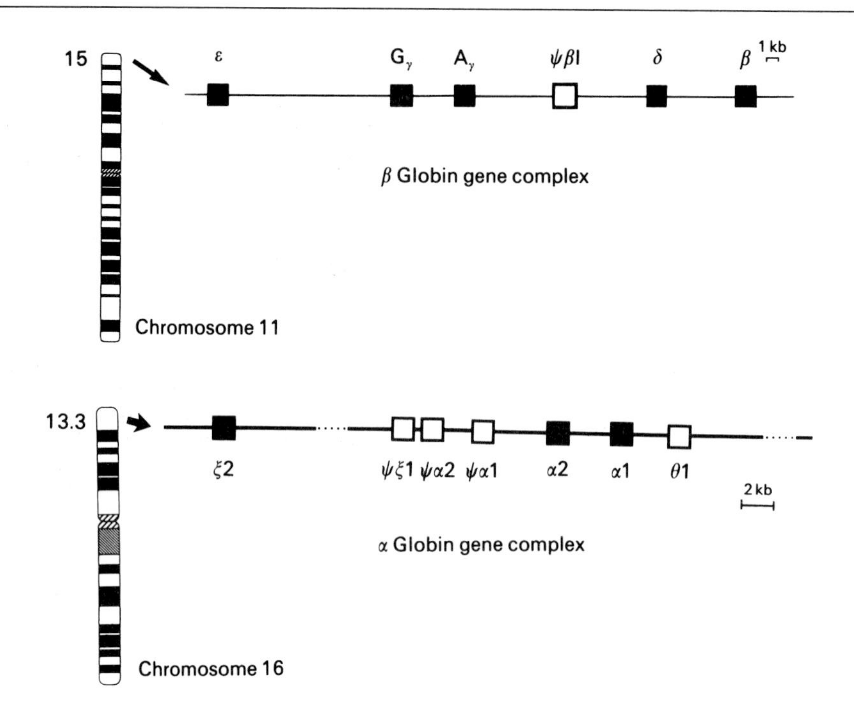

Fig. 1.2 α and β globin gene clusters.
Functional genes are indicated as ■ and non-functioning genes (called pseudogenes) as □. On the short arm of chromosome 11 at band position 15 is found the β globin gene complex. There is one gene which is active during embryonic life (ε); two which are fetal-life specific (Gγ, Aγ) and two are expressed in adult life (δ, β). The α globin complex is on the short arm of chromosome 16 at band 13.3. There are a lot more genes situated in this complex but many are non-functional. The embryonic/fetal gene is ξ and the two adult genes are α2 and α1. .... marks the position of a DNA polymorphism (see Chs 2, 3, 8 for further discussion of polymorphisms).

clone the human α and β globin genes. This was achieved in the late 1970s.

Following cloning, the α and β globin genes were able to be characterised to a much greater depth than would have been possible by protein analysis. From the DNA studies it became evident that the globins belonged to a family of closely related genes that had *evolved* by duplication. Cell fusion studies localised a cluster of such genes on chromosome 16 (α globin cluster) and a second cluster on chromosome 11 (β globin cluster) (Fig. 1.2).

Phenotypic versus genotypic comparisons

DNA mapping, cloning and sequencing of natural mutants for the various thalassaemias were extensively undertaken during the 1980s and 1990s. Correlating phenotypes (clinical and laboratory features) with genotypic (DNA) changes enabled the functional *significance of mutations* to be determined. Thus, the globin genes became useful models for understanding genetic disorders which had arisen from mutations in single genes.

The thalassaemias were also shown to be valuable models to study how the activity of genes was *regulated*. This knowledge would have relevance to all eukaryotic genes. The globin genes were also interesting because they displayed a developmental switch. This switch was present in the form of separate embryonic, fetal and adult-specific genes which were operational during specific periods of development (Table 1.3). As will

Table 1.3 Globin gene switching in development.
Different genes are operational during specific periods of development resulting in sequential changes in the globin subunits of haemoglobin and hence

	Organ		
	Yolk sac	Fetal liver	Adult marrow
α Genes	α,ξ	α	α
β Genes	ε	γ	β, δ
Haemoglobins	$\xi_2\varepsilon_2$ $\alpha_2\varepsilon_2$	$\alpha_2\gamma_2$	$\alpha_2\beta_2$ $\alpha_2\delta_2$

Box 1.3 Identification of regulatory genes by mutation characterisation

The thalassaemia syndromes illustrate how knowledge of spontaneously occurring genetic mutants can lead to a greater understanding of more fundamental biological issues. For example, what are important regulatory sequences found in association with an eukaryotic gene? Information about this has come from a rare thalassaemia which was characterised at the DNA level and shown to have an extensive deletion which started 5′ (upstream) of the β globin gene and extended 100 kb beyond the ε, γ and δ genes.

(a) The intact β gene was non-functional which was puzzling since sequence analysis failed to disclose any mutation in this gene. Furthermore, the β gene was transcriptionally active using in vitro assays.

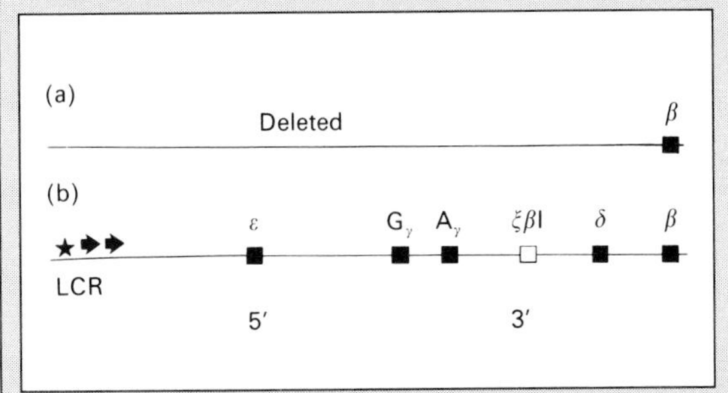

(b) Molecular characterisation identified a region 10.5 kb upstream of the ε globin gene called the locus control region (LCR). This has now been shown to be a potent regulatory region known as an *enhancer*. The LCR exerts a powerful effect on the globin genes enabling high levels of gene expression which is specific to red blood cells alone. Deletion of the LCR in the above thalassaemia prevented the remaining (normal) β globin gene from functioning. Subsequently, a second LCR was located 3′ to the β globin gene. When human globin genes are introduced into transgenic mice, the LCR permits their expression to be position independent. The finding of the LCR has identified a key regulatory element in the eukaryote genome. This enhancer is likely to play an important role if a gene therapy approach is to be taken in the thalassaemias.

be shown in Chapter 3, information gained from characterising mutations in the globin genes has now provided insight into the types of mechanisms which control gene expression (Box 1.3).

FURTHER READING

Collins F S 1992 Positional cloning: let's not call it reverse anymore. Nature Genetics 1: 3–6

King R C, Stansfield W D 1985 A dictionary of genetics. Oxford University Press, Oxford (for historical information)

Watson J D, Tooze J, Kurtz D T 1983 Recombinant DNA. A short course. W H Freeman, New York, p242–249

Weatherall D J 1980 Of some common inherited anemias: the story of thalassemia. In: Wintrobe M M (ed) Blood, pure and eloquent. McGraw-Hill, New York, p373–414

2

MOLECULAR TECHNOLOGY

DNA

Ingredients

Production of DNA requires the provision of: (1) nucleated cells (2) enzymes to break up cell membranes and proteins (3) chemicals to separate proteins from nucleic acids.

Sources

DNA has a number of properties which are exploited in the laboratory. DNA in all cells of an organism is identical in its sequence. Therefore, obtaining a specimen of tissue for DNA studies is relatively simple since 10 ml of blood usually suffice. This yields approximately 250 μg of DNA. Isolation of DNA is straightforward. Nuclei are first separated from cellular debris by enzymatic means. DNA is then separated from proteins by using chemicals such as phenol and chloroform.

Structure/function

Two properties of DNA are of particular relevance in molecular technology. The first is the genetic code which is present in the form of nucleotide triplets called *codons* (Table 2.1). This means that the signal for individual polypeptides is coded by different triplet combinations. For example, the codons for a polypeptide such as glycine-serine-valine-alanine-alanine-tryptophan will read: –GGT–TCT–GTT–GCT–GCT–TGG–. Similarly, the positions where to start and where to end a polypeptide are clearly defined by the triplets ATG (start) and TAA or TAG or TGA (stop). Point mutations (single changes in the nucleotide bases) in any of the above codons can lead to genetic disease (see Detection of point mutations p. 29).

The second significant property of DNA is that it is made up of two strands (Fig. 2.1). Each strand has a sugar phosphate backbone linked from the 5′ and 3′ carbon atoms of deoxyribose. At the end of a strand is either a 5′ phosphate group, the 5′ end, or a 3′ phosphate group, the 3′ end. One strand of DNA (sense-strand) contains the genetic information in a 5′ to 3′ direction in the form of the four nucleotide bases adenine (A), thymine (T), guanine (G) and cytosine (C). Its partner strand (antisense strand) has the complementary sequence, i.e. A pairs with T; G with C and vice versa. For example, the double-stranded DNA sequence for the polypeptide described above will be:

Sense strand: 5′- ATG (start)–GGT–TCT–GTT–GCT–GCT–TGG–TAA (stop)-3′
Antisense strand: 3′- TAC–CCA–AGA–CAA–CGA–CGA–ACC–ATT-5′

In biological terms, the double-stranded DNA structure is essential for replication to ensure that each dividing cell receives an identical DNA copy.

Table 2.1 The genetic code.
Nucleotides code in sets of three, or triplets, for individual amino acids. The triplets or codons are shown as they appear in DNA (T = thymine, C = cytosine, A = adenine and G = guanine). In mRNA, T is replaced by U (uracil). The code is degenerate, i.e. there can be more than one codon per amino acid. The genetic code is read from left to right, i.e. TTT = phe (phenylalanine); TCT = ser (serine); TAT = tyr (tyrosine)

First nucleotide [5′]	Second nucleotide T	C	A	G	Third nucleotide [3′]
T	Phe	Ser	Tyr	Cys	T
T	Phe	Ser	Tyr	Cys	C
T	Leu	Ser	STOP	STOP	A
T	Leu	Ser	STOP	Trp	G
C	Leu	Pro	His	Arg	T
C	Leu	Pro	His	Arg	C
C	Leu	Pro	Gln	Arg	A
C	Leu	Pro	Gln	Arg	G
A	Ile	Thr	Asn	Ser	T
A	Ile	Thr	Asn	Ser	C
A	Ile	Thr	Lys	Arg	A
A	Met	Thr	Lys	Arg	G
G	Val	Ala	Asp	Gly	T
G	Val	Ala	Asp	Gly	C
G	Val	Ala	Glu	Gly	A
G	Val	Ala	Glu	Gly	G

Other amino acids are Cys = cysteine; Trp = tryptophan; Leu = leucine; Pro = proline; His = histidine; Gln = glutamine; Arg = arginine; Ile = isoleucine; Met = methionine; Thr = threonine; Asn = asparagine; Lys = lysine; Val = valine; Ala = alanine; Asp = aspartic acid; Glu = glutamic acid; Gly = glycine

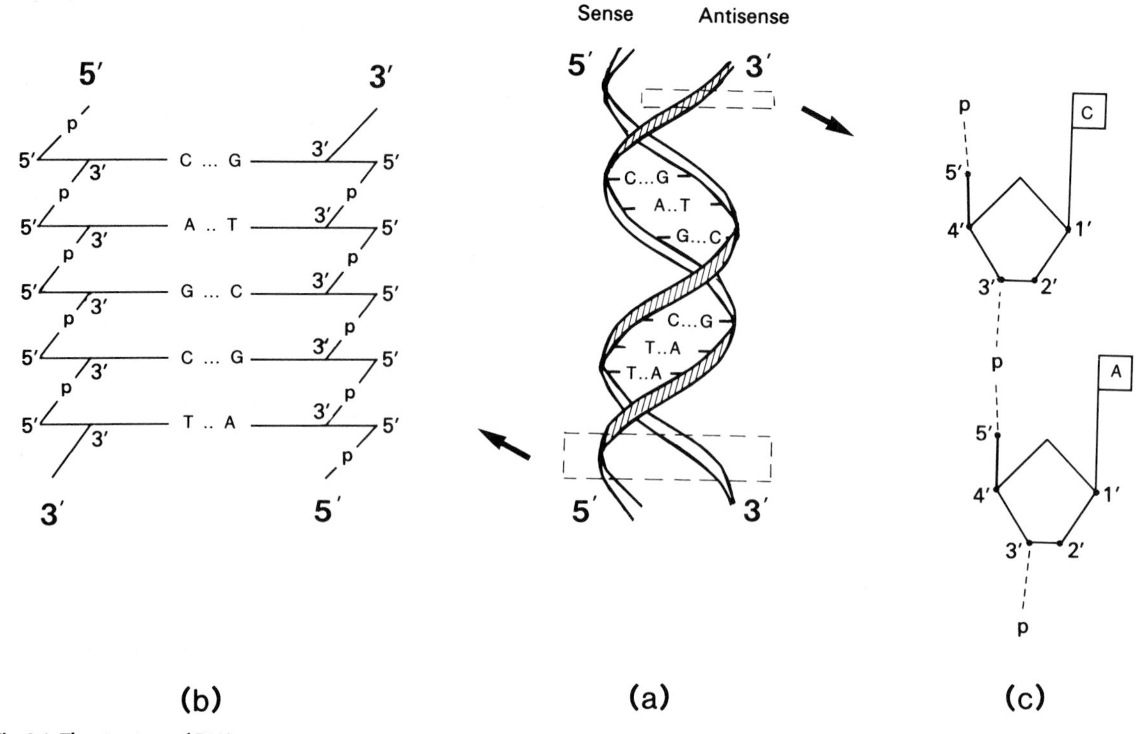

Fig. 2.1 The structure of DNA.
(**a**) A schematic drawing of the DNA double-helix. The sugar phosphate backbone is shown as two strands which run in opposite directions: the sense strand 5′ → 3′ and the antisense 3′ → 5′. The four nucleotide bases are indicated as A,T,G,C (note the complementary pairing of A/T, T/A, C/G and G/C. (**b**) The two strands are held together by hydrogen bonds between the nucleotide bases (two hydrogen bonds between A/T and three between G/C). (**c**) An expanded view of two nucleotides. Each nucleotide has three essential components: a sugar containing five carbons (●), a phosphate backbone (P) and the nucleotide base (□). 1′ to 5′ refers to the position of the carbons on the deoxyribose sugar. The direction for transcription is 5′ → 3′.

From the above example, it can be seen that the genetic code needs to be read from the sense strand. Hence, transcription to give the complementary mRNA sequence is taken from the antisense strand so that the single-stranded mRNA will have the sense sequence present. This was previously mentioned in Chapter 1 when describing antisense RNA and is discussed further in Chapter 7.

DNA probes

The double-stranded structure of DNA is exploited in making and utilising probes. These comprise single-stranded segments of DNA which have the complementary nucleotide sequences to bind a segment of DNA which is also single-stranded. For example, if the single-stranded target has the sequence: 5′-GGTTACTACGTXXX-3′ the single-stranded DNA probe will be 3′-CCAATGATGCAXXX-5′. The *specificity* of a probe thus resides in its *nucleotide sequence*. Since double-stranded DNA is held together by hydrogen bonds, it is relatively easy to make both DNA probe and target DNA single-stranded, e.g. by boiling or treating with sodium hydroxide. Once cooling occurs or the pH is neutralised, the complementary DNA strands will reanneal, i.e. reform into double, base-paired strands. Reannealing will occur between the following combinations: DNA probe + DNA probe; target DNA + target DNA and DNA probe + target DNA. If the DNA probe is radiolabelled with ^{32}P, then the DNA probe + target DNA hybrids can be detected by autoradiography after a procedure such as DNA mapping (Fig. 2.2; see also p. 18). DNA probes are of three types: cDNA, genomic and oligonucleotide (Box 2.1). Plasmid vectors are now available which allow the production of *RNA probes*. These can be used in a similar way to that described above for

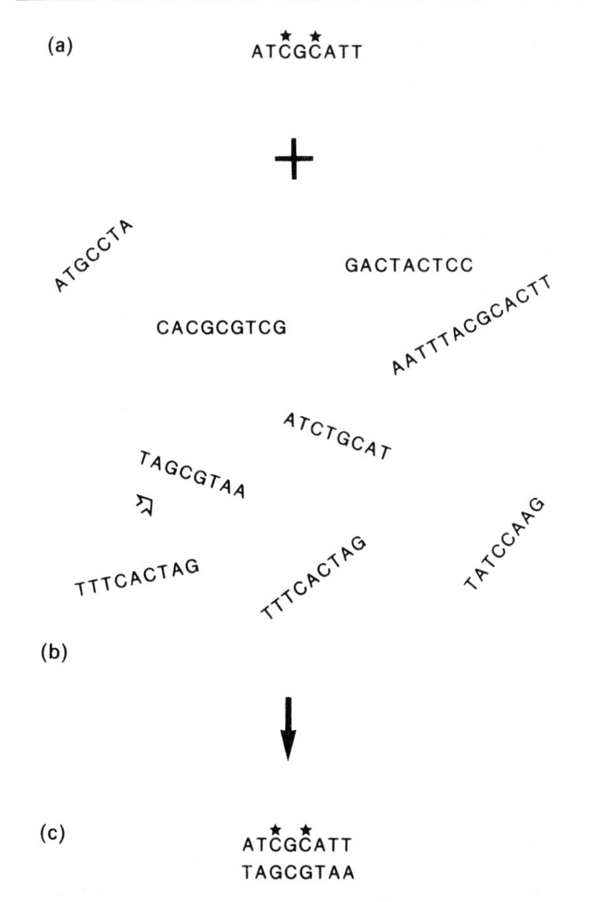

Fig. 2.2 Utilising a probe to identify a specific DNA segment.
Probes are single-stranded segments of DNA or RNA which will bind to a complementary sequence in DNA which is also single-stranded. The specificity of a probe resides in its nucleotide sequence.
(**a**) Nucleotides (e.g. cytosine) in the DNA probe are radiolabelled with ^{32}P (denoted by ★). The probe is then made single-stranded.
(**b**) Target DNA is digested into small fragments with a restriction endonuclease and then made single-stranded. In the mixture will be a fragment (indicated by ⇒) whose sequence is complementary to the probe, i.e. the base pairing of adenine with thymine and guanine with cytosine is matched between target and probe.
(**c**) Radiolabelled probe and target DNA bind together or anneal. Non-specific binding of the probe to other areas of the genome can be inhibited by the type of hybridisation (annealing) conditions used and a washing step following hybridisation. The washing uses high temperature and low salt to break non-specific joining of the probe with other segments of DNA.

Box 2.1 Types of DNA probes

A *cDNA* probe is derived from mRNA by using reverse transcriptase. Thus, it represents DNA sequences which are only present in exons. On the other hand, *genomic probes* can be derived from any segment of DNA, i.e. they comprise any combination of flanking sequences, exons and introns. A genomic probe may contain 'anonymous' DNA sequences, i.e. segments of DNA that are not genes and may not even have a known chromosomal location. The above two types of probes vary in size from a few hundred base pairs to a number of kilobases. *Oligonucleotide* probes are much smaller (20–30 base pairs) and are synthesised by automated means according to the required DNA sequence. Probes are usually labelled with ^{32}P using techniques such as random primer labelling, nick translation or end-labelling. To date non-radiolabelled probes have been of limited value because sensitivity with these probes is less than that obtained with the corresponding ^{32}P labelled probes. This has now changed with the ever increasing utilisation of amplified DNA (obtained through the polymerase chain reaction) and the substitution of colorimetric methods with chemiluminescent substrates. Oligonucleotides are unsatisfactory as probes if total genomic DNA is used because the small size of the probes produces non-specific background hybridisation. However, oligonucleotides are very effective if hybridised against amplified DNA. DNA probes can be purchased commercially or obtained through central processing facilities for a nominal fee, e.g. ATCC–the American Type Culture Collection–lists an extensive catalogue of DNA probes. In the UK, the Human Genome Mapping Project Resource Centre in Harrow offers a similar service.

DNA probes or in the identification of RNA species (see also Expression vectors p. 33).

DNA probes have a variety of names, which can lead to confusion. To create uniformity, a nomenclature for DNA loci against which a number of related DNA probes can hybridise has been devised. For example, the locus D19S51 means human chromosome 19 segment 51. Thus, probe pTD3-21 (p = plasmid; TD = T Donlon, the scientist who prepared the probe) is more usefully identified by its official name of D15S10 (human chromosome 15 segment 10).

RNA

Structural differences between RNA and DNA relevant to molecular technology include the single-stranded nature of RNA and its utilisation of uracil in place of thymine. RNA which is sought in most instances is mRNA. In contrast to DNA, RNA is less robust and isolation techniques require the addition of chemicals to ensure that any RNAase enzymes which may be present are inactivated to avoid degradation of RNA.

Another difference between RNA and DNA lies in the former's tissue-specificity. Thus, the relevant mRNA can only be isolated from a tissue which is transcriptionally active in terms of the target protein. For example, reticulocytes, the red blood cell precursors, would contain predominantly erythroid-specific mRNAs (i.e. mRNAs for the α and β globin genes). The reticulocyte would be an inappropriate source for neuronal-specific mRNA. The tissue-specificity requirement limited the use of mRNA until fairly recently. Now it has been observed that mRNA production in cells which are easy to access, such as peripheral blood lymphocytes, can be 'leaky', i.e. there is transcription of mRNA species which are not directly relevant to the lymphocytes' function. These mRNAs are found in minute amounts but the amplification potential of the polymerase chain reaction (see p. 26) can be utilised to isolate rare mRNA species.

In terms of recombinant DNA technology, mRNA has an important size advantage over DNA because mRNA contains only the essential genetic data resident in exons without the superfluous information found in introns and flanking sequences.

DNA MAPPING

Ingredients

DNA mapping requires the provision of: (1) DNA, (2) restriction endonucleases, (3) electrophoresis, (4) Southern transfer, (5) DNA probes and (6) autoradiography.

Restriction endonucleases

The unique property of restriction enzymes, i.e. their ability to recognise specific base sequences, usually four to six nucleotides in length, is a key element in DNA mapping. How restriction enzymes do this has been discussed in Chapter 1. At any DNA locus, there will be a number of sites which are recognised by restriction enzymes. DNA fragments produced following digestion with restriction enzymes will make up a map for that region. The restriction map may be represented by one or more restriction fragments or a composite of many. Disruption of the map will indicate an alteration in DNA sequence. A change in only one of many restriction sites occurs when there is a discrete modification such as a point mutation affecting a single nucleotide base. This may indicate a genetic disorder or more likely it is a neutral mutation which has given rise to a DNA polymorphism. An alteration in more than one restriction site usually indicates that a structural

Table 2.2 Cleavage sites and some properties of restriction endonucleases.
Fragments with sticky ends, i.e. a few bases not paired, are particularly useful in cloning where they anneal with greater efficiency compared with blunt-ended fragments

Enzyme	Cleavage site	Properties
*Eco*RI	5′ G▼AATTC 3′ 3′ CTTAA▲G 5′ ↓ G AATTC CTTAA + G	Sticky ends; 6 base pair recognition
*Hpa*II	5′ C▼CGG 3′ 3′ GGC▲C 5′ ↓ C CGG GGC + C	Sticky ends; methylation sensitive; 4 base pair recognition
*Rsa*I	5′ GT▼AC 3′ 3′ CA▲TG 5′ ↓ GT AC CA + TG	Blunt ends; 4 base pair recognition
*Not*I	5′ GC▼GGCCGC 3′ 3′ CGCCGG▲CG 5′ ↓ GC GGCCGC CGCCGG + CG	Sticky ends; recognises 8 base pair sequence (i.e. a rare cutter enzyme and so is useful in PFGE)

▼▲ indicate the site of digestion; PFGE = pulsed field gel electrophoresis

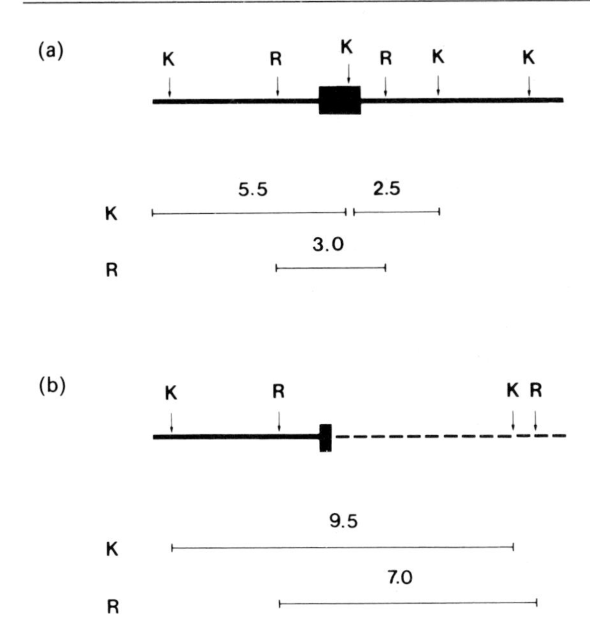

Fig. 2.3 How a restriction enzyme map can define a specific DNA locus.
Restriction enzymes recognise a specific base sequence at which they will cleave DNA. A change in the DNA sequence will disrupt the site(s) of cleavage, altering the map of fragments. A small change, such as a point mutation, will only affect one site; a larger change such as a deletion will affect more than one restriction enzyme cleavage site. A DNA probe is selected to hybridise against a gene (■). Two restriction endonucleases (designated K and R) are used to digest DNA in the vicinity of this gene.
(**a**) The restriction enzyme digests give 5.5 kb and 2.5 kb bands with enzyme K and a single 3.0 kb band with R. This combination of restriction fragments in association with the probe used provides a 'gene map' for this particular locus.
(**b**) Any rearrangement around this locus, such as a deletion, will alter the restriction enzyme pattern since the novel DNA sequence found in association with the deletion (-------) will have different restriction endonuclease recognition sites, i.e. enzyme K now gives a 9.5 kb fragment and enzyme R a 7.0 kb fragment.

rearrangement, e.g. deletion, has taken place (Fig. 2.3).

Restriction enzymes which recognise a four base pair sequence will digest DNA frequently whilst the rare cutting enzymes will give larger fragments (Table 2.2). Restriction enzymes can produce overlapping (sticky) ends following digestion or clean-cut (blunt) ends. The former are particularly useful if a fragment of DNA is to be cloned. Some restriction enzymes will not digest DNA if cytosine residues are methylated. This property is useful in distinguishing a segment of DNA which is *hypomethylated* compared to DNA which has been methylated. The former finding may provide a clue that the region is transcriptionally active.

Southern blotting

DNA fragments generated by restriction endonucleases are separated into their different sizes by electrophoresis in agarose gels. DNA fragments are then made single-stranded following treatment with sodium hydroxide. Single-stranded DNA fragments are transferred from agarose to a more robust medium such as a nylon membrane. The transfer step is called Southern blotting. DNA is now ready for hybridisation to a DNA probe which is labelled with ^{32}P or, less commonly, with a non-radioactive marker.

Non-specific hybridisation of the probe to the membrane or to sequences which might have imperfect homology to that found in the probe is prevented in a number of ways: (1) prehybridising the membrane with DNA from another species (heterologous DNA, e.g. salmon sperm DNA) and (2) following the hybridisation step, the membrane is washed under stringent conditions (e.g. high temperature such as 65°C and a low salt

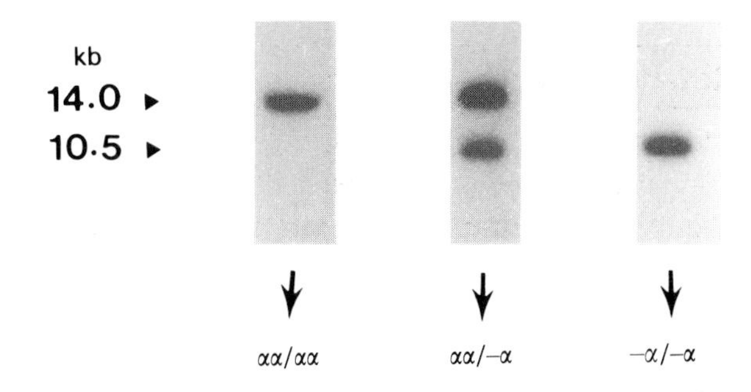

Fig. 2.4 Analysis of α thalassaemia by using a DNA probe: autoradiograph illustrating a DNA restriction map.
The DNA probe is specific for the α globin gene locus on chromosome 16 and the restriction enzyme used is called *Bam*HI. The normal (also known as wild-type) α globin gene structure at this locus is represented by a 14 kb band since there is one *Bam*HI recognition site on either side of the duplicated α globin genes (see Fig. 1.2). Loss through a deletion of one of the two α globin genes (such as occurs in one form of α thalassaemia) brings the two *Bam*HI restriction sites closer together and the 14 kb band becomes 10.5 kb. Thus, this DNA map shows from left to right: normal genotype since αα/ represents one allele and αα/ the second; heterozygous α^+ thalassaemia because there are both the normal 14 kb and mutant 10.5 kb bands. The latter represents deletion of one of the two genes, i.e. −α/ and homozygous α^+ thalassaemia because only the 10 kb fragment is present.

solution). The stringent wash breaks hydrogen bonds which may have formed by chance between random complementary base pairs. The end result will be binding of the radiolabelled probe to its specific target DNA fragment which can be detected by autoradiography (Fig. 2.4).

Controls for DNA mapping include normal DNA which has been processed under identical conditions and DNA markers which enable the size of the hybridised fragment to be determined. A DNA map may require a number of different restriction endonuclease digestions before interpretation can be made. The upper limit of resolution for DNA mapping is approximately 30–40 kb (kb = kilobase; 1 kb = 1000 nucleotide base pairs). Differences in restriction fragments 100–200 base pairs in size can be detected by conventional DNA mapping.

RNA may also be 'mapped' although restriction endonucleases are not used since these enzymes digest double-stranded nucleic acid. RNA 'mapping' involves an estimation of the RNA size. Another difference between RNA and DNA is that RNA must be prepared from transcriptionally active tissues. Electrophoresis of RNA is also undertaken in a denaturing gel to prevent secondary structures from forming during electrophoresis. Transfer of RNA from an agarose denaturing gel to nitrocellulose membranes is known as 'Northern' blotting.

Specific RNAs can be identified by using radiolabelled DNA or RNA probes. Detection and quantitation of low abundance mRNAs is possible through a technique called S1 mapping or RNAase protection. In essence this means that RNA probes which will anneal with a corresponding mRNA species are less liable to breakdown by single-strand specific enzymes such as SI nuclease or RNAase. The 'protection' resulting from formation of a double-stranded structure, i.e. the RNA probe + mRNA hybrid, provides indirect evidence for the presence of mRNA.

DNA polymorphisms

An important application of DNA mapping involves the identification of DNA polymorphisms. This refers to variations in restriction fragment length sizes which occur secondary to changes in the DNA caused by:

1. *Point mutations* which fortuitously delete or add restriction endonuclease recognition sites. This polymorphism is called an RFLP (**r**estriction **f**ragment **l**ength **p**olymorphism). An RFLP is biallelic, i.e. the polymorphism comprises either a large or small fragment.

2. *Insertions* between two restriction endonuclease recognition sites of a segment of DNA which is composed of tandem repeats. The repeats can be of any number and so the end result is a multiallelic polymorphism which is of variable length (called VNTR: **v**ariable **n**umber of **t**andem **r**epeats) (Fig. 2.5).

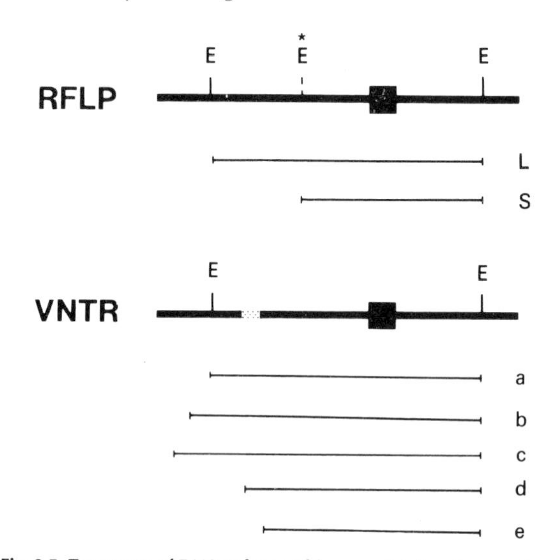

Fig. 2.5 Two types of DNA polymorphisms: *r*estriction *f*ragment *l*ength *p*olymorphisms (RFLPs) and *v*ariable *n*umber of *t*andem *r*epeats (VNTRs).
DNA polymorphisms are caused by changes in the nucleotide sequence which result in changes in the fragment length pattern, or polymorphic map, produced after digestion of the DNA with restriction enzymes. A DNA probe will hybridise against a DNA segment marked (■) and DNA is digested with a restriction enzyme 'E'. RFLPs are caused by point mutations affecting a single site. They are biallelic: there are two options, large (L) or small (S), which depend on whether or not the polymorphic restriction fragment site (*) is absent or present respectively. In contrast, the multiallelic VNTR has the potential to be more polymorphic (and so more informative) since the changes in the E-specific restriction fragment are brought about by the insertion of a variable number of repeat units at the polymorphic site (· · ·). Thus, the number of polymorphic restriction fragments generated is potentially much greater (e.g. the arbitrary sizes indicated by a–e). Because of their greater intrinsic variability, VNTRs are usually more 'informative' in polymorphism studies since there is more chance that heterozygous patterns will be detected at any one locus.

For convenience, a single base change is called a DNA polymorphism if it occurs at a frequency of 1% or more within a population and is inherited along Mendelian lines. DNA polymorphisms have proven to be invaluable markers to distinguish the two alleles at a locus (for example, wild-type versus mutant alleles). DNA polymorphisms enable a gene locus to be detected *indirectly* by allowing co-segregration between a phenotype (which can be normal or abnormal) and a particular DNA polymorphism to be followed within the context of a family study. It is not even necessary to have identified a gene or its underlying defects to utilise DNA polymorphisms. Predictive estimations concerning disease status of individuals on the basis of their DNA polymorphism patterns are now frequently undertaken (examples may be found in Chs 3, 4). DNA polymorphisms also provide patterns which are useful in forensic analysis (see Ch. 8). There are few diagnostic or research studies which do not utilise these invaluable markers.

Pulsed-field gel electrophoresis

Pulsed field gel electrophoresis, a modification of gene mapping, was described by Schwartz and Cantor in 1984. This technique has increased the upper limit of resolution for DNA mapping from kilobases to megabases (Mb = megabase; 1 Mb = 1000 kb or 1 x 10^6 base pairs). Technology developments which have made this possible include: (1) the isolation of restriction endonucleases which cleave DNA into much larger fragments than is possible with the more frequently used restriction enzymes and (2) the application of a non-homogeneous and interrupted electric field.

DNA for pulsed field gel electrophoresis needs to be prepared in a special way since it is essential to avoid random shearing which will break DNA into fragments which are relatively small. Rare cutting restriction enzymes then digest DNA into fragments ranging in size from 100–2000 kb. Application of a constant and homogeneous electric current, such as that used for conventional gene mapping, will not distinguish large DNA fragments (e.g. 100–500 kb) since the forward motion brought about by the electric charge and the drag due to friction as the DNA passes through agarose pores does not allow differential migration of large molecules. In pulsed field gel electrophoresis, this differential is obtained by periodically altering the orientation of the electric field. Each time this happens, large

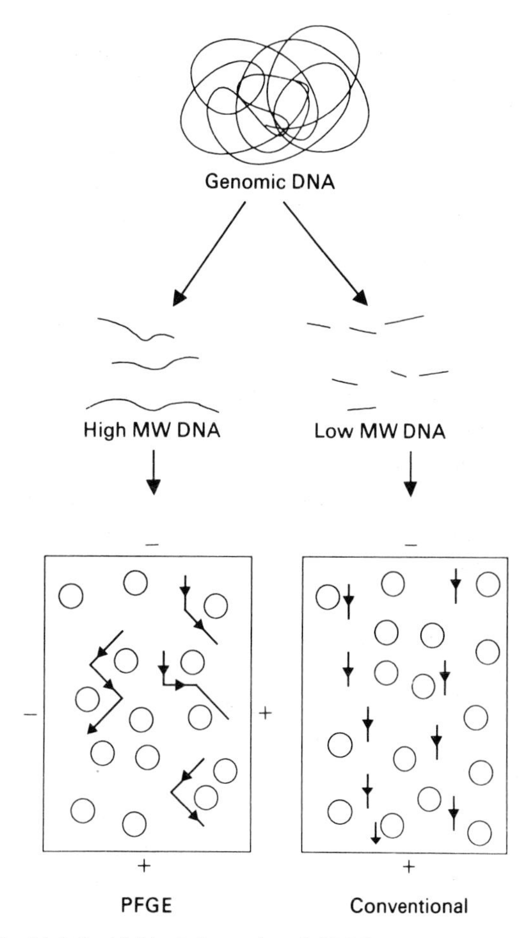

Fig. 2.6 Pulsed field gel electrophoresis (PFGE).
This technique enables much larger fragments of DNA to be separated which migrated together in conventional electrophoresis. Together with the rare cutter enzymes this has enabled megabase fragments of DNA to be isolated and cloned. DNA for PFGE needs to be of high molecular weight (MW). The high MW DNA molecules are separated from each other by altering the orientation of the electric field. DNA is negatively charged and so will move towards the positive electrode through the agarose particles (○). The snake-like movement of the large particles, because there is a periodic change in polarity of the electric field, separates them. Their reorientation time to a new electric field is dependent upon the size of the fragment. The smaller DNA fragments in conventional electrophoresis move in one direction.

molecules need to reorientate and then find a new path through the gel matrix. The reorientation time will be dependent on the size of the fragment (Fig. 2.6). Restriction fragments up to 10 Mb in size can be distinguished by pulsed field gel electrophoresis. This technique is increasingly being used to construct physical maps of DNA loci. Large fragments distinguished by pulsed field gel electrophoresis can also be isolated from gels and cloned.

CLONING DNA

Ingredients

Cloning of DNA requires: (1) DNA fragmented into specific size ranges, (2) a vector which will accept DNA to be cloned, (3) a host to allow propagation of vector and its inserted DNA and (4) a method to screen a DNA library for the clone of interest.

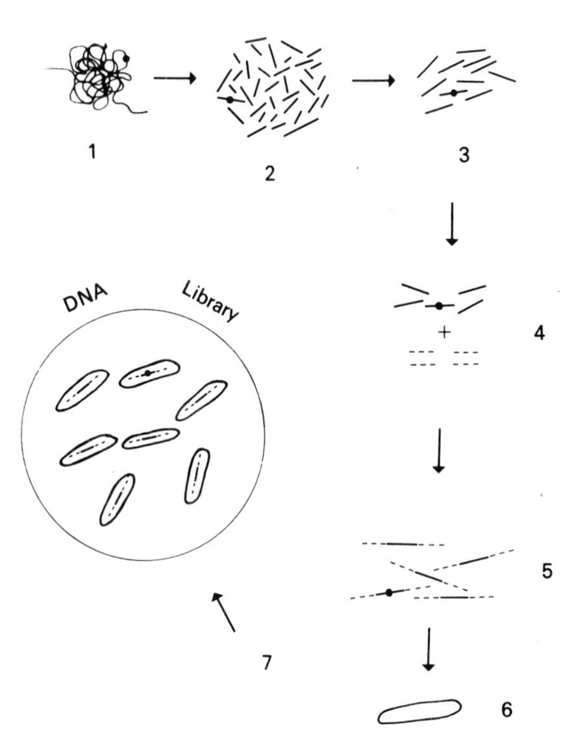

Fig. 2.7 Steps involved in cloning.
1. Genomic DNA with target DNA depicted as (●). **2**. DNA is randomly fragmented, by physical methods or restriction enzymes, into many pieces. **3**. The DNA fragments are size-selected. **4**. Appropriately-sized DNA is then joined, using the enzyme DNA ligase, to the DNA vector (- - -) such as a plasmid or a phage. **5**. There will be many ligated DNA segments–hopefully with the target gene being included in the ' library '. **6**. Vector plus inserts are taken up into a host such as *Esch. coli* (transformation). **7**. Vector, inserts and host are plated out onto an agarose plate where screening can be undertaken to identify the correct clone (see Fig. 2.8 for a description of how a library can be screened). The relevant colony can then be used to grow up a large number of copies of the DNA concerned.

Cloning strategies

The principles behind cloning are summarised in Figure 2.7 and a summary of cloning vectors is given in Table 2.3. The first step is to decide on the size of DNA to be cloned. Thus, a limited segment of DNA needs to be cloned if the gene being sought is small (for example, the globin genes). On the other hand, trying to clone a large gene (for example, cystic fibrosis) or finding a gene somewhere on a chromosome will mean a different approach is necessary, i.e. the largest possible DNA segments will need to be cloned. Once size is known, the appropriate vector system required to carry the cloned DNA fragment can be

Table 2.3 Vectors available for DNA cloning.
Vectors will carry segments of DNA into a host cell where it can be both identified and produced in larger amounts. The vector chosen will depend on the size of the DNA to be cloned

Vector	DNA insert size	Features
Plasmid	< 10 kb	Technically simple but relatively inefficient
Phage	~ 20 kb	Most conventional approach; size of inserts is limited
Cosmid	~ 40 kb	Larger inserts possible
P1 phage	< 100 kb	Recent vector; in between cosmids and yeast artificial chromosomes; still being evaluated
YAC	~ 300 kb*	Complex to make and difficult to screen; large inserts possible

YAC = yeast artificial chromosome. * average insert sizes for the most recent YACs from a resource centre, CEPH, are approximately 1.2 Mb

selected. Random shearing by physical methods or restriction endonucleases will break DNA into fragments. Target DNA will be found within one or more of these fragments.

Cloning with plasmid vectors

The simplest cloning system involves a plasmid vector. In this situation genomic DNA and plasmid are both digested with the same restriction endonuclease creating, if possible, fragments with unpaired bases at each end (sticky ends). Mixing the two together in the presence of DNA ligase will allow ligating to occur so that each individual plasmid will have one fragment of genomic DNA. Using electrical or calcium chloride shock, it is possible to get the plasmids plus their cloned inserts to be taken up (in a process called transformation) by *Esch. coli*. The host *Esch. coli* divides and is then plated out as a lawn on an agarose plate. If all goes well there should be a broad representation of DNA from most of the genome (including the target DNA) inserted into the plasmids. Thus, a *library* of DNA fragments has been produced. The next step involves screening of the library to find the target DNA. For screening, a replica of the colonies on the agar plate is obtained by placing the plate against a nylon membrane. DNA which has been transferred to the nylon by this contact is then made single-stranded and can be screened for target DNA with the appropriate probe. When a positive colony is found it is traced back to the relevant colony on the agar plate and isolated (Fig. 2.8).

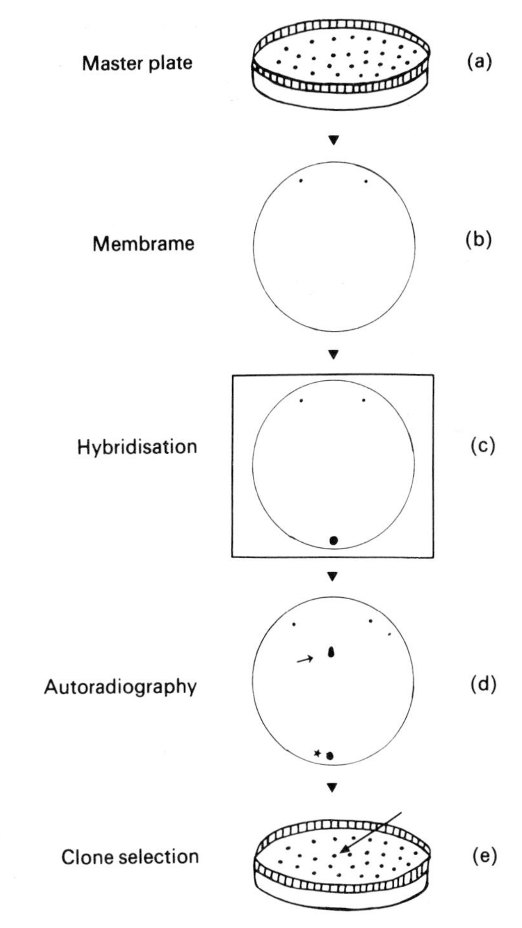

Fig. 2.8 Screening a DNA library.
(**a**) Master plate of agarose containing a lawn of *Esch. coli*. The position of clones will be seen as colonies (plasmid clones) or plaques (phage clones). (**b**) The master plate is overlain with a nylon membrane. Portions of the colonies/plaques will be transferred to the nylon membrane. Radioactive markers are used to orientate the master plate in respect to the nylon membrane(••). (**c**) DNA in the nylon membrane is made single-stranded and hybridised against a radiolabelled DNA probe which is specific for the DNA to be cloned. (**d**) Autoradiography indicates the position of the positive clone (→) where probe and target DNA have joined (★ = positive control). (**e**) From the master plate the relevant colony/plaque can be selected from the autoradiography pattern which is aligned correctly with assistance from the radioactive markers.

Cloning with bacteriophage vectors

The next level of complexity for cloning is the bacteriophage ('phage') vector. This has the advantage over plasmids in that it can accept a larger DNA insert and is overall a more efficient way to clone. The same steps as described for the plasmids are undertaken except that phage plus its DNA insert need to be prepared into an infectious unit (called 'packaging') for insertion into *Esch. coli*. Phage vector cloning will produce plaques (areas of lysis) wherever a phage infects an *Esch. coli*. These plaques on the agar plate represent individual clones. A disadvantage of using phage as a vector is poor yield of target DNA once it has been isolated. This is usually overcome by finding the relevant DNA in a phage library and then subcloning that DNA into a plasmid vector.

Cloning with yeast artificial chromosomes

To clone larger DNA fragments requires more sophisticated vectors. These include cosmids, P1 phage or yeast artificial chromosomes (abbreviated to YACs). The last are autonomously replicating stable chromosomes with selectable markers suitable for yeast. They are particularly valuable since large DNA fragments, e.g. 300–500 kb in size, can be cloned. DNA plus vector are inserted into a eukaryotic (yeast) host rather than the prokaryotic (bacterial) host described above. One major advantage in cloning with yeast artificial chromosomes is that the inserts (DNA fragments) are so large that it is very likely target DNA will be cloned within a single fragment.

The larger cloned fragment also make it easier for directional movement along a chromosome (called 'chromosome walking') to occur. This will be discussed further under positional cloning. To isolate a gene, it is necessary to have overlapping, contiguous segments of DNA (called 'contigs') to assist in chromosome walking (this is discussed further below and in Chapter 3 under cystic fibrosis). The larger the contigs the easier is 'the walk'. It should also be noted that some segments of the genome are difficult to clone and so phage or cosmid libraries may be inadequate to provide contigs over those regions. The ability to clone into yeast artificial chromosomes is very useful in terms of the above considerations.

There are technical problems associated with yeast artificial chromosomes. The libraries are difficult to make. To get around this a number of laboratories have become resource facilities. Recently, yeast artificial chromosome libraries with inserts which are much larger than previously described have become available through CEPH (Centre d'Etude du Polymorphisme Humain) one of the resource centres. Screening of YAC libraries can also be difficult, but this has improved with the availability of the polymerase chain reaction (see p. 26). Finally, DNA inserts, particularly the larger ones, are frequently unstable, i.e. they can recombine or delete.

Functional and positional cloning

Haemophilia, a genetic bleeding disorder, provides an example of what is meant by *functional* cloning. In this case, a disease has been sufficiently characterised to enable identification of an abnormal end-product (the factor VIII coagulation protein). From this the relevant gene can be cloned and mapped at the DNA level (Fig. 2.9). Cloning has resulted in a wider range of diagnostic tests, an increased knowledge of the disorder's pathogenesis and alternative therapeutic approaches (discussed further in Chs 3, 7). However, the situation found in most genetic disorders, e.g. Duchenne muscular dystrophy, is quite different. In these circumstances, the underlying defect remains unknown and there can be no further progress with understanding of the disease or even the possibility of diagnosis. The molecular strategy known as *positional* cloning (initially called *reverse genetics* but a more appropriate description is positional cloning) can now be adopted.

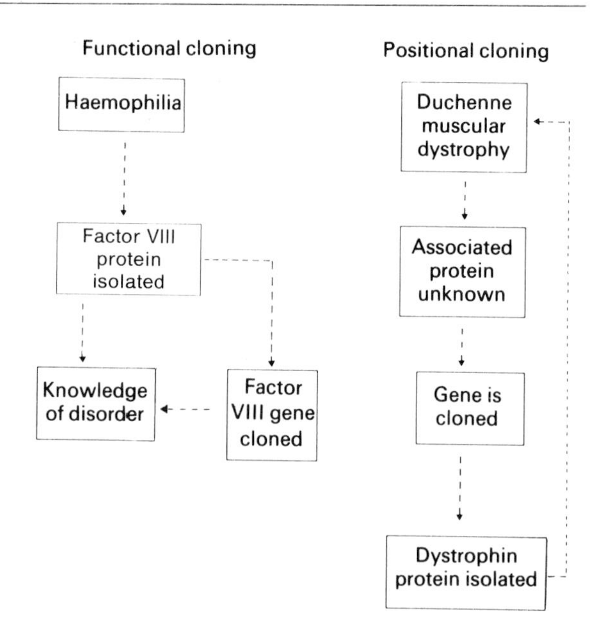

Fig. 2.9 Functional and positional cloning.
The modern molecular approach to the study of genetic diseases illustrates the differences between the functional cloning strategy, used when the abnormal gene product is known, and the more recently described positional cloning, used when the underlying defect is unknown. In the latter method clues to the likely locus are obtained from clinical studies or candidate genes are postulated from a knowledge of the general clinical area involved in the disorder. Testing of probes chosen in this way is made to determine if there is linkage to the genetic defect and a genetic map is made using affected families.

The first step in positional cloning is to obtain, if possible, a clue as to the likely locus or chromosome involved. This comes from case reports or observations in which chromosomal rearrangements have been noted to occur in association with the clinical picture. An alternative approach is to look with DNA markers derived from 'candidate genes'. For example, a good candidate gene for a heart muscle disorder would be myosin, a component of muscle (see Ch. 3, familial hypertrophic cardiomyopathy for further discussion). The above steps are very helpful in narrowing the search for DNA markers which are essential for the positional cloning strategy to progress. However, the preliminary steps to identify a locus are not crucial for positional cloning. Random testing of DNA probes for linkage to a disease locus is tedious and risky but has been successful on a number of occasions as illustrated by Huntington disease and adult polycystic kidney disease (see Ch. 3).

Once a DNA marker (usually a DNA polymorphism) has been linked to a genetic defect it is necessary next to construct genetic and physical maps of that locus. The *genetic map* is made by looking at DNA polymorphisms within affected families. The closer the polymorphism is to the gene, the fewer will be the recombinations (breaking and rejoining of the DNA) that are observed (see Ch. 3 for a more detailed description of recombination). Eventually, a polymorphism associated with the gene itself will produce no recombination events. *Physical maps,* which are based on actual measurements e.g. kb or Mb, are also constructed. The availability of pulsed field gel electrophoresis, fluorescent in situ hybridisation and yeast artificial chromosomes has helped in the construction of physical maps. These new techniques make positional cloning a more realistic strategy for isolating genes (see Chs 1 and 3 for examples of how positional cloning has been utilised). From clones contained within the yeast artificial chromosomes, genes can be identified and these are then further characterised until the correct one is found. Information derived from this enables a diagnostic test to be developed and the function of the underlying genetic disorder to be determined. The many successes of the positional cloning approach have already been mentioned in Chapter 1.

SEQUENCING DNA

Ingredients

For successful sequencing it is necessary to have: (1) cloned DNA, (2) DNA (oligonucleotide) primers, (3) DNA polymerase, (4) the four nucleotide bases (A,T,G,C), (5) the four nucleotide bases in modified form as dideoxy bases, (6) ^{35}S -labelled base and (7) electrophoresis.

Sequencing DNA with the dideoxy method

The majority of genetic disorders are caused by point mutations and less frequently by small discrete deletions. In these circumstances, DNA mapping alone is unlikely to detect an abnormality unless, by chance, the mutation deletes or creates a recognition site for a restriction endonuclease. This occurs infrequently and so sequencing of individual nucleotide bases may be required. The methodology for sequencing DNA has evolved rapidly since the first descriptions in the mid-1970s by Sanger, Maxam and Gilbert. Two procedures are now available. One utilises chemical cleavage of DNA and the second is the more popular enzymatic or dideoxy chain termination method (Fig. 2.10).

The dideoxy chain termination method has undergone a number of modifications which have improved resolution, e.g. utilisation of the isotope ^{35}S rather than ^{32}P for radiolabelling; the isolation of a more efficient polymerase enzyme (Sequenase™) and better gel electrophoresis techniques to increase the length of readable DNA sequence. Generation of cloned, single-stranded DNA required for sequencing has been facilitated with the development of a special phage vector (M13) or plasmids containing M13 replication origins. An alternative to cloned DNA for sequencing is DNA which has been amplified by the polymerase chain reaction. Sequencing with this product will be described in the following section.

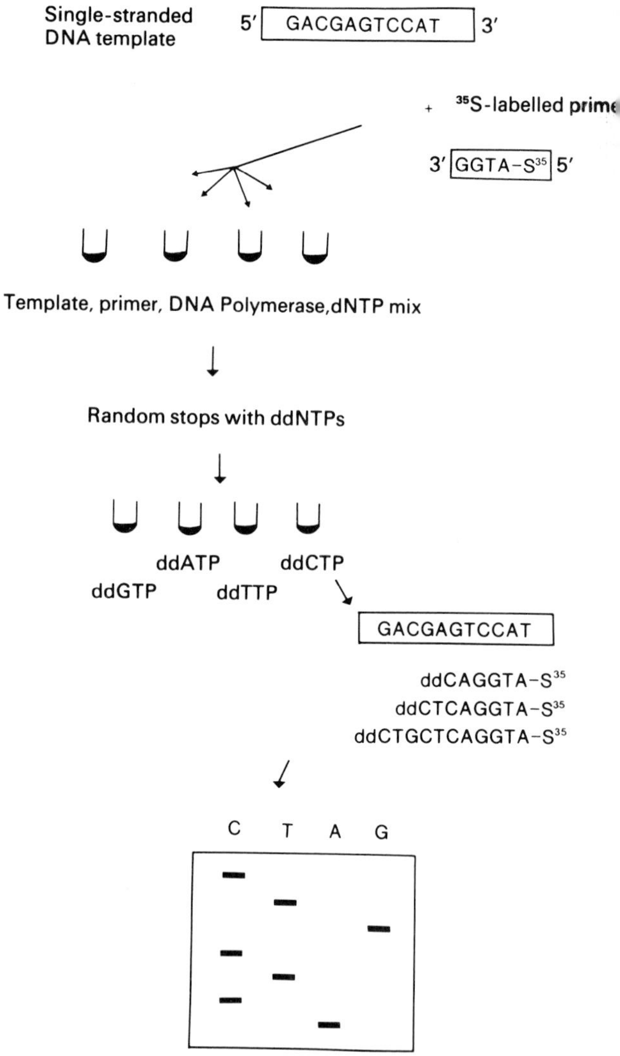

Single-stranded DNA template 5' GACGAGTCCAT 3'

+ ^{35}S-labelled prime

3' GGTA-S^{35} 5'

Template, primer, DNA Polymerase,dNTP mix

Random stops with ddNTPs

ddGTP ddATP ddTTP ddCTP

GACGAGTCCAT

ddCAGGTA-S^{35}
ddCTCAGGTA-S^{35}
ddCTGCTCAGGTA-S^{35}

C T A G

Fig. 2.10 DNA sequencing with the dideoxy chain termination method.
Single-stranded DNA template and radiolabelled single-stranded primer are prepared. The two are allowed to anneal. In the presence of DNA polymerase and a mixture of the four deoxynucleotides (dGTP, dATP, dTTP, dCTP) there is extension from the primer/template double-stranded site. Random stops in the extension are then produced by adding to each tube one of the dideoxynucleotides (ddNTP). As illustrated, ddCTP will produce random stops wherever there is a cytosine nucleotide. The remaining three dideoxynucleotides will do likewise in their individual reactions. The end result is a mixture containing variable lengths of extended DNA segments. Each mixture is electrophoresed in the gel track corresponding to the dideoxynucleotide added, e.g. G = ddGTP etc. The DNA sequence is read from bottom to top. In this example the sequence reads: ACTCGTC which represents DNA sequence 3' to 5' following annealing between GGTA (primer) and its complementary sequence in the template (CCAT).

POLYMERASE CHAIN REACTION

Ingredients

The polymerase chain reaction requires: (1) DNA (or RNA), (2) oligonucleotide primers, (3) *Taq* polymerase, (4) a mixture of the four nucleotide bases and (5) temperature cycling apparatus.

Utility

The polymerase chain reaction (abbreviated to PCR) utilises a DNA extension enzyme (polymerase) which can add nucleotide bases once a template is provided (Fig. 2.11). There are three basic steps in the polymerase chain reaction. (1) Denaturation of double-stranded DNA into its single-stranded form. (2) Joining of oligonucleotide primers to both ends of a target sequence. The oligonucleotides are constructed so that they are complementary to target DNA or RNA sequence and this gives the polymerase chain reaction its specificity. Oligonucleotide primers are present in excess and so will not be limiting in the subsequent amplification steps. (3) Addition of the four nucleotide bases and a polymerase. *Taq*

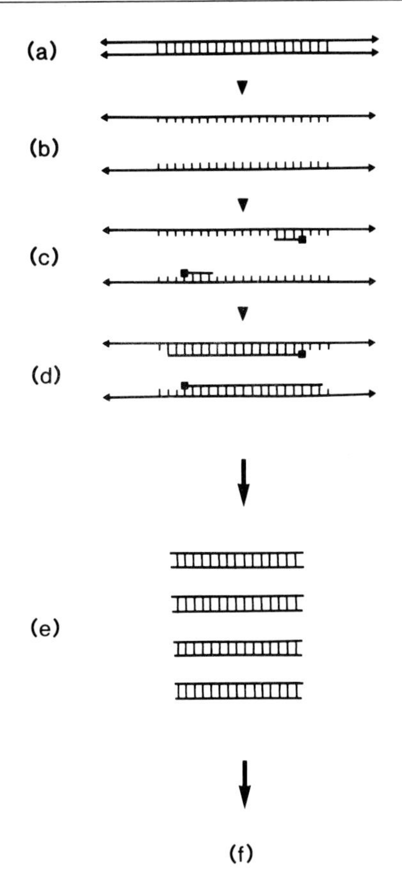

Fig. 2.11 The polymerase chain reaction.
This process allows amplification of a targeted DNA sequence by using a DNA extension enzyme (polymerase) to make new copies of the sequence. Oligonucleotide primers are constructed to bind to the required DNA sequence initiating synthesis of the specified area. (**a**) Double-stranded target DNA. (**b**) The first step in DNA amplification is denaturation to produce single-stranded DNA. (**c**) DNA primers (–■) anneal to both the single-stranded target DNA segments through complementary base pairing. (**d**) Once the primers anneal, the *Taq* polymerase enzyme allows extension from the primers by adding the corresponding nucleotide bases. At the end of one cycle, there is twice the amount of target DNA which is again double-stranded. (**e**) Denaturation, annealing and extension steps (**b**–**d**) are repeated in the second cycle. At the end of this cycle there is four times the amount of target DNA present. (**f**) Cycles are repeated to produce (in theory if the process is 100% efficient) 2^n times the amount of template DNA (where n = number of cycles).

polymerase is used since it is relatively heat resistant and so the step for denaturation can be incorporated into the overall cycle without interfering with the polymerase activity. One polymerase chain reaction cycle comprises the above three steps. After a cycle, each of the single-stranded DNA target segments has become double-stranded through the polymerase's activities. The cycle is then repeated and each time a

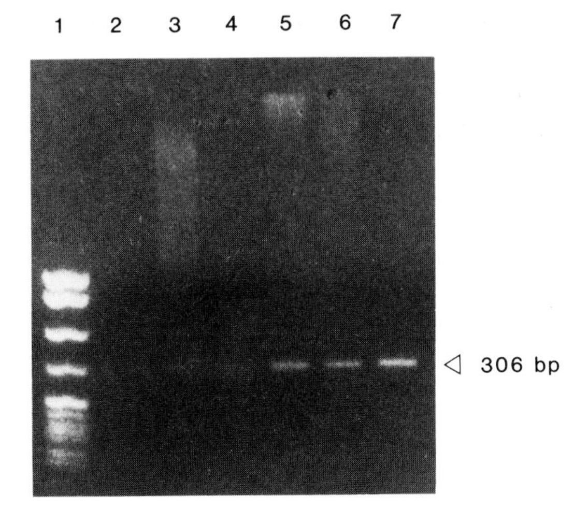

Fig. 2.12 Amplified DNA can be visualised directly by staining with ethidium bromide.
This intercalates between the bases and fluoresces under ultraviolet light. Lane 1 has a DNA size marker. Lanes 2–7 show increasing concentrations of DNA 306 base pairs in size which have been amplified from exon 5 of the human aldolase B gene (see also Fig. 4.6).

new target segment of DNA is synthesised. Theoretically, the number of templates produced equals 2^n, i.e. after 20 cycles of amplification there should be 1×10^6 templates.

Amplified DNA products can be visualised in a number of ways. The simplest makes use of electrophoresis and staining of DNA with ethidium bromide which intercalates between the nucleotide bases and fluoresces under ultraviolet light (Fig. 2.12). Radiolabelled nucleotides can also be incorporated into the polymerase chain reaction steps and the amplified products visualised by autoradiography. Amplified DNA products can be transferred to nylon membranes by Southern blotting and detected by their probe hybridisation patterns. Colorimetric assays have been used with the polymerase chain reaction. These depend on oligonucleotide primers which have been modified to allow them to be labelled with chemicals such as biotin or fluorescein.

A feature of the polymerase chain reaction is its exquisite *sensitivity* so that even DNA from a single cell can be amplified. The ability of the polymerase chain reaction to amplify small numbers of target molecules has been put to use in detecting 'illegitimate transcription'. As described earlier, mRNA is tissue-specific except for

Box 2.2 Applications of the polymerase chain reaction

DNA amplification by the polymerase chain reaction (PCR) is used in the diagnosis of many genetic disorders. The polymerase chain reaction enables a segment of DNA to be screened for polymorphisms or point mutations. In the infectious diseases, amplification can detect DNA from microorganisms which are present in too few numbers to be visualised or whose growth characteristics are such that there will be a delay in diagnosis. In the long term the sensitivity of the polymerase chain reaction will play a role in the forensic scenario where tissue left at the scene of the crime is small in amount or degraded. Research applications of the polymerase chain reaction are numerous. Amplification has enabled cloning, direct sequencing and the creation of mutations in the DNA segments. Investigation of DNA/protein interactions by a process called 'footprinting' is possible. Quantitation of DNA by using the polymerase chain reaction can be undertaken although it requires a number of inbuilt control reactions to allow for variability in amplification. Non-radioactive labelling becomes feasible if used in conjunction with amplified DNA. Novel automated procedures for amplifying DNA on a large-scale are described and will enable rapid and sensitive screening for a number of defects or polymorphisms. Multiplex PCR refers to the simultaneous amplification of a number of DNA segments using a combination of multiple oligonucleotide primer sets. RNA may also be studied along the same lines as described for DNA. An additional step, which incorporates the enzyme reverse transcriptase, enables RNA to be first converted to cDNA from which amplification is then able to proceed.

some 'leakiness' which occurs in cells such as the lymphocyte. Thus, mRNA specific for muscle tissue in disorders such as Duchenne muscular dystrophy and the hereditary cardiomyopathies has been characterised by amplification of mRNA from lymphocytes (see Box 3.8 for further discussion). Sensitivity and specificity of DNA amplification can be increased further by use of *nested primers*, i.e. a second round of amplification utilising oligonucleotide primers which are derived from DNA sequences within the segment amplified during the first round. The polymerase chain reaction is rapid and can be automated with a 30-cycle procedure taking approximately 2–3 hours to complete. Applications of this technique are extensive with new modifications or innovations

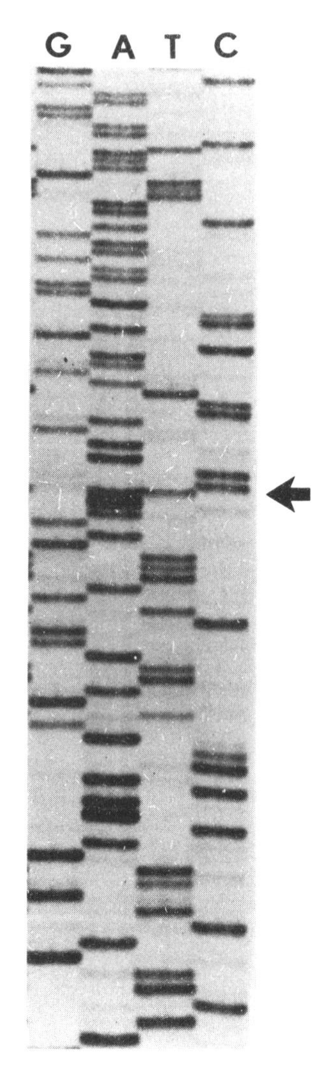

Fig. 2.13 Direct sequencing of double-stranded DNA amplified with the polymerase chain reaction.
The four gel lanes marked GATC indicate the relevant nucleotide in the sequence. Reading from the bottom of the gel the DNA sequence in this example is: ATCTTGACTGTTGA etc. An arrow marks where two lanes (A and T) have a band present in the same position. Therefore, because this is double-stranded DNA sequencing, the simultaneous appearance of bands in the A and T tracks means one allele has adenine and the second has thymine. This may represent a DNA polymorphism or a point mutation producing a genetic defect.

being constantly described (Box 2.2). Single-stranded or double-stranded DNA generated by the polymerase chain reaction can be used directly for DNA sequencing (Fig. 2.13). Automated sequencing of single-stranded amplified DNA is now available utilising fluorochromes in place of ^{35}S.

Problems

Two significant disadvantages of the polymerase chain reaction are the necessity to have a DNA sequence to allow synthesis of oligonucleotide primers and the relative ease with which contamination can occur. The latter may come from extraneous DNA such as other samples or the operator. The commonest source for contamination is the previously amplified products. Many strategies have been described to avoid contamination which is one reason why the polymerase chain reaction has not been utilised in more of the routine diagnostic laboratories.

The sequence fidelity of amplified products is an additional consideration when assessing the usefulness of DNA amplification since in vitro DNA synthesis is an error-prone process. The error rate associated with *Taq* DNA polymerase activity is in actual fact low and has been estimated to be about 0.25%, i.e. one misincorporation per 400 bases over 30 cycles.

DETECTION OF POINT MUTATIONS

Ingredients

Detection of point mutations requires: (1) cloned or amplified DNA, (2) electrophoretic separation of DNA fragments and (3) methods to distinguish DNA changes involving one to a few nucleotide bases.

Allele-specific oligonucleotides

The ultimate way to detect a point mutation is by sequencing DNA. However, this is a lengthy procedure and need only be done the first time to identify and then characterise a specific mutation. Thereafter, more rapid means are required to detect the same mutation in other specimens. The detection of point mutations in DNA has relevance to a number of areas in medicine. They include: genetics, to identify genetic disorders; oncology, to look for mutations in cancer-producing genes and microbiology, for epidemiological purposes.

Known point mutations can be identified if, by chance, they occur at a restriction endonuclease recognition site. Thus, the point mutation produces a novel restriction enzyme fragment. Unfortunately, this is an uncommon event. An alternative way to detect point mutations is to hybridise with *allele-specific oligonucleotides*, i.e. oligonucleotide probes which are synthesised so that one probe will hybridise to the wild-type (normal) sequence and the second probe will hybridise to the mutant DNA. Hybridisation patterns will define the underlying genotypes, e.g. mutant probe alone–homozygous for that mutation; wild-type probe alone–normal; both mutant and wild-type probes–heterozygote (Fig. 2.14).

Oligonucleotides as DNA probes have widespread application in the diagnosis of genetic disorders (see Ch. 3) and have proven useful in the screening of DNA libraries. Because of their small size, oligonucleotide probes have a limited function when total genomic DNA is used. In this circumstance, there is considerable background hybridisation. On the other hand, amplified DNA is an ideal target against which oligonucleotide probes can hybridise. The probes can be end-labelled with ^{32}P or linked to chemicals, enzymes or fluorochromes.

Screening DNA to look for point mutations

In the past few years the necessity to 'screen' amplified DNA products to identify which one is likely to contain a mutation has assumed increas-

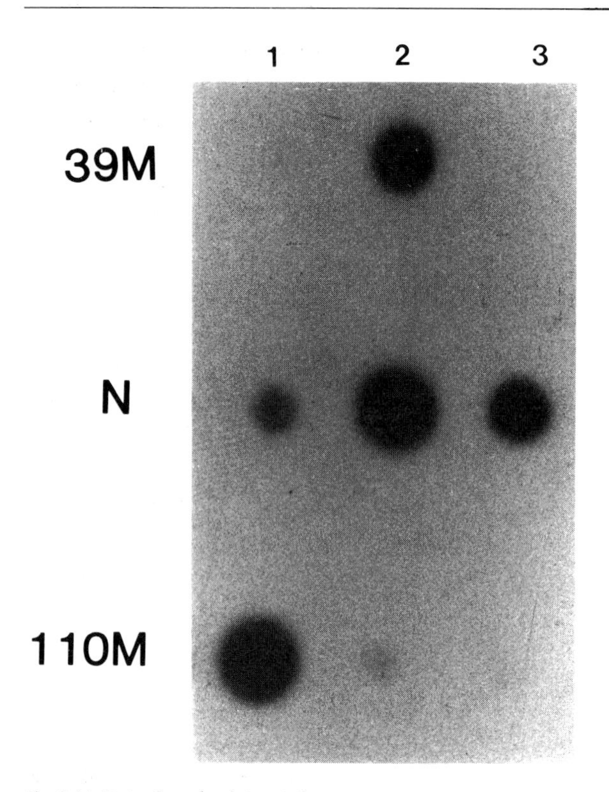

Fig. 2.14 Detection of point mutations.
Allele-specific oligonucleotide (usually abbreviated to 'ASO') probes are synthesised so that one probe hybridises to the normal (wild-type) sequence and the second to the mutant. ^{32}P labelled ASO probes hybridise to amplified, single-stranded DNA which is fixed onto nylon membranes in the form of dots (hence the name dot blots). There are three specimens numbered 1–3. DNA has been amplified in the region of the β globin gene which covers two common Mediterranean mutations: codon 39 and IVS1,110 which are located in close proximity. Specimen 1 hybridises only with the 110M (mutant) oligonucleotide probe, i.e. the individual in question is homozygous for this mutation. Specimen 2 hybridises to the codon 39 mutant oligonucleotide *and* the normal (N) oligonucleotide. This individual is a heterozygote for the codon 39 mutation. Specimen 3 only hybridises to the normal oligonucleotide and so does not have the codon 39 or IVS1,110 mutations. The last result cannot exclude another mutation which is located outside the amplified region of DNA. There is some cross-hybridisation with the normal oligonucleotide probe (N) in specimen 1. This can occur because the oligonucleotide probes differ by only one base. However, the relative intensities indicate that this is only cross-hybridisation.

ing importance. This will be illustrated in Chapter 3 by reference to Duchenne muscular dystrophy, haemophilia and familial hypertrophic cardiomyopathy. In these examples, there are a number of point mutations which cause the underlying disease but unfortunately the genes are too large to make sequencing a practical diagnostic strategy. One approach to this problem involves the amplification of cDNA in a number of segments. For example, the cDNA itself might be 2 kb in size, but it is amplified in ten overlapping segments each of which might be 0.2 kb in size. The ten segments are next screened to see which has a

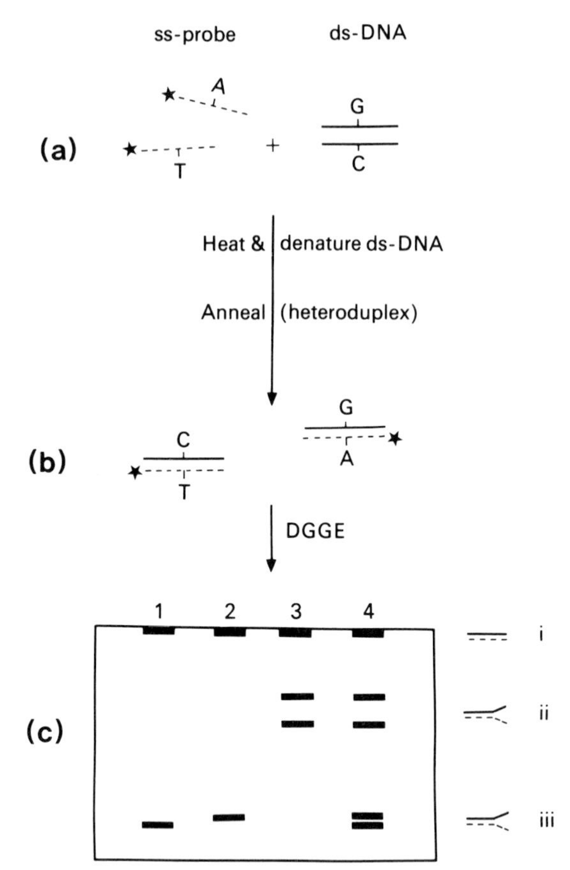

Fig. 2.15 Screening for point mutations in DNA by denaturing gradient gel electrophoresis (DGGE).
In this technique point mutations are detected as points of mismatch in double-stranded DNA made with wild-type sequence probes. Such mismatches will cause separation of the strands under denaturation resulting in altered gel mobility. There are a number of variations for the DGGE technique. A simplified summary is illustrated here. Radiolabelled single-stranded (ss) DNA probe for a wild-type sequence (*) is mixed with double-stranded (ds) target DNA. (**a**) Single-stranded ^{32}P-probe is mixed with double-stranded target DNA which contains a point mutation (A to G and so T to C in the antisense strand). The mixture is heated to make the target DNA single-stranded. (**b**) As the temperature falls, probe and target DNA will anneal and form heteroduplexes which contain the mismatched base pairs (A + G; T + C). (**c**) DNA is electrophoresed through a denaturing gradient increasing from i to iii. In position i, DNA will be in the double-stranded form. As the denaturing gradient increases, melting domains where mismatched bases occur will be affected and branched structures form. These reduce the DNA's mobility. Lanes 1, 2 illustrate the mobility of wild-type and mutant homoduplex controls respectively. In lane 3 the heteroduplexes formed will migrate less since they are unstable because of the mismatched bases. In lane 4 there is a mixture of homoduplexes and heteroduplexes.

point mutation. That amplified fragment alone is sequenced. The end result is sequencing of 0.2 kb rather than 2.0 kb. cDNA is used in preference to genomic DNA because the former is much smaller in size since it is only comprised of exons. These are also more likely to have mutations compared to introns (see Ch. 3 for examples).

Three methods have been developed to *screen* DNA for point mutations. They are: denaturing gradient gel electrophoresis (commonly abbreviated to DGGE); single-strand conformation polymorphism (SSCP) and the chemical cleavage of mismatch method (CCM). The first two depend on the mobility shifts of single-stranded DNA which has one or more nucleotide base mismatches. The last relies on chemical reactivity if there is a mismatch between nucleotide bases. The above methods do not identify the actual changes in DNA sequence but they provide a comparative means by which differences (e.g. mutations, polymorphisms etc.) can be detected rapidly.

Denaturing gradient gel electrophoresis, as a way to distinguish a segment of DNA containing a single base pair mutation from its corresponding wild-type segment, was first described in 1983 by Fischer and Lerman. In this method, target DNA (genomic, cloned or amplified) is annealed to a radiolabelled single-stranded DNA probe. If target DNA contains a point mutation, heteroduplexes (two strands of DNA in which there is a base-mismatch) between the target and probe will form. DNA is then electrophoresed through an increasing linear denaturant gradient (for example by adding urea, formamide or having a temperature gradient). At a certain denaturant level, a region of the double-stranded DNA will become single-stranded thereby producing a branched structure. This becomes entangled in the gel matrix pores and so retards mobility. The normal mobility will be shown by the migration pattern of the wild-type homoduplex which is run in parallel with the above. Thus, variations in nucleotide bases will be distinguished by different mobilities since the denaturing domain is dependent on the underlying nucleotide sequence (Fig. 2.15).

CHROMOSOME ANALYSIS

Ingredients

Chromosomal analysis requires provision of: (1) karyotype, (2) DNA probe, radiolabelled or conjugated to fluorochromes and (3) cell fusion procedure using human and non-human cells.

Karyotype

A karyotype describes an individual's chromosomal constitution. It was only in 1956 that the human diploid chromosome number was shown to be 46. During the 1970s, methods were developed to distinguish bands within individual chromosomes. Each of the 44 human autosome chromosomes and the X or Y sex chromosomes can now be counted and characterised by banding techniques. The most common is called G-banding and involves trypsin treatment of chromosomes followed by staining with Giemsa. G-banding produces a pattern of light and dark staining bands for each chromosome. The light bands (euchromatin) are more likely to be associated with genes while DNA comprising the darker bands (heterochromatin) is rich in repetitive sequences (Fig. 2.16).

The banding patterns, the size of the chromosome and the position of the centromere enables the accurate identification of each individual chromosome. The number of bands which can be detected per haploid chromosome set with G-banding varies with the technique being used. An average number is 550. Each band is estimated to contain approximately $5–10 \times 10^6$ base pairs of DNA. One suggestion is that there is in the vicinity of 100 genes per band. However, this will vary considerably since some genes are small (e.g. globin genes are approximately 1 kb in size) and others are very large (e.g. Duchenne muscular dystrophy gene is over 2000 kb in size). Some

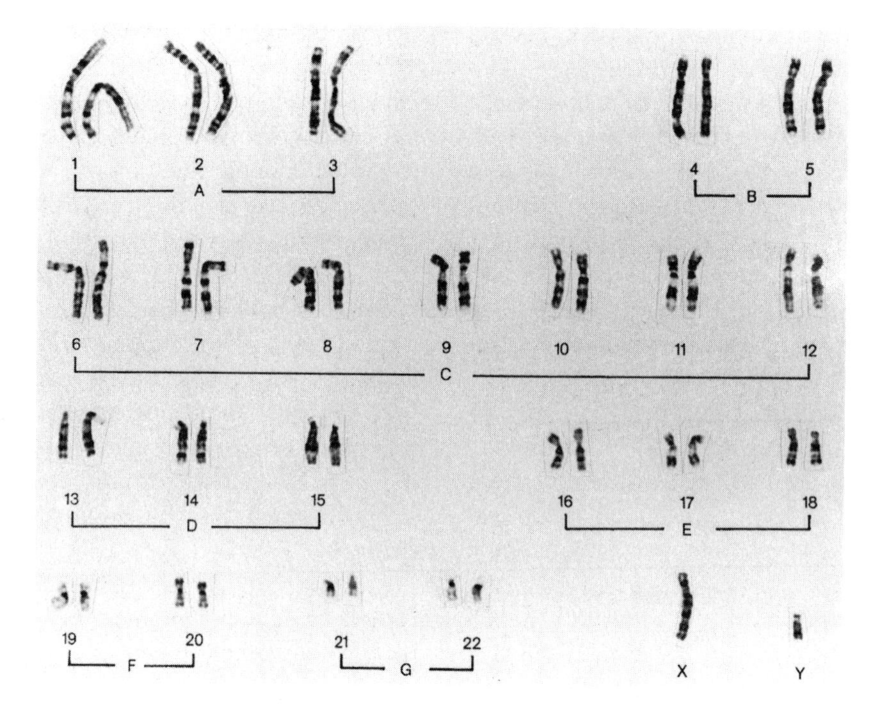

Fig. 2.16 A normal human karyotype (46,XY) which illustrates G-banding.
A normal individual has 44 autosomal (non-sex) and two sex chromosomes. These can be counted and characterised by staining techniques which produce bands. Note the light and dark bands on the chromosomes. (Karyotype provided by Dr A Smith, Department of Medical Genetics, Children's Hospital Camperdown.)

chromosomes appear to be more gene-rich than others.

Cytogenetic nomenclature is derived from the 1985 International System for Human Cytogenetics recommendations. The centromere, a constricted portion of the chromosome where the chromatids are joined, divides the chromosome into a short arm designated as 'p' (for petit) and the long arm or 'q'. Each arm is divided into regions which are marked by specific landmarks. Regions comprise one or more bands. Regions and bands are numbered from the centromere to the telomere (chromosome end) along each arm (Fig. 2.17). Therefore, each band will have four descriptive components. For example, the cystic fibrosis locus on chromosome 7q31 defines a band involving chromosome 7, on the long arm at region 3 and band 1. Additional information is available by higher resolution banding techniques which enable sub-bands to be identified. In the case of the cystic fibrosis locus this becomes

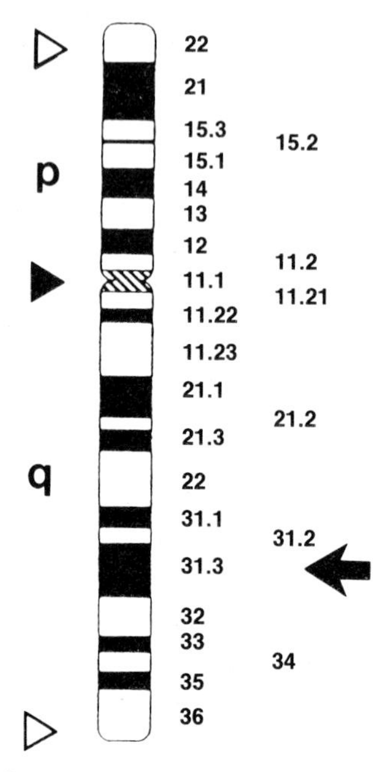

Fig. 2.17 Banding patterns for human chromosome 7.
The individual bands are designated by numbers. The short and long arms are shown by p and q respectively; the centromere by a filled triangle and the telomeres by open triangles. An arrow marks position 31.3.

7q31.3 where the .3 defines the sub-band (Fig. 2.17).

In situ hybridisation

Even greater resolution than was possible by chromosomal banding became available through the development in the late 1970s and early 1980s of in situ hybridisation. Radiolabelled DNA probes were now able to define the chromosomal localisation of single copy DNA sequences in metaphase chromosomes. In situ hybridisation can be used to assign a chromosomal location for DNA probes. In the past few years the emergence of non-isotopic in situ hybridisation, particularly fluorescence in situ hybridisation (abbreviated to FISH), has greatly enhanced the utility of this technique. The potential to use a number of DNA probes each labelled with a different fluorochrome in the same procedure means that separate loci can be identified, comparisons can be made and relationships to the centromere and telomeres established. In the long term, fluorescence in situ hybridisation will have an important role, perhaps more so than pulsed field gel electrophoresis, as the method for mapping and ordering DNA probes along a chromosomal segment. Since chromosomes are more extended in interphase compared with metaphase, the application of fluorescence in situ hybridisation during interphase will resolve even further the location of DNA probes and the characterisation of chromosomal rearrangements, such as microdeletions in genetic or malignant disorders.

Somatic cell hybrids

Human cells (e.g. fibroblasts) can be fused with tumour cells from other animals (e.g. rodents) using an agent such as the sendai virus. After fusion there are both human and non-human chromosomes in the hybrid cells. Subsequently, chromosomes, particularly the human ones, will be gradually lost from the fused cells until after a few generations the cells become more stable. Hybrids can be selected for and their stability improved by using selectable markers so that hybrids with human chromosomes are more likely to survive following culture in special medium. Each of the stable hybrids can then be propagated as an individual cell line. The cell lines will differ from each other in the numbers of human chromosomes which they have retained. Karyotypes of the cell lines will identify which human chromosomes are present. These cell lines can then be used for mapping purposes, e.g. a DNA probe is available but its location in the genome is unknown. This probe can be hybridised to a panel of somatic cell hybrids. Some hybrids will be positive and others negative following hybridisation. From this it will be possible to determine the chromosomal origin for the DNA probe, e.g. somatic cell line No.1 has human chromosomes 1,3,7; line No.2 has human chromosomes 1,5,8,19; line No.3, human chromosomes 1,12,15,18,21 and line No.4, human chromosomes 3,4,6,19. If the DNA probe hybridises to lines 1,2,3 but not 4 it is likely that the probe is located on chromosome 1.

EXPRESSION OF RECOMBINANT DNA

Ingredients

Expression of recombinant DNA requires: (1) cloned DNA, (2) an expression vector and (3) microinjection techniques.

In vitro expression

The most basic expression vector is a plasmid which has an origin of replication, a selectable DNA marker to allow it to be detected, and a multiple cloning site into which the gene to be expressed is inserted. The vector with the inserted gene is then transfected into a host cell such as *Esch. coli* and allowed to replicate. Selection, usually via an antibiotic resistance gene, ensures that only *Esch. coli* with the plasmid insert will remain viable (Fig. 2.18). In the prokaryotic system,

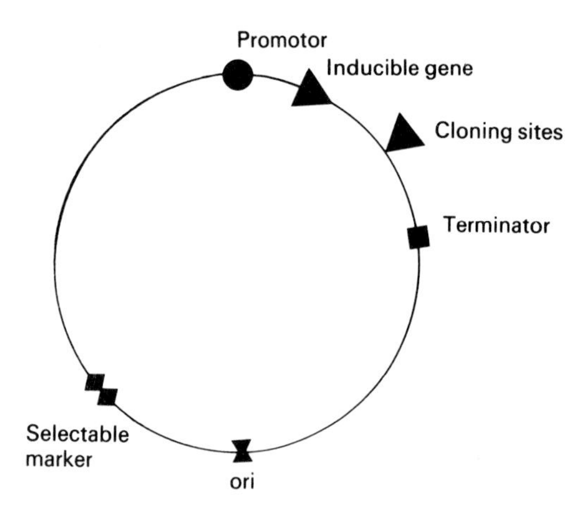

Fig. 2.18 An idealised bacterial expression vector.
The vector has an origin of replication (ori) which allows it to replicate autonomously. There is a selectable marker (usually antibiotic resistance) which gives a growth advantage to the host (e.g. *Esch. coli*) which contains the vector. The bacterial promotor ensures that transcription will occur. An inducible bacterial-specific gene is useful since it allows expression to be controlled. The DNA sequence to be expressed is inserted into the multiple cloning site. Transcription is stopped by placing a termination signal downstream of the multiple cloning site. A fusion product–protein from the inducible gene plus protein from the cloned gene–is expressed which protects the non-bacterial protein from degradation. These two components need to be separated and the expressed protein purified from the bacterial products.

bacterial proteins are expressed but eukaryotic-derived products will be degraded unless they are fused to a bacterial protein. Expressed protein is isolated, purified and able to be used as a therapeutic agent. More sophisticated in vitro expression systems utilise yeast, insect or mammalian-derived vectors. In the non-bacterial expression systems, protein products can undergo post-translational changes. These can be significant for biological activity (see Ch. 7 for a description on how expression vectors are utilised to produce drugs and vaccines).

In vitro expression systems are also used to test gene function particularly that involving promoter regions. For example, the enzyme chloramphenicol acetyl transferase can be included in an expression plasmid. It is then possible to insert eukaryotic promoter sequences 5′ to the chloramphenicol acetyl transferase gene. The activity of these promoters can be tested by noting their effect on the production of the above substance. RNA transcripts can be produced from plasmid vectors which have special promoters located 5′ to their cloning site. The promoters are recognised by DNA-dependent RNA polymerases and so produce RNA rather than DNA. RNA transcripts formed in this way can be used as RNA probes or to identify and quantify mRNA production in vitro.

In vivo expression

Fertilised oocytes contain two pronuclei. Into one of these a gene, in the form of cloned DNA, can

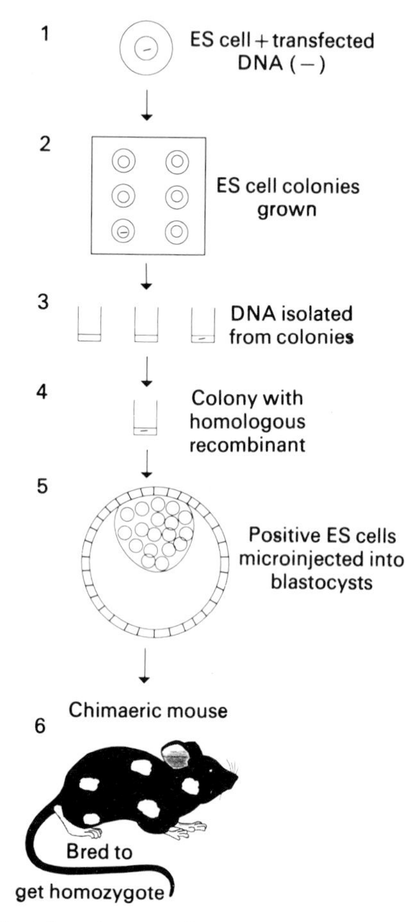

Fig. 2.19 Embryonic stem cells (ES cells) used for in vivo expression of recombinant DNA.
This method produces transgenic mice which can be used to test the function of genes in vivo. 1. ES cells are transfected with foreign DNA. 2. Colonies of ES cells are grown. 3. DNA is isolated from pools of colonies. 4. The colony which has DNA integrated into the correct position in the genome by homologous recombination is identified with the polymerase chain reaction. 5. ES cells which have the homologously recombined DNA are injected into mouse blastocysts. 6. Using different coloured mice as sources of ES cells and blastocysts will enable chimaeras to be distinguished. If the germline is also chimaeric, it will be possible to obtain a homozygote animal by appropriate matings.

be microinjected using a finely drawn pipette. The injected (foreign) DNA becomes randomly integrated as multiple copies in a head to tail tandem arrangement in oocyte DNA. Expression of foreign DNA will occur if there are sufficient copy numbers and the environment into which the DNA has become integrated is suitable. Transgenic animals (animals, usually mice, with the foreign DNA) can be detected by screening their DNA with Southern blotting or through DNA amplification by the polymerase chain reaction. Expression of the transgene can be altered if an inducible promoter is included in the gene construct, e.g. the metallothionein promoter is inducible in the presence of heavy metals which can be added to the animal's drinking water.

Transgenic mice provide very useful in vivo expression systems to test the function or significance of genes, gene sequences or promoters. Disadvantages of the technology include the relatively inefficient production of transgenics (e.g. approximately 20% of microinjections will give a viable transgenic animal); the inability to control where the gene will integrate, and, frequently, low expression of the transgene. To overcome the integration problem, *embryonic stem cells* are becoming increasingly more popular since it is possible to utilise *homologous recombination* to target the gene into its correct position in the chromosome. Embryonic stem cells which have the appropriately integrated gene are identified. These are microinjected into blastocysts to produce chimaeric animals. These represent a model closer to the in vivo situation since inserted DNA is now in its normal location in the genome (Fig. 2.19). Further discussion of embryonic stem cells and homologous recombination is given in Chapters 7 and 10.

FURTHER READING

Chelly J, Concordet J P, Kaplan J C, Kahn A 1989 Illegitimate transcription: transcription of any gene in any cell type. Proceedings of the National Academy of Sciences, USA 86: 2617–2621

Cotton R G H 1992 Detection of mutations in DNA. Current Opinion in Biotechnology, 3: 24–30

Davies K E (ed) 1988 Genome analysis–a practical approach. IRL Press, Oxford

Erlich H A, Gelfand D, Sninsky J J 1991 Recent advances in the polymerase chain reaction. Science 252: 1643–1651

Ferguson-Smith M A 1991 Putting the genetics back into cytogenetics. American Journal of Human Genetics 48: 179–182

Gelehrter T D, Collins F S 1990 Principles of medical genetics. Williams & Wilkins, Baltimore

Mathew C G (ed) 1991 Protocols in human molecular genetics. In: Methods in molecular biology. Humana Press, New Jersey, vol 9

Naylor J A, Green P M, Montandon A J, Rizza C R, Giannelli F 1991 Detection of three novel mutations in two haemophilia A patients by rapid screening of whole essential region of factor VIII gene. Lancet i: 635–639

Old J M, Higgs D R 1983 Gene analysis. In: Weatherall D J (ed) Methods in hematology, vol 6. The thalassemias, Churchill Livingstone, Edinburgh, p 74–102

Schlessinger D 1990 Yeast artificial chromosomes: tools for mapping and analysis of complex genomes. Trends in Genetics 6: 248–258

Vosberg H-P 1989 The polymerase chain reaction: an improved method for the analysis of nucleic acids. Human Genetics 83: 1–15

3

MEDICAL GENETICS

INTRODUCTION

Genetic diseases can be classified into four categories:

1. Single-gene disorders
2. Multifactorial disorders
3. Chromosomal abnormalities
4. Somatic cell disorders, a more recently defined group.

The first two will be discussed in the present chapter and somatic cell disorders in Chapter 6.

The 1992 version of McKusick's 'Mendelian inheritance in man' has over 5500 entries describing genetic loci. Between 2000 and 3000 of those with Mendelian inheritance lead to clinical disorders that are autosomal dominant in more than 70% of cases and autosomal recessive in approximately 20%. Autosomal disease is the result of abnormalities in the 22 pairs of non-sex chromosomes. Disorders that are X-linked occur in less than 10%. The majority of the above involve single genes. Another group, known as the multifactorial disorders, represent traits which occur as a result of interactions between environmental factor(s) and multiple genes. The contribution of genetic disease to ill-health has been assessed in a Canadian survey of 1 million consecutive live-births. This study found that, before they reached the age of 25, approximately 53/1000 live-borns developed some disorder which had an identifiable genetic component (Table 3.1).

Table 3.1 Genetic contributions to ill-health before the age of 25. Results of a Canadian study of 1 million consecutive live-births. The total number affected were approximately 53/1000 live-borns (Baird et al 1988). Autosomal disease is the result of abnormalities affecting the 22 pairs of non-sex chromosomes. In dominant disorders the disease is produced if only one of the two alleles at a particular locus is mutated. In recessive disorders, both alleles must be non-functional for the disorder to occur

Category	Group	Number/1000 with genetic components
Single-gene disorders	Autosomal recessive	1.4
	autosomal dominant	1.7
	X-linked	0.5
Multifactorial disorders	–	46.4
Chromosomal anomalies	–	1.8
Undefined	–	1.2

A number of key words need definition. Different forms of a gene at a locus are called *alleles*. The *haplotype* refers to a set of closely linked DNA markers at one locus which are inherited as a unit. The *genotype* is the genetic (DNA) make-up of an organism. In the present context, genotype would also refer to the genetic constitution of alleles at a specific locus, i.e. the two haplotypes for a particular locus. The *phenotype* reflects the recognisable characteristics determined by the genotype and its interaction with the environment. An individual is *homozygous* if both alleles at a locus are identical and *heterozygous* if the alleles are different. *Autosomal* inheritance involves traits which are encoded for by the 22 pairs of human autosomes. *X-linked* inheritance refers to genes located on the X-chromosome. The products of both normal (wild-type) alleles at a particular locus need to be non-functional in a *recessive* disorder, e.g. cystic fibrosis. On the other hand, a *dominant* disorder results if only one of the two wild-type alleles is mutated, e.g. Huntington disease.

In this chapter, genetic diseases will be described in terms of the above inheritance patterns. Emphasis will be placed on how a greater understanding of their pathogenesis has evolved from the applications of molecular technology.

THALASSAEMIAS – MODELS FOR MOLECULAR GENETICS

Haemoglobinopathies are inherited disorders of haemoglobin. They are classified into the thalassaemias and the variant haemoglobins. An example of the latter is sickle cell haemoglobin.

Table 3.2 Clinical and molecular classification of the thalassaemias.
Thalassaemias are inherited disorders of haemoglobin resulting from insufficient production of normal α or β chains

Classification	Genotypes*	Clinical consequences	Molecular defect(s)
β Thalassaemia	β^T/β^A	Mild anaemia	Point mutations
	β^T/β^T	Severe anaemia	
α Thalassaemia	–α/αα	Nil, not usually detectable	Deletions
	–α/–α – –/αα	Mild anaemia	
	–α/– –	Mild to severe anaemia (HbH disease)	
	– –/– –	Fatal (Hb Bart's hydrops fetalis)	
HPFH	Variable	Nil, not usually detectable	Point mutations or deletions

* Genotypes: ^{T} Thalassaemia genes, ^{A} normal genes. – = deletion. Normal genotype for the β globin cluster is β^A/β^A and for the α globin cluster αα/αα.

The haemoglobinopathies represent one of the commonest single gene disorders. Estimates from the World Health Organization predict a carrier rate of 7% by the year 2000. The high frequency for β thalassaemia and sickle cell heterozygotes reflects the protection that these disorders provide against malaria. Clinical and molecular classifications for the thalassaemias are given in Table 3.2.

Linkage analysis

Before the availability of the polymerase chain reaction, it was difficult to detect individual point mutations in genes if mutations were multiple or heterogeneous in type. For example, there are in excess of 200 different mutations which produce β thalassaemia. Thus, an *indirect* strategy which identified the segregation of DNA *polymorphisms* in members of a family was developed. The β thalassaemias became one of the earliest models to demonstrate the utility of what is called *linkage analysis*.

DNA polymorphisms associated with the β globin gene cluster were first described in 1978. Today, there are many polymorphisms identified giving many variations in the size of DNA fragment produced following digestion with restriction enzymes (Fig. 3.1). The great advantage of a DNA polymorphism is that it provides an *indirect* marker for a particular locus or gene. A minimum requirement for a DNA linkage study is a family unit in which there is a key individual who is completely normal or is homozygous-affected. The key person is essential to allow assignment of the polymorphic markers to the wild-type (normal) or mutant alleles. DNA from each of the relevant family members is digested with the restriction endonuclease which is known to be polymorphic for a particular locus. After Southern blotting, the restriction enzyme digest is hybridised to the relevant DNA probe (Fig. 3.2, see also Ch. 2) for further analysis.

In some cases, DNA polymorphisms are not informative, i.e. family members are homozygous for the polymorphic alleles which means that the normal and mutant alleles are unable to be distinguished. In this situation, additional DNA polymorphisms are sought until informative ones are found. Informativeness in linkage studies has been improved with the use of more variable polymorphisms, for example, VNTRs (**v**ariable **n**umber of **t**andem **r**epeats), minisatellites or microsatellites

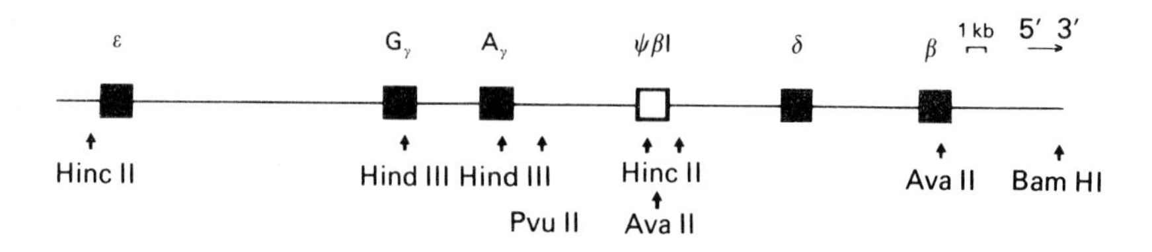

Fig. 3.1 The β globin gene cluster and its associated restriction fragment length polymorphisms (RFLPs).
The ↑ indicate the various RFLP sites and the names of the restriction endonuclease enzymes which digest DNA at these loci to give fragments of varying lengths.

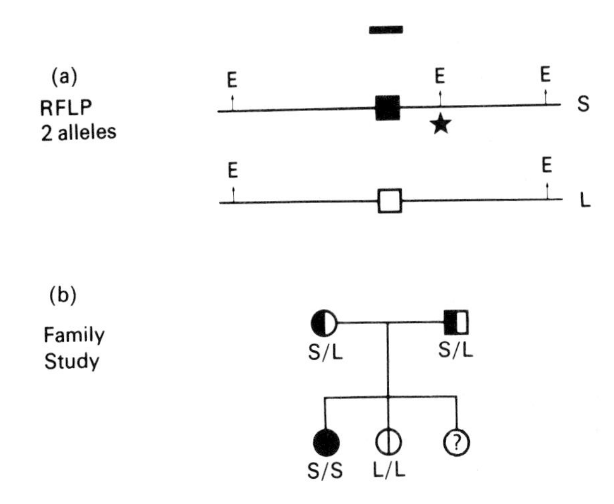

Fig. 3.2 The application of DNA polymorphisms in linkage analysis. A mutation in the DNA causes a change in the restriction enzyme cleavage site resulting in changes in the fragment lengths of DNA produced. Using a key individual who is completely normal or has both alleles affected (homozygous) the polymorphisms can be linked to the disorder and used as markers. (**a**) A polymorphic site (★) where an additional restriction cleavage occurs which can be detected using a DNA probe (solid bar). The two allelic genes are shown by ■ and □. Digestion with a restriction enzyme (E) gives either a small (S) or large (L) band depending on whether the polymorphic site is present (S) or absent (L). (**b**) A pedigree is drawn in which both parents are informative for the polymorphic marker since they are heterozygotes (S/L). An affected offspring is homozygous S/S thereby showing that an autosomal recessive disorder in this family co-segregates with the 'S' marker. This is confirmed by a normal child who is homozygous 'L'. Subsequent offspring can have their genetic status identified on the basis of this polymorphic marker.

(the different types of polymorphisms will be discussed in greater detail in Ch. 8). The number of alleles in the group of multiallelic polymorphisms are greater than the biallelic RFLPs (**r**estriction **f**ragment **l**ength **p**olymorphisms) as illustrated in Figure 2.5 and the examples given in Chapter 8.

Studies based on a polymorphic linkage analysis strategy have a number of disadvantages. These include: (1) the requirement for a family study, (2) non-paternity can lead to an erroneous pattern and (3) recombination is possible. The last is a function of the distance between a polymorphic marker and the gene of interest. Although there are many exceptions, a *physical* distance of 1 Mb (Mb = megabase or 1×10^6 bp) is roughly equivalent to a *genetic* distance of 1 cM (cM = centiMorgan). 1 cM indicates a 1% recombination potential (i.e. in 100 meioses there will be one recombination event between the DNA polymorphism and the target DNA of interest). Thus, genetic distances are measured by family studies which allow the closeness of an association between a DNA polymorphism and the DNA of interest to be assessed.

The three problems described above make DNA polymorphic linkage analysis an unsatisfactory way to detect β thalassaemia mutations since now there are alternative strategies available, e.g. the polymerase chain reaction (see Fig. 2.14). However, DNA polymorphisms and linkage analysis remain a basic component in most positional cloning strategies as will be seen from the examples which follow. DNA polymorphisms are also essential to diagnose disorders for which the gene has not been isolated, e.g. adult polycystic kidney disease.

Gene interactions

γ Globin genes and β thalassaemia or sickle cell disease

Individuals who have the ability to produce an excess of fetal haemoglobin (HbF) (see Ch. 1) will have milder forms of β thalassaemia and sickle cell disease. Homozygotes with a rare thalassaemia called deletional hereditary persistence of fetal haemoglobin have no β globin gene activity (just like β thalassaemia) because their β globin genes are deleted. However, their clinical phenotypes are mild compared to homozygous β thalassaemia. This occurs because there is greater γ globin gene activity and so there are increased amounts of HbF in the blood.

At the molecular level, there is considerable interest in characterising the 'mutants of nature' such as hereditary persistence of fetal haemoglobin. An understanding as to why HbF output is elevated in these rare disorders will be invaluable in planning therapeutic strategies which are based on HbF (the therapeutic implications of HbF are discussed further in Ch. 7).

Gene regulation

As well as showing how genes might interact to

change the severity of genetic diseases, the thalassaemia syndromes have proven to be good models for the study of gene regulation in the eukaryote. The approach followed in the first instance has again centred around the experiments of nature, i.e. human genetic disorders whose changed clinical phenotypes are examined to determine, by molecular techniques, which elements in their genes are responsible for the altered phenotype. From this, the normal regulatory components can be deduced.

The increased output of HbF seen in the deletional hereditary persistence of fetal haemoglobin disorders may occur by a number of mechanisms. The first involves a loss of putative inhibitors which are normally present in the deleted segment. Alternatively, the deletions may enable juxtaposition of the γ globin genes to normally distant 3′ regulatory elements e.g. enhancer sequences, which are then able to increase the γ globin gene's activity. Which of these two mechanisms is the more significant remains to be determined. Other, as yet undefined, mechanisms may also be operational (Fig. 3.3).

Another group of disorders related to the above is called *nondeletional* hereditary persistence of fetal haemoglobin. In this condition, the levels of elevated HbF are lower and the molecular defects are point mutations. It is noteworthy that the point mutations associated with nondeletional hereditary persistence of fetal haemo-

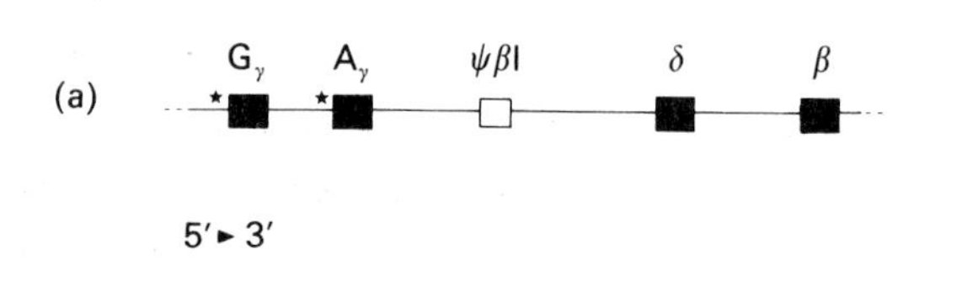

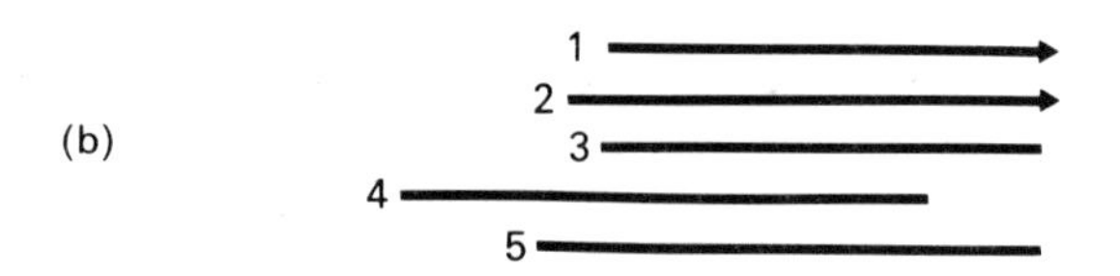

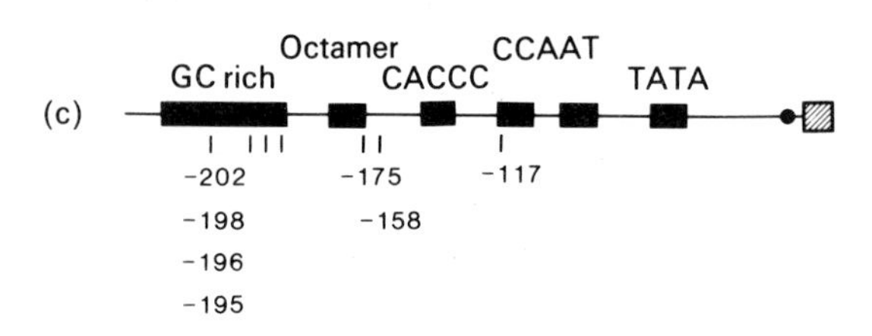

Fig. 3.3 Molecular mechanisms involved in fetal haemoglobin (HbF) regulation.
(**a**) The normal β globin gene complex; ■ = functional genes; □ are pseudogenes; ★ defines the γ globin gene promotor regions. (**b**) Five examples of deletional hereditary persistence of fetal haemoglobin (HPFH) are given: 1 and 2, African black; 3, Italian; 4, Haemoglobin Kenya; 5, Indian HPFH. The arrow indicates that the 3′ end continues for an indeterminate distance. (**c**) The γ globin gene promotor region (G_γ and A_γ in a). Six loci (including a duplicated CCAAT box) which are important in transcription are identified and the positions of the various nondeletional HPFH mutations are shown. The (●) defines the gene's Cap site (initiation site for protein synthesis) (the −202 to −117 mutations are numbered from the Cap which is +1). The start of the gene's translation (the ATG start codon) is indicated by a hatched box.

Box 3.1 Genetic control of fetal haemoglobin synthesis

Different mechanisms have evolved for dealing with the low oxygen environment of embryonic and fetal life. For example, in the primates and ruminates there are separate high oxygen affinity fetal haemoglobins which are active in middle to late uterine life. The α-like and β-like globin gene clusters are arranged in order of their developmental expression with two switches occurring during embryonic and fetal life (Fig. 1.2, Table 1.3). The important switch in the present context involves HbF (fetal haemoglobin) which is produced in utero. At approximately 6 months after birth, HbF is replaced by adult haemoglobin (HbA). This explains why patients with homozygous β thalassaemia are normal when born (since the genes for HbF are normal) but become ill at about the time that HbF is replaced by the defective HbA. Despite a lot of research, the molecular basis for HbF switching remained elusive until transgenic mice containing human globin genes were made. From these animals, preliminary data are beginning to emerge which suggest that switching may involve a number of components including DNA binding proteins which interact with DNA regulatory elements such as the *locus control region* situated on either side of the β globin gene cluster (Box 1.3). In the near future the complete regulatory mechanism involved in HbF switching will be defined.

globin occur within the gene's immediate 5′ flanking region which has an important part to play in the regulation of transcription (Fig. 3.3). The regulatory regions associated with the human Gγ and Aγ globin genes are a GC-rich region, an octamer sequence, the CACCC, CCAAT and TATA boxes. It has been shown that these DNA motifs are important for binding of proteins to DNA. Various binding proteins have been isolated. These can be ubiquitous (found in all cells) or specific to the red blood cells.

Further work is required to define the molecular mechanisms involved in the hereditary persistence of fetal haemoglobin disorders. The complex interactions possible between DNA/DNA, DNA/proteins and proteins/proteins will enable the molecular controls of eukaryote genes to be understood in greater depth. Data are also starting to emerge from studies of the physiological HbF switch mentioned earlier in Chapter 1 to illustrate potential ways in which proteins can modulate the expression of genes (Box 3.1). As well as explaining the pathogenesis of some genetic diseases, knowledge from the HbF models will identify potential targets for future therapeutic strategies.

AUTOSOMAL RECESSIVE DISORDERS

CLINICAL FEATURES

The appearance of an autosomal recessive disorder in a pedigree gives rise to a *horizontal* rather than vertical pattern. This occurs because affected individuals tend to be limited to a single sibship and the disease is not usually found in multiple generations (Fig. 3.4). Males and females are affected with equal probability. In specific populations with rare autosomal recessive traits, *consanguinity* can be demonstrated in affected families. The usual mating pattern which leads to an autosomal recessive disorder involves two heterozygous individuals who are clinically normal. From this union, there is a one in four (25%) chance that each offspring will be homozygous-normal or homozygous-affected for that trait or mutation. There is a two in four (50%) chance that offspring will themselves be carriers (heterozygotes) for the trait or mutation. Two thirds (66%) of the *normal* children will be heterozygous carriers like their parents.

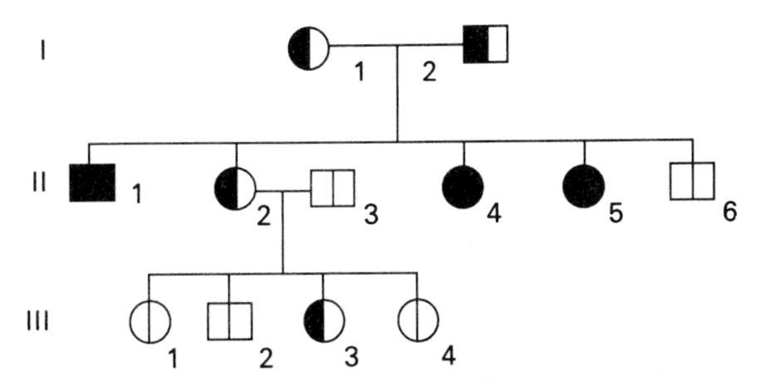

Fig. 3.4 Idealised pedigree for an autosomal recessive disorder. Females are represented by ○ ; males by □. Carriers of the genetic trait are indicated as half-filled circles or squares. Affected as ● or ■. Note the horizontal distribution for the affected (see Fig. 3.9).

The inheritance patterns described above are not always present, particularly in communities where the number of offspring are few. In these instances, the genetic trait or mutation may appear to be *sporadic* in occurrence. Therefore, the finding of a negative family history in the autosomal recessive disorders should not be ignored since the genetic trait or mutation can still be transmitted to the next generation, particularly if the mutant gene occurs at a high frequency in the population e.g. cystic fibrosis.

CYSTIC FIBROSIS

Cystic fibrosis is the most common autosomal recessive disorder in Caucasians (Box 3.2). It affects approximately one in 2000–2500 live births with a carrier rate in northern Europeans of one in 20 to one in 25. The high incidence of cystic fibrosis remains unexplained. Several hypotheses have been proposed, but as yet they remain unproven. The hypotheses include: an increased mutation rate, genetic drift, multiple loci, repro-

Box 3.2 Cystic fibrosis: an autosomal recessive disorder

Cystic fibrosis is a multisystem disorder of children and adults characterised by an abnormality in exocrine gland function which manifests as chronic respiratory tract infections and malabsorption. The first comprehensive description of cystic fibrosis was given in 1938. Approximately 1 in 2000 newborns of northern European descent are affected although considerable differences exist within ethnic groups, e.g. the incidence in oriental Hawaiians is 1 in 90000. Complications associated with cystic fibrosis include: (1) respiratory – chronic bacterial infections which eventually lead to respiratory and cardiac failure; (2) gastrointestinal tract – 10% of newborn infants with cystic fibrosis present with obstruction of the ileum (meconium ileus). More than 85% of children show evidence of malabsorption due to exocrine pancreatic insufficiency which requires dietary regulation and supplementation with vitamins and pancreatic enzymes. Rare gastrointestinal tract complications include biliary cirrhosis. (3) Infertility which affects 85% of males and also females although to a lesser extent. Long-term outlook for cystic fibrosis has improved dramatically over the years although the median survival is still only 25 years. The availability of specialised cystic fibrosis clinics has enabled a multidisciplinary approach to follow-up and treatment of this disorder. Support groups, for example the various cystic fibrosis associations, have ensured that families are aware of recent developments which have followed from the cloning of the cystic fibrosis gene.

ductive compensation and selective advantage of the heterozygote carrier.

After asthma, cystic fibrosis is the commonest cause of chronic respiratory distress in childhood and is responsible for the majority of deaths from respiratory disease in this age group. Clinical features of the disease are related to the thick tenacious secretions which can manifest by intestinal obstruction in the newborn (called meconium ileus), pancreatic insufficiency and chronic respiratory infections in childhood. Despite intensive treatment for these, affected individuals, until recently, were unlikely to survive beyond their 3rd decade.

Although first described as a clinical entity in the 1930s, the pathogenesis remained elusive despite clinical, electrophysiological and other conventional approaches to study this disorder. The only clue to the underlying defect in cystic fibrosis related to the elevated chloride in sweat. In the mid-1980s, chloride ion conductance across the apical membranes of respiratory epithelial cells or sweat ducts was shown to be decreased. Whether this defect represented an abnormal chloride channel or aberrant control of a normal channel was unknown. This remained the state of knowledge until 1989 when the cystic fibrosis gene was isolated by the recombinant DNA strategy of *positional cloning*.

Positional cloning to identify the cystic fibrosis gene

Linkage analysis

The first step in positional cloning is *chromosomal localisation* of the mutant gene. The clue for this often comes from cytogenetic studies which identify a chromosomal rearrangement such as a deletion or translocation associated with the clinical phenotype. For example, the isolation by positional cloning of the Duchenne muscular dystrophy gene and the neurofibromatosis 1 gene was made easier by finding a deletion of Xp21.2 (Duchenne muscular dystrophy) and a number of balanced translocations involving chromosome 17q11.2 (the neurofibromatosis 1 locus).

Initial attempts at chromosome localisation in cystic fibrosis were unsuccessful. This delayed isolation of the gene since a blind 'trial and error' approach was required to determine which DNA polymorphic markers would co-segregate with the cystic fibrosis defect within pedigrees. In contrast to the relatively simple linkage analysis steps described for the β thalassaemias, the procedures involved in demonstrating co-segregation between an *unknown* locus and DNA polymorphisms are more complex. Larger pedigrees and an increased number of DNA markers are now essential. Analyses of the linkage data require sophisticated computer programs.

In 1985, linkage of cystic fibrosis to DNA markers on the long arm of chromosome 7 at band 31 was shown. Multiple markers were tested until it was possible to narrow the region containing the gene to approximately 1.5 Mb (1.5×10^6 base pairs). The two closest markers were designated KM-19 and XV-2c. These were used in 1st trimester prenatal diagnoses for cystic fibrosis although there was an inbuilt error rate (less than 1%) through recombination between these markers and the actual cystic fibrosis gene. A second disadvantage of the linkage approach was the requirement for an individual *affected* with cystic fibrosis. This person was essential to allow assignment of polymorphisms to the normal or mutant alleles. Without such a person, the family linkage study was not possible. Thus, couples having their first child and who were at increased risk for cystic fibrosis were unable to have prenatal diagnosis in this way. Even if the affected person were available, not all families would turn out to be informative for the DNA polymorphisms (Fig. 3.5).

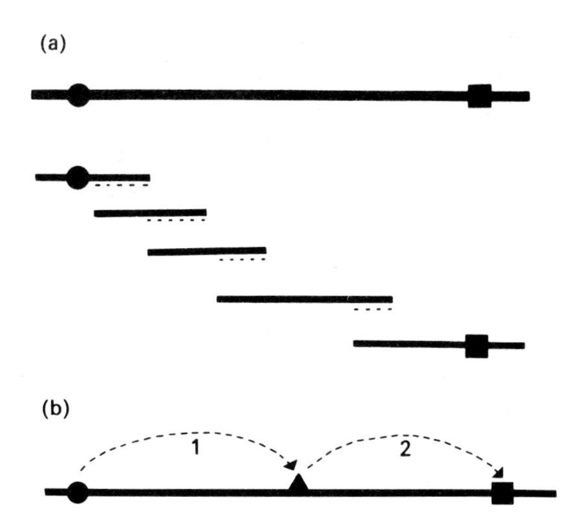

Fig. 3.6 Chromosome walking and chromosome jumping.
Identifying a gene within a targeted region of DNA can be achieved by mapping overlapping clones from the region so that DNA markers can be ordered in both their orientation and distance from the gene locus. The closer a marker is to the gene of interest the fewer recombinations occur. (**a**) Illustrates a number of contigs or overlapping sections of DNA, which cover a region between the starting DNA (●) and the target sequence (■) in a chromosome walk. - - - - indicates the segment which can be used as a DNA probe to isolate overlapping clones. (**b**) In comparison to walking, there are only two chromosome jumps required to reach the target DNA. A DNA segment (▲) between the start and the target allows a second jump to occur. Intermediate DNA can be bypassed by jumping from one segment to another.

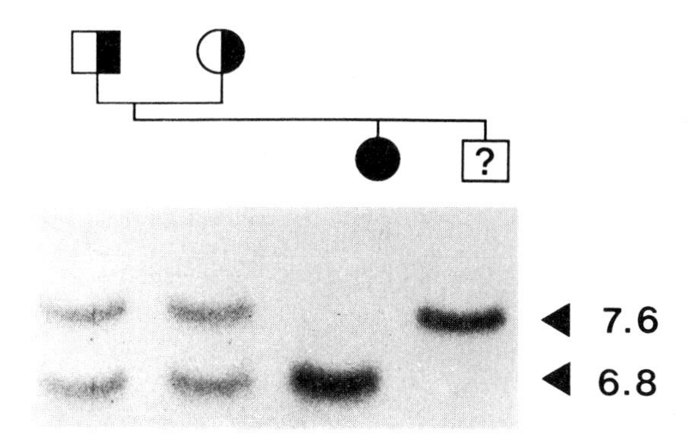

Fig. 3.5 DNA polymorphic patterns associated with the cystic fibrosis locus.
The probe is KM-19 and the restriction enzyme is *PstI* which gives RFLP polymorphic fragments of 7.6 kb or 6.8 kb. The parents who carry the cystic fibrosis defect are heterozygous for the polymorphism since they have both fragments and so are informative. Their affected child is homozygous for the 6.8 kb marker which indicates that within this family the cystic fibrosis defect co-segregates with the 6.8 kb restriction enzyme fragment. The second offspring has an unknown carrier status (?). However, DNA polymorphisms indicate that this individual is homozygous-normal (with approximately a 1% error rate to allow for recombination).

Chromosome walking

To identify the actual cystic fibrosis gene required *chromosome walking*. Cloned segments of DNA can be ordered by using the overlapping sections as probes to identify the adjacent segment. In other words, overlapping clones for the 1.5 Mb cystic fibrosis region needed to be ordered so that DNA markers and their orientation as well as distance from the putative cystic fibrosis locus could be determined (Fig. 3.6). As indicated earlier, distance in genetic terms is measured by the number of meiotic recombinations (breakage and rejoining) which occur between a DNA marker and the clinical phenotype. The fewer recombination events, the closer the marker is to the gene of interest. Once the correct orientation is determined (by showing that the genetic distance for each marker is progressively reduced) chromosomal walking allows unidirectional progress until the gene of interest is reached. Two disadvan-

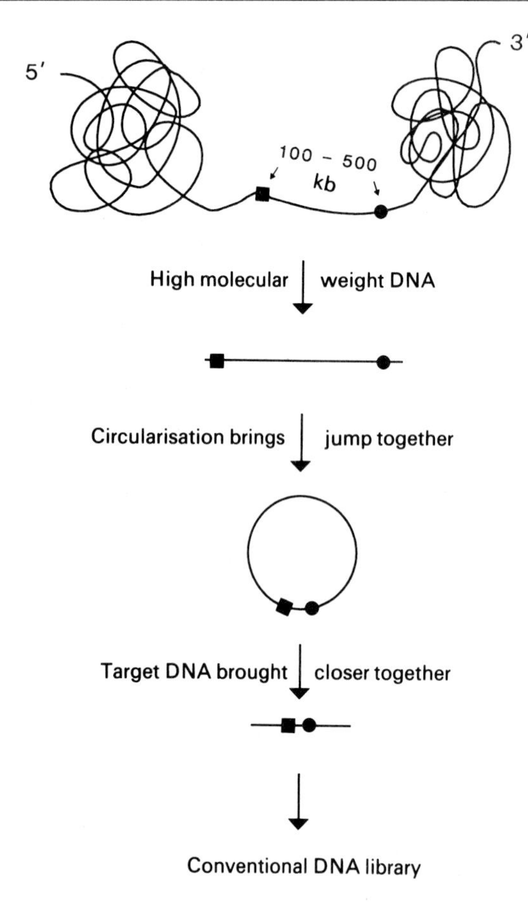

Fig. 3.7 Construction of a DNA jumping library.
Two DNA sequences in total genomic DNA are depicted (■ and ●) which are about 100–500 kb apart. To 'walk' from one to the other would be difficult. The alternative is to 'jump' from one to the other by first isolating large segments from total genomic DNA. One of these will contain the two sequences within the one segment. All the DNA segments are circularised. This enables the sequences of interest to come into close proximity. A much smaller segment of DNA now contains the two sequences. A conventional library is prepared (see Fig. 2.7). This library will contain many jumping fragments. The one of interest will be identified by screening the library. If the ■ sequence is used as a probe it will be possible to identify within the library a segment which contains both ■ and ●. In this way, a jump has been obtained along a large chromosomal region.

tages of chromosome walking include: (1) it is a slow, tedious process since only short distances can be transversed at any one time and (2) the occurrence of regions in the genome which cannot be cloned and thus will interrupt the chromosome walk. *Chromosome jumping*, an alternative and more efficient strategy to chromosome walking was developed and proved to be very successful in cystic fibrosis. In this technique a DNA segment between the start and the target is used to link the two. Intermediate DNA does not have to be characterised (Fig. 3.6).

Chromosome jumping

The ways to go about chromosome jumping are illustrated in Figures 3.6 and 3.7. The advantage of this strategy is that linked probes, which can be located large distances apart, e.g. 100–500 kb, are generated. Probes produced in this way allowed the extensive 1.5 Mb region of the cystic fibrosis locus to be analysed more rapidly than would otherwise have been possible by chromosome walking. Unclonable regions were bypassed. The probes generated from jumping libraries were used to screen conventional phage or cosmid libraries for *candidate* genes, features of which are summarised in Table 3.3.

Table 3.3 Features which would indicate that a segment of DNA contains a candidate gene.
Probes generated from jumping libraries can be used to screen conventional libraries for candidate genes

Molecular findings	Significance
Deletions or gene rearrangements	A consistent finding in affected individuals but not in normal controls would suggest a possible functional role
Cross-hybridisation to primate, rodent and DNA from other species	Evolutionary conservation of DNA sequences would suggest functional significance (procedure is called a zoo blot)
Identification of CpG islands	Regions of DNA which are rich in hypomethylated cytosine followed by the base guanine are frequently found 5′ to vertebrate genes
Identification of open reading frames	Computer programs enable DNA sequences to be scanned for stop codons. If these are not present, the DNA sequence has the potential to encode for a gene
Identification of mRNA transcripts	Putative candidate genes identified through chromosome walking or jumping are more likely to be significant if a corresponding mRNA sequence can be isolated by Northern blotting and/or from a cDNA library. The DNA probe in this instance would be derived from the candidate gene

Finding the right gene

The search for the cystic fibrosis gene eventually narrowed to a region of DNA approximately 0.5 Mb in size. This area contained a handful of genes. Clues which suggested that one of these was the cystic fibrosis gene included: (1) there was conservation of DNA sequence across a number of species, i.e. the gene has an important function. (2) Northern blotting showed that mRNA from this gene was present in tissues connected with cystic fibrosis, i.e. lung, pancreas, intestine, liver and sweat glands.

Once cloned, the putative cystic fibrosis gene was demonstrated to code for a protein of 1480 amino acids. The gene was large, comprising 27 exons located over 250 kb of DNA. The mRNA transcript was 6.5 kb in size. Final proof of the gene's identity came with the demonstration of mutation(s) which correlated with the cystic fibrosis phenotype. The first mutation to be found involved a 3 bp deletion in exon 10. This resulted in loss of the amino acid phenylalanine at residue 508 (the mutation is called ΔF508 mutation). This was a causative mutation rather than a neutral polymorphism since it was consistently found in cystic fibrosis patients but not in the normal population.

When the cystic fibrosis gene was being sought by positional cloning, it was predicted that the number of underlying defects would be few. This has now been proven incorrect with more than 200 different mutations being described. The commonest is the ΔF508 mutation which is present in over 70% of cystic fibrosis chromosomes from northern Europeans. An additional 20 or so mutations occur at a frequency of only 1–5% in most Caucasian populations. The remainder occur sporadically.

After the gene is cloned

Decoding

An increasingly common consequence of molecular strategies such as positional cloning will be the identification of genes or gene-like sequences for which a function needs to be found. This has been called 'decoding' by Collins. The function of a gene can be decoded in a number of ways (Box 3.3). The 168 kDa protein encoded for by the cystic fibrosis gene (the gene is also called CFTR or cystic fibrosis transmembrane conductance regulator) was shown to have considerable similarity to a family of membrane associated ATP-dependent transporter proteins which are found conserved throughout evolution since they are present in a wide range of species including bacteria, *Drosophila* and mammals. These proteins are involved in the active transport of substances such as ions and small proteins across membranes. Common structural findings in the above transporter proteins include one or two hydrophobic transmembrane domains and one or two nucleotide-binding folds for ATP attachment from which the energy for transport is

Box 3.3 Determining the function of a gene

The function of a gene or DNA segment can be *decoded* in a number of ways. (1) An expression vector containing the gene of interest is constructed and from this a recombinant protein produced. Antibodies are raised against this protein. By in situ hybridisation it is possible to determine which tissues produce the protein. At the cellular level, the location and distribution of the protein can provide a clue to its function. An example of how this helped decoding is seen with Duchenne muscular dystrophy which is discussed in the section on X-linked disorders. (2) A computer search is undertaken of DNA and protein databases to look for homology, i.e. similarity. There are three outcomes from this search. The gene's function is determined since the database lists another substance which has been described previously and to which the DNA or protein in question has 100% homology. Alternatively, database searches produce a complete blank. Considerable work will now be required to identify the gene's function. The final option results in the finding of partial homology with one or more substances in the databases. This provides some clues from which function can be sought. This occurred with cystic fibrosis.

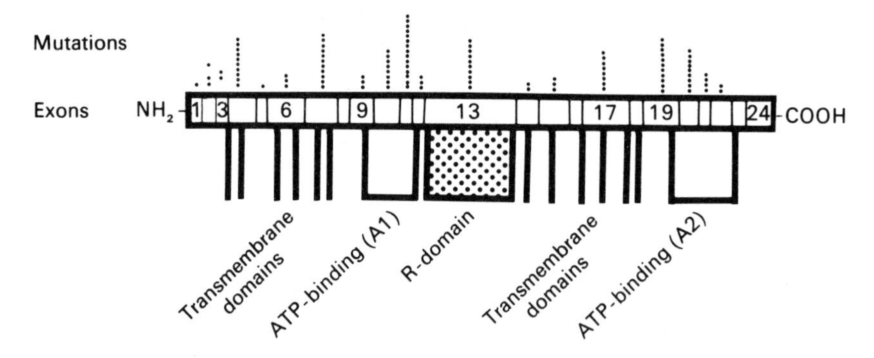

Fig. 3.8 The cystic fibrosis gene.
The diagram shows its exons and above them the location of mutations (the number and position are indicated by ●). There are 24 exons illustrated but the total number is 27 since introns were recently detected dividing exon 6 (so that there are now exons 6a and 6b), exon 14 (exon 14a and exon 14b) and exon 17 (exon 17a and exon 17b). An indication of the areas of the putative cystic fibrosis protein sub-structures thought to be coded for by the exons is given (Tsui et al 1992).

obtained. CFTR has the above features and as well there is a highly charged central R domain (R for regulatory) which is considered to be involved in phosphorylation by protein kinase. The various components which make up the cystic fibrosis protein are depicted in Figure 3.8. There is now good experimental evidence that CFTR codes for a *chloride ion channel*. Activation of this channel can occur by cyclic AMP or following phosphorylation by protein kinase. The latter involves the R domain and may work through a conformational change which then allows the passive flow of chloride ions. Whether CFTR has other functions remains to be determined.

Diagnosis of cystic fibrosis

Diagnosis of an individual affected by cystic fibrosis is possible by sweat testing, i.e. local sweating is induced and the chloride ion content measured. Those with cystic fibrosis have an elevated chloride level. Disadvantages of the sweat test are its inability to detect heterozygotes (carriers) and it is not suitable for prenatal diagnosis. Once DNA polymorphisms became available it was possible to utilise a linkage study approach to undertake prenatal diagnosis and carrier testing within a family unit.

Since 1989, identification of the ΔF508 mutation by DNA amplification has become the method of choice for laboratory detection. Diagnostic applications include: prenatal diagnosis, carrier testing and helping to distinguish those disorders which resemble cystic fibrosis but have atypical features (Box 3.4).

As indicated previously, the ΔF508 mutation affects approximately 70% of the cystic fibrosis chromosomes in northern Europeans with a lower frequency in other groups, e.g. 50% in southern Europeans. The multiplicity of mutations makes detection of all cystic fibrosis defects an unrealistic proposal with present technology. Therefore, DNA-based diagnostic tests incorporate ΔF508 and a limited number of other mutations (e.g. six) which are selected on the basis of their prevalence in each population. This enables approximately 80–90% of the cystic fibrosis mutations to be detected. Tissues which have been used with the polymerase chain reaction include blood, hair follicles, chorionic villus, blood spots such as those taken from neonatal heel pricks (Guthrie spots), cells shed in amniotic fluid (amniocytes) and buccal cells which can be obtained from mouth washes.

The potential of the polymerase chain reaction to be automated and so screen for the ΔF508 defect on a widespread basis has produced a controversy, i.e. whether there should be population-based (random) cystic fibrosis screening. The protagonists point out the importance that this knowledge would have on future reproductive decisions. Those against random population testing indicate that benefits and risks of population

Box 3.4 Diagnostic applications of DNA testing

Two cases involving a cystic fibrosis-related problem are described to illustrate the applications of DNA testing. In the first, the consultand (II_3) requested that her carrier status be determined since she was planning to become pregnant. The consultand has a cousin with cystic fibrosis and a brother (II_2) who is being treated for pancreatic malabsorption. Sweat tests in the latter are equivocal. After DNA testing, the cousin with cystic fibrosis is shown to be homozygous for the ΔF508 mutation. This mutation is not present in the consultand's mother, her spouse and brother. Two facts emerge from the DNA results: (i) the risk for cystic fibrosis in future offspring of the consultand and her spouse is very low and (ii) cystic fibrosis is an unlikely explanation for pancreatic malabsorption in the consultand's brother. N = normal.

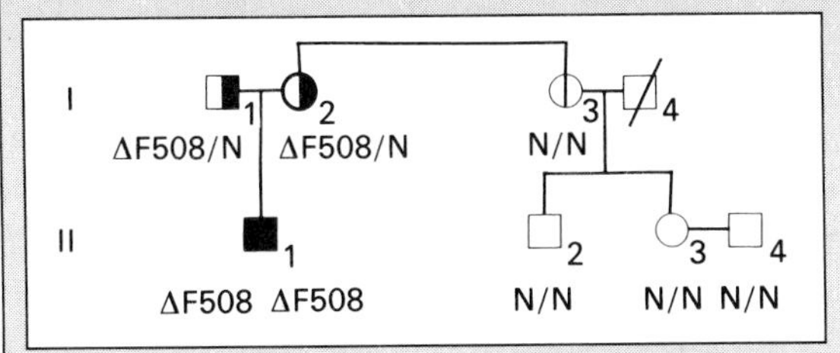

The second case involves a pregnant woman who is an obligatory carrier of the cystic fibrosis defect since she has an affected child (→). The story is more complicated in that it is not known which of two men is the father in the present pregnancy and the mother requests that neither is tested. DNA studies of the affected child and her mother show that the former is a double-heterozygote for the ΔF508 and the G551D mutations. The mother has the less frequently found G551D defect. Cystic fibrosis is excluded in the fetus since DNA from a chorion villus sample does not have the G551D defect. In this circumstance the genetic status of the potential fathers is irrelevant since cystic fibrosis is excluded in the fetus. N = normal.

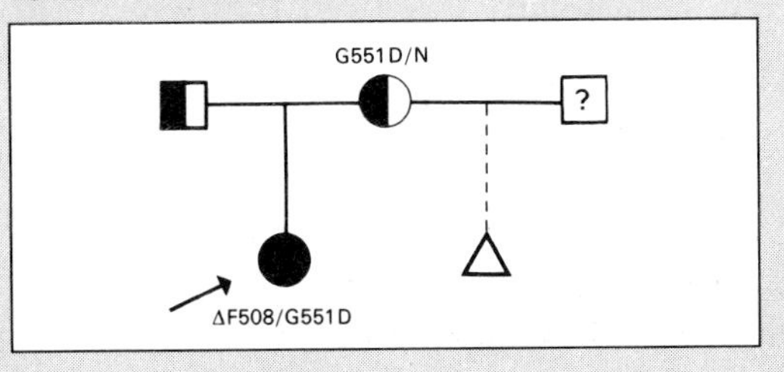

screening are uncertain unless 90–95% of carriers can be detected which is presently not an economical or realistic goal in cystic fibrosis (see Ch. 9 for a further discussion of screening).

Future directions

Considerable effort has gone into the correlation of genotypes (DNA defects) with phenotypes (clinical features) since there is both genetic and clinical heterogeneity in cystic fibrosis. For example, normal pancreatic function is often found in mild forms of the disorder. Molecular defects which are associated with pancreatic sufficiency can now be identified. In these circumstances, prognosis for cystic fibrosis is considerably improved and mutations usually result in missense codon changes. The molecular defects in those with pancreatic insufficiency (i.e. the cystic fibrosis is of the severe type) are more likely to involve the ΔF508 deletion which is located within the first ATP binding site (Fig. 3.8). Similarly, meconium ileus is frequently found in the newborn with the ΔF508 deletion. Other mutations associated with severe phenotypes produce premature stop codons, frameshifts or splicing defects. However, the story is not that simple since exceptions are found.

Research into the molecular defects, other genetic components such as immune responsiveness and the elucidation of interacting environmental factors will enable a more accurate assessment of prognosis in children with cystic fibrosis. Both in vitro and in vivo studies are under way to characterise further the role played by the CFTR and the consequences of mutations in this gene. A recent important development has been the production of a mouse model of cystic fibrosis using the strategy of homologous recombination where an inserted gene is targeted to its correct position (see Chs 2, 10). Knowledge gained from the above studies will help to rationalise the therapeutic regimens and reduce the fear associated with cystic fibrosis. A form of gene therapy to treat this disorder is presently being considered (discussed further in Ch. 7).

AUTOSOMAL DOMINANT DISORDERS

CLINICAL FEATURES

The characteristic feature in a pedigree with autosomal dominant inheritance is a *vertical* mode of transmission. This appearance comes from the fact that the disorder is apparent in every generation of the pedigree. Both males and females are affected. Offspring of an affected are at 50% risk (Fig. 3.9). There are a number of additional features which need to be considered when dealing with autosomal dominant disorders. The following become important in counselling.

Sporadic cases occur and these become increasingly more common as the mutation in question interferes with fertility. This may represent a secondary effect of the disorder or because death results before reproductive age is reached. For example, it is estimated that in achondroplasia, an autosomal dominant form of dwarfism, 50% of cases are spontaneous mutations since the disorder indirectly reduces the individual's reproductive capacity in terms of finding a partner. In contrast, Huntington disease per se does not affect reproduction. Thus, the finding that sporadic cases of Huntington disease are rare is not surprising.

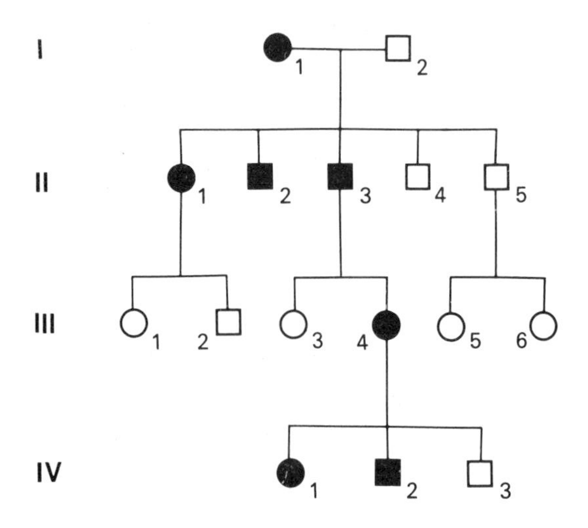

Fig. 3.9 Idealised pedigree for an autosomal dominant disorder. Affected individuals are indicated by ■ and ●. Note the vertical disease pattern (cf. Fig. 3.4) with the disease apparent in every generation and affecting males and females.

Penetrance is an all-or-nothing phenomenon that describes the clinical expression of a mutant gene in terms of its presence or absence. Thus, an individual carrying a mutant gene may not express the clinical phenotype, i.e. the condition is non-penetrant. From family studies it is possible to determine the number of obligatory heterozygotes for a mutant allele. If seven out of ten heterozygotes show the clinical phenotype, the disorder is described as being 70% penetrant. That is, there is a 70% probability that an individual carrying a mutant gene at a certain age will have the clinical phenotype. This aspect of autosomal dominant disorders is discussed further under Huntington disease. Apart from spontaneous mutations and death before onset of symptoms, penetrance is an additional mechanism which would account for affected offspring having an apparently normal parent.

Expressivity refers to the severity of the phenotype. There are genes which can produce apparently unrelated effects on the phenotype or act through involvement of multiple organ systems. This is called *pleiotropy*. Such genes often show variable expressivity. An example of this is Marfan syndrome which has autosomal dominant inheritance and involves connective tissues in the skeletal system, the eye or the heart. Individuals with Marfan syndrome have any combination of manifestations which can also be present in different degrees of severity. Such variability can occur even within families in which it is presumed the same mutant allele is present. To date, the underlying basis for expressivity has not been defined but it is thought to represent either gene/environment or gene/gene interactions. A recent molecular mechanism producing somatic instability (variation in the gene in different cell lines in an individual) may provide another explanation for expressivity (see Fragile X syndrome below).

Severity of an autosomal disorder may also be influenced by the sex of the transmitting parent or the sex of the affected person. The former will be discussed in this chapter under 'Imprinting' (p. 66). An example of the latter is otosclerosis (a cause of deafness in adults due to overgrowth of bone in the ear) in which the female to male ratio is approximately 1.8 to 1. The reason for this is unknown. One hypothesis suggests that affected males may have a selective disadvantage compared to females and this selection is having its effect prenatally.

HUNTINGTON DISEASE

Huntington disease is a neurodegenerative disorder with autosomal dominant inheritance. It usually manifests with progressive movement disorder (typically chorea), psychological disturbance and dementia. Onset of symptoms is typically in the mid-thirties and there is complete penetrance by the age of 80. Because of the relatively late appearance of Huntington disease, reproductive and many other life decisions have been made before knowledge of genetic status is known. Offspring of affected individuals have a 50% risk of inheriting the disease.

As a model for molecular genetics, Huntington disease contrasts with cystic fibrosis in a number of ways. First, conventional investigative approaches to obtain some understanding of the pathogenesis of Huntington disease were uniformly unsuccessful. Therefore, presymptomatic diagnosis (diagnosis prior to the onset of clinical manifestations) was unavailable. In the early 1980s, the only hope for an advance in our knowledge of Huntington disease rested with recombinant DNA techniques. A second feature to emerge from the Huntington disease example is the difficulties which can occur when positional cloning is used even when the best DNA resources are available.

Positional cloning to identify the Huntington disease gene

An important consideration in the decision to utilise positional cloning to study Huntington disease was the availability of a number of very large pedigrees. These would be very useful for linkage analysis. Since cytogenetic data had not indicated the likely chromosomal location for Huntington disease, a trial and error approach was required to look for DNA polymorphisms linked to the disease phenotype. In 1983, a DNA marker name G8 (locus D4S10) located on the short arm of chromosome 4 at position 4p16.3 was found to co-segregate with Huntington disease. Subsequently, G8 was estimated to have

Box 3.5 Familial hypertrophic cardiomyopathy: investigation using candidate genes

Familial hypertrophic cardiomyopathy (FHC) is an autosomal dominant disorder affecting heart muscle. It can lead to sudden death in adolescents and young adults. It is a common autopsy finding in young athletes who die suddenly. Clinical examination and tests such as echocardiography are unable to detect all affected individuals particularly the young. Cytogenetic studies had not identified a chromosomal location for FHC. Thus, positional cloning to find the FHC gene would need to be undertaken blind in a similar way to that required for Huntington disease. However, a short-cut existed since potential *candidate genes* existed for FHC even before any DNA studies were undertaken. These were genes that encoded for muscle proteins particularly those localised to the heart. Examples of such genes included actin and myosin. Using DNA probes associated with these candidate genes, linkage studies were undertaken. In 1990, linkage was established between the β isoform of the cardiac heavy chain myosin gene on chromosome 14 and FHC. Since then it has been shown that a number of individuals with FHC have point mutations involving the above gene. However, this is not the entire story since there are families who do not show linkage to the chromosome 14 locus or where mutations cannot be detected in the β myosin heavy chain gene. Therefore, like the adult polycystic kidney disease example, FHC is heterogeneous at the DNA level and is likely to involve other genes. These will need to be sought by positional cloning using linkage analysis and the more difficult strategy of chromosome walking.

a recombination potential of approximately 4%, i.e. the marker was situated approximately 4 Mb from the actual gene.

Since 1983, chromosomal walking has been utilised to look for the actual disease gene. Initial linkage data suggested that the Huntington disease locus was located between G8 and the telomere for the short arm of chromosome 4. By 1991, this telomeric region had been fully characterised. However, few genes were found and there were no DNA rearrangements detectable in association with the disease phenotype. Reassessment of linkage data was then undertaken and it became evident that the earlier interpretations were incorrect and the Huntington disease gene was more likely to be situated closer to the G8 marker than the telomere in a region approximately 2.5 Mb in size. One factor which complicated isolation of the Huntington disease gene was the presence of a 'hot spot' (where mutations occur at unusually high frequencies) for recombination near the D4S10 locus. Thus, genetic distances were more difficult to calculate and in effect this led to chromosome walking in the wrong direction! Intensive study of the 2.5 Mb locus for candidate genes (there are likely to be 50 or more for this amount of DNA) was undertaken and this led to isolation of the putative Huntington disease gene in 1993. The difficulties in isolating the Huntington disease gene can be contrasted to another autosomal dominant disorder called familial hypertrophic cardiomyopathy. Here positional cloning was simplified by the availability of candidate genes (Box 3.5).

Box 3.6 Predictive testing for Huntington disease

The following pedigree shows three family members who have Huntington disease. One is deceased and two are alive. The proband is indicated by →. In the second generation there is also one elderly normal relative who is aged 80 (II_2) and a spouse (II_4). The consultand is III_2 and she has requested predictive testing for Huntington disease (her a priori risk is 50%). Her two siblings have donated blood to help establish the inheritance patterns for the three DNA polymorphisms used in the study. This is called determining the 'phase' of the polymorphisms. Alleles for polymorphisms No.1 are 14 or 9 kb; alleles for polymorphisms No.2 and No.3 are designated 'a' or 'b'. Sibling III_3 is particularly useful in identifying phase because she is homozygous for the three polymorphisms, i.e. she has only the 14 kb band for marker No.1; the 'b' allele for marker No.2 and the 'a' allele for marker No.3. Therefore, her two haplotypes can only be 14-b-a and 14-b-a (a haplotype is defined as the inheritance of two or more linked alleles at the one locus). It follows from this that the 14-b-a haplotype must have been inherited from each of this individual's parents. With this information it is possible to construct haplotypes for the consultand, her parents and perhaps more distant family members. In this family, the Huntington disease gene co-segregates with haplotype 9-b-b. This is determined by identifying which haplotype is shared by the two affected individuals. It is also helpful that the elderly sister of an affected individual is normal and she does not have the 9-b-b pattern. Therefore, the consultand's risk can be reduced to approximately 5% (the small error reflects the risk from recombination) because she does not have 9-b-b. Her siblings have not requested predictive testing although the laboratory data would indicate that III_3 is also at low risk. Results in III_1 are more difficult to interpret since haplotype 14-b-b does not fit in with the overall patterns. Either there has been non-paternity or a recombination event has occurred. More DNA studies are required to determine which of the two is likely.

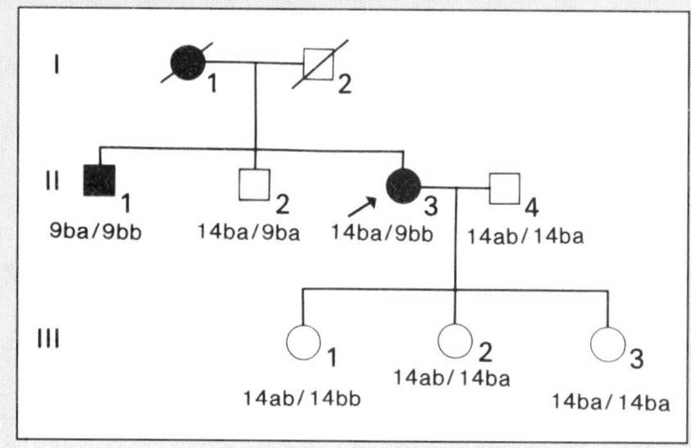

Benefits to emerge from molecular studies

Predictive testing

Although the Huntington disease gene had not been isolated until early 1993, the molecular approaches taken since 1983 have made available a diagnostic test for this disorder. By using DNA polymorphisms linked to the Huntington disease locus it has been possible to undertake predictive testing within the confines of a family unit (Box 3.6).

It is important to emphasise again that there are prerequisites before linkage analysis is possible. For example, the family unit has to be sufficiently large with either known affected and/or normal individuals. These key family members are critical to indicate which polymorphisms are co-inherited with the Huntington disease and which go with the normal phenotype. A computer program produces a numerical risk estimate. The program takes into account the recombination potential for the various probes (since the DNA markers are located at some distance from the actual gene) as well as including a correction factor to allow for penetrance in individuals who are clinically normal. For example, at ages over 60 years there remains the possibility, estimated to be about 10%, that an at-risk individual who is clinically normal can still develop Huntington disease.

Individuals with a family history of Huntington disease now have an opportunity through predictive testing to alter their a priori risks by DNA studies. However, two important issues have emerged from the Huntington disease predictive testing programs which are being undertaken in many centres. First, key individuals can be lost through death (including suicide) and this has prevented a number of families from having access to predictive testing. The concept of a *DNA bank*, which will be discussed below, has assumed increasing importance to avoid this problem.

A second consideration relates to the comprehensive clinical, counselling and support facilities which are necessary in a predictive testing program. This has major resource implications. It should also be noted that results from DNA testing have not always proven to be beneficial to the overall welfare of a family. The potential ethical/social issues resulting from DNA testing for Huntington disease will be discussed further in Chapter 9. Prenatal detection or prenatal exclusion for this disorder is also possible by DNA testing and will be described in Chapter 4.

Future directions

A feature of Huntington disease has been the difficulty in finding true sporadic cases. Thus, it has been proposed that the mutation rate at the Huntington disease locus is low. Perhaps the defect has arisen only once in history. However, more recent evidence would suggest that the multiple haplotypes (i.e. DNA markers for two or more closely linked alleles (forms of the same gene) at one locus which are usually inherited as a unit – see Box 3.6) found in association with Huntington disease are unlikely to have arisen by simple genetic events such as recombination. If this is correct, the haplotype data would indicate there are multiple mutations involved. This dilemma will soon be resolved now that the actual gene(s) has been cloned.

At present, nothing is known about the pathogenesis of Huntington disease. Thus, current therapy is empirical, e.g. drugs for depression. One of the priorities once any gene(s) is isolated is identification of its protein product(s). Following from this will come more rational therapeutic measures to treat or even prevent a disorder such as Huntington disease from developing.

ADULT POLYCYSTIC KIDNEY DISEASE

This example will be used to illustrate how heterogeneity at the clinical and molecular levels can produce diagnostic predicaments.

Clinical features

There are many causes for polycystic kidney diseases. These disorders affect children or adults and can be sporadic or genetic in occurrence.

Because of this heterogeneity, diagnosis is difficult. One form of this disorder, adult polycystic kidney disease (abbreviated to PKD1), has been intensively investigated at the molecular level. Adult polycystic kidney disease is characterised by bilateral enlargement of the kidneys as a result of multiple cysts. Cysts may be found in other locations (e.g. pancreas, liver, spleen). Hypertension is an important complicating factor and there is progressive renal failure. Adult polycystic kidney disease comprises approximately 10% of patients who are awaiting renal transplantation for end-stage renal failure. Affected individuals usually present for medical advice in their 4th or later decades of life. Presymptomatic detection of those at risk is possible by showing renal cysts with ultrasonography or nuclear imaging techniques. The accuracy of these studies is of limited value in childhood when renal cysts may not have developed fully. Moreover, simple cysts are found in approximately 10% of the normal population and can make interpretation of imaging studies difficult.

Box 3.7 Detection of adult polycystic kidney disease

The consultand (who is also the proband) has requested confirmation that his three children (aged 12, 14, 16 years) have adult polycystic kidney disease. The latter diagnoses were made a few years earlier on the basis of abdominal ultrasound examination. The consultand has the disease, one of his two sisters has undergone renal transplantation and the second sister has chronic renal failure and is awaiting transplantation. The pedigree illustrated below is an abbreviated version of a much larger family study which showed linkage of the adult polycystic kidney disease defect to the α globin gene complex on chromosome 16p13.3. DNA polymorphism results are listed in the pedigree and indicate that the disease locus co-segregates with DNA marker 2.2. in the consultand and his siblings. Therefore, one of the consultand's offspring (II_1) has DNA evidence for the disease but the other two lack this marker and have a low probability for the disease. Again, the major source of error would reflect the recombination risk for the DNA probe used in the study. Review of the original ultrasound pictures confirmed that the reports for the last two children were probably incorrect.

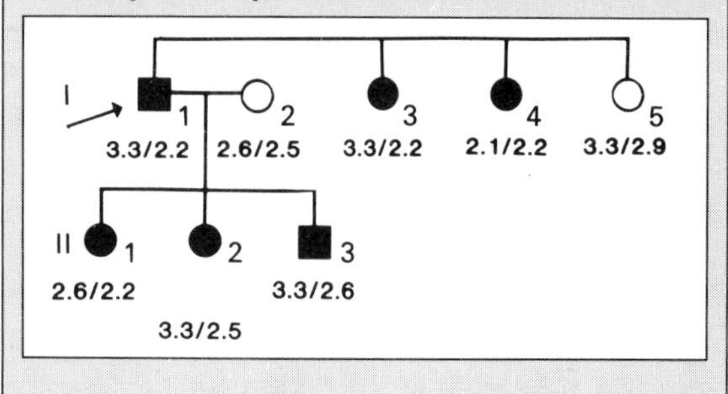

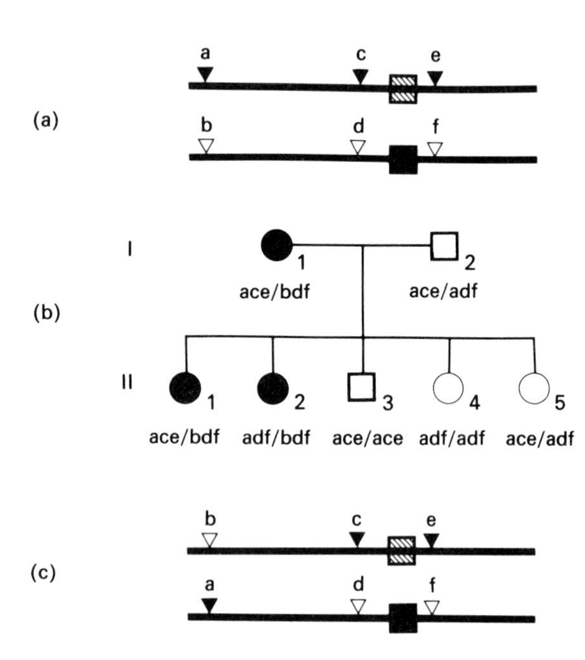

Fig. 3.10 Detecting recombination using flanking DNA markers in the adult polycystic kidney disease locus (PKD1).
It is assumed that in this family the PKD1 defect involves the chromosome 16p locus. (**a**) The three polymorphic markers and their alleles for the PKD1 locus are a or b; c or d; e or f. The hatched box is the normal gene; the solid box is the mutant. (**b**) The pedigree illustrates the segregation patterns for the above three polymorphisms. I_1 (female) has PKD1. Two of her children (II_1,II_2) are affected and so they allow the mutant-specific haplotype to be identified as bdf/ since this is what the three have in common. The only male offspring (□) has not inherited the maternal bdf/ haplotype which is consistent with his normal phenotype at age 50 years. The remaining two siblings are a problem. The adf/adf genotype in II_4 is not possible. It does not indicate non-paternity since the adf/ paternal haplotype is present. This is an example of recombination which has occurred somewhere between the a/b and the c/d loci (shown in panel **c**). The mutant-specific haplotype has now become adf/. Therefore, II_4 has inherited the PKD1 mutation. Individual II_5 is even more difficult to assess since the ace/ haplotype may have come from either parent and so it is not possible to determine whether haplotype adf/ is paternal in origin or a maternal recombinant. The latter cannot be excluded.

DNA diagnosis

DNA detection of adult polycystic kidney disease became possible in 1985 when linkage between the disease phenotype and a DNA polymorphism located on the end of the short arm of chromosome 16 (16p13.3) near the α globin gene locus was shown. This polymorphic marker was called 3′αHVR and it enabled both presymptomatic and prenatal detection of adult polycystic kidney disease with a recombination risk of 1–5% (Box 3.7). Today, there are closer markers available including polymorphisms located on either side of the putative gene locus. Flanking markers are useful in these circumstances because they make it more likely that recombination events will be detected (Fig. 3.10).

Nearly 3 years after linkage of the adult polycystic kidney disease locus to 3′αHVR was established, it became apparent that a second locus for the disease must be present. This conclusion was reached on the basis of linkage studies which failed to show co-segregation with the chromosome 16-specific probes in a few families. Thus, *genetic heterogeneity* was identified. The adult polycystic kidney disease example illustrates the importance of careful evaluation of DNA markers which rely on the indirect, linkage analysis approach. This type of assessment will give errors if genetic heterogeneity is involved.

Because of the risk (albeit low) that at least one other locus apart from chromosome 16 is implicated in adult polycystic kidney disease, it has been recommended that family studies should be extensive enough to allow statistical confirmation that there is actual linkage between the chromosome 16 specific DNA markers and the clinical phenotype. Once linkage is confirmed, the data can be used to predict clinical status. However, this is not possible in small pedigrees and so the utility of the DNA test becomes limited and must be considered in terms of what other diagnostic approaches are available. Hopefully, linkage studies will become unnecessary once the actual gene(s) is cloned and specific mutations, rather than indirect linkage analysis, can be sought.

X-LINKED DISORDERS

CLINICAL FEATURES

X-linked disorders result from abnormal gene function associated with the X-chromosome. Males, who have only one X chromosome (i.e. they are *hemizygous*), will fully express an X-linked disorder. On the other hand, females, who have two X chromosomes, will be carriers of the defect in the majority of cases. Although females have two X chromosomes to the male's one, products from this chromosome are quantitatively similar in both sexes. Therefore, one of the two X chromosomes in females must be inactive.

Lyonisation (named after M Lyon) describes the random X- inactivation of one of the two female X chromosomes. This occurs during early embryonic development. Because of the early onset and randomness of the process, female carriers of X-linked disorders can demonstrate variable amounts of the gene product which will depend on the proportion of normal or mutant X chromosomes remaining functional. The majority of the X chromosome is inactivated although there are some segments which escape inactivation. The molecular basis for X inactivation is unknown. Methylation may play some role (methylation is discussed further in the section on the fragile X syndrome). Since only one X chromosome is functional in females it is difficult to understand why Turner syndrome (a female phenotype associated with poorly developed gonads, sexual immaturity and a chromosomal constitution 45,X) occurs. To explain this finding it has been proposed that normal development of ovarian function requires both X chromosomes.

A typical pedigree illustrating X-linked inheritance is shown in Figure 3.11. The X-linked pedigree has an oblique character through in-

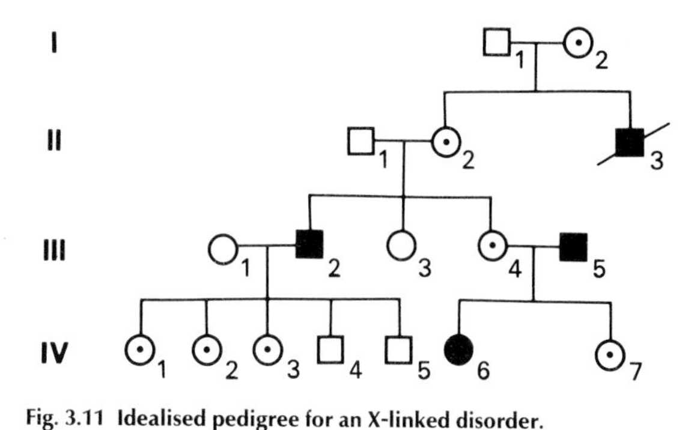

Fig. 3.11 Idealised pedigree for an X-linked disorder.
Females are represented by ○ and males by □. Female offspring of affected males are obligatory carriers. IV$_6$ is an affected female since both her parents carry the X-linked disorder. Affected males are shown by ■; carrier females by a dot within ○; individual II$_3$ is deceased. Because the disease is X-linked, a male cannot transmit the disorder to his sons (to whom he contributes a Y chromosome). Construction of pedigrees are key steps in the understanding of genetic diseases and how they are transmitted (see also mitochondrial DNA inheritance and imprinting which reinforce the value of a pedigree).

volvement of uncles and nephews related to the female consultand. The usual mating pattern involves a heterozygous female carrier and a normal male. Each son has a 50% risk of being affected by inheriting the mutant maternal allele. Similarly, each daughter has a 50% chance of inheriting the mutant maternal allele but will remain unaffected since she has her father's normal X-chromosome. Male to male transmission is not seen but may appear to occur if the trait is sufficiently common that by chance the mother also carries the mutant gene. An example of this would be glucose-6-phosphate dehydrogenase deficiency in those of black African origin. Approximately 10–20% of blacks in the United States are carriers or hemizygous for this defect.

Females can be symptomatic carriers or develop X-chromosome related disorders in a number of ways. (1) If a disproportionate number of normal X chromosomes has been inactivated. This can be a chance event or following a translocation between an X chromosome and an autosome. In the latter situation, X inactivation appears to be non-random since the normal X chromosome is preferentially inactivated. However, this may represent a selective process as cells with the normal X inactivated are least imbalanced and so will have a survival advantage. (2) If the female is hemizygous (Turner syndrome or 45,X). (3) Females affected with X-linked disorders may result following inheritance from both parents of a frequently occurring gene, e.g, glucose-6-phosphate dehydrogenase deficiency. (4) The recently defined heritable unstable DNA sequence which is described below under the fragile X syndrome.

Just as for autosomal dominant conditions, the frequency of *spontaneous mutations* in the X-linked disorders needs to be considered, particularly when counselling females who are potential carriers. Haemophilia does not interfere with the reproductive capacity of the affected individual. In contrast, Duchenne muscular dystrophy is usually fatal in the 2nd decade of life. Therefore, spontaneous mutations occurring in the latter disorder would be greater in numbers and correspondingly the proportion of females who are carriers will be less.

HAEMOPHILIA

The clinical, laboratory and molecular features of the haemophilias are summarised in Table 3.4. In

Table 3.4 Clinical, laboratory and molecular features of the haemophilias.
These are coagulation disorders with distinctive features

	Haemophilia A	Haemophilia B (Christmas disease)
Frequency	1 in 10 000 males	1 in 50 000 males
Defective protein	Clotting factor VIII: complex protein which circulates bound to von Willebrand factor. Produced in the liver	Clotting factor IX: serine protease produced in the liver
Clinical	Prolonged bleeding spontaneously or after minor trauma involving joints, muscles, subcutaneous tissues and organs	As for haemophilia A
Genetics	X-linked, 10–30% spontaneous mutations	As for haemophilia A
Gene structure	26 exons over 186 kb Chromosomal location Xq28	8 exons over 34 kb Chromosomal location Xq26-q27

the context of molecular medicine, the haemophilias will be used as models to illustrate the difficulties which can arise in identifying carriers particularly when the X chromosome is involved. Much has been written about positional cloning strategies to detect genes. The haemophilias also illustrate the value of functional cloning.

Carrier detection in X-linked disorders

Protein assays

Protein levels for coagulation factor VIII (deficiency of which produces haemophilia A) and coagulation factor IX (deficiency gives haemophilia B) demonstrate a broad normal range in blood. Because of random X-inactivation, the levels of factors VIII and IX can vary considerably in females who are carriers of haemophilia. This scatter makes an accurate assessment of carrier status difficult if the subject being tested demonstrates a normal or borderline result for the coagulant protein (Fig. 3.12). This may reduce the individual's a priori risk but does not provide definitive proof of her carrier status. In addition to X-inactivation there are physiological fluctuations seen with the coagulation factors, e.g. pregnancy (or taking the oral contraceptive) at which times the baseline levels for coagulation factors can increase. Finally, there is the not infrequent problem of assessing whether an affected relative is an example of a spontaneous mutation rather than the transmission of a haemophilia defect within a family. This occurs when there is a family history of only one male with haemophilia.

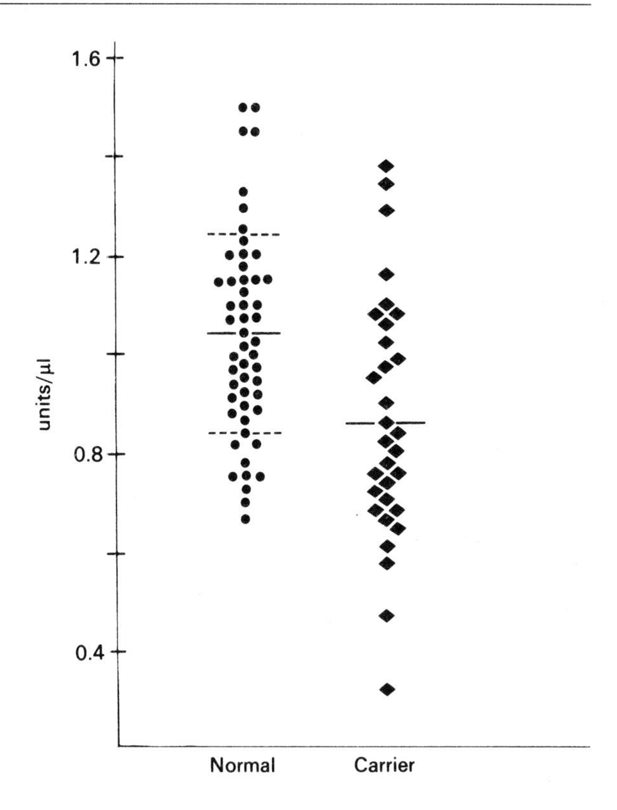

Fig. 3.12 The levels of factor VIII as measured by its coagulant activity in a normal population and in obligatory female carriers for haemophilia A.
Levels of factor VIII have a broad normal range in blood. The levels in carriers can also vary considerably because of random inactivation of the X chromosome. This makes accurate assessment of carriers difficult. The mean values are indicated as well as the standard deviation for the normal values. There is considerable overlap between normal and factor VIII coagulant levels in carriers. Better discrimination can be obtained by measuring the ratio of factor VIII coagulant to factor VIII antigen although there is also overlap with these ratios.

DNA linkage analysis

Testing for DNA mutations has advantages over proteins assays: (1) access to DNA is unlimited whereas an abnormal protein may not be easy to obtain and (2) DNA is not affected by physiological fluctuations, unlike proteins which may demonstrate marked variations in their normal level. The former is not a problem in haemophilia in which a blood sample suffices. The latter is an important consideration. Since the majority of defects associated with the haemophilias are point mutations it has been easier to utilise an indirect DNA linkage approach for diagnosis. A number of DNA polymorphisms have been described which are located within (intragenic) and in close proximity to (extragenic) the factor VIII and factor IX genes. These polymorphisms enable DNA diagnosis (prenatal or carrier) in up to 70–80% of families (Fig. 3.13). Intragenic polymorphisms have the advantage that recombination is unlikely to occur since these markers are located within the gene.

The disadvantages of DNA testing must also be considered. The DNA polymorphic approach requires key family members (who may be de-

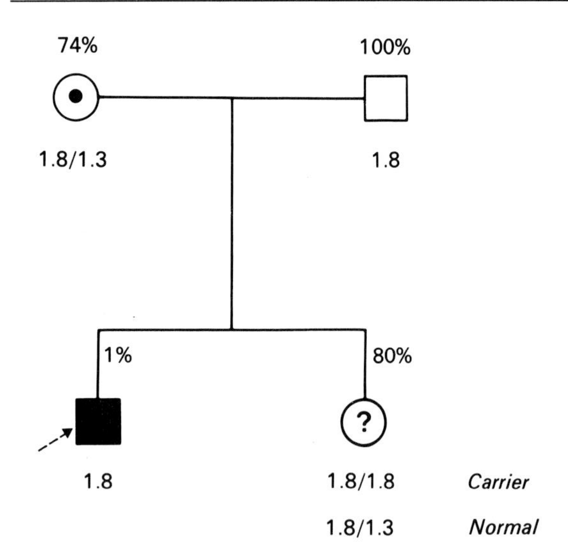

Fig. 3.13 Utility of the DNA linkage analysis approach in detecting female carriers for haemophilia.
The consultand (⊙), the mother of the child with severe haemophilia B (factor IX level is 1%) is an obligatory carrier because she also has an uncle who is affected (not shown in the pedigree). Factor IX coagulant levels are given as percentages in the pedigree (normal is > 50%). Factor IX levels for the mother and daughter are within the normal range (74% and 80% respectively) but this does not exclude the carrier state because of random X-inactivation in females. From the DNA polymorphisms, it is evident that the haemophilia B defect co-segregates with the 1.8 kb DNA polymorphism since this is the marker present in the haemophiliac. Therefore, the daughter's carrier status can be determined on the basis of which DNA polymorphism she inherits from her mother, i.e. if the daughter is homozygous for the 1.8 kb marker (she will always inherit one 1.8 kb marker from her father) she is a carrier. If the daughter has both the 1.8 and the 1.3 kb markers then the latter must have come from her mother. The daughter cannot be a carrier since the 1.3 kb polymorphism is a marker for the normal maternal X chromosome. The proband in this family is the haemophiliac child (→) who has only one polymorphic marker compared to his female relatives, i.e. he is hemizygous.

ceased or unavailable). It is very difficult to determine if mutations are spontaneous events. Germline mosaicism, where an individual has two or more cell lines of different chromosomal content derived from the same fertilised ovum, cannot be excluded. This is discussed further under the topic of non-traditional inheritance (p. 65). An additional problem associated with the DNA linkage approach in haemophilia B is the effect that linkage disequilibrium (preferential association of linked markers) can have on the informativeness of polymorphisms. For example, there are five biallelic DNA polymorphisms associated with the factor IX gene (Fig. 3.14). Some of these polymorphisms are inherited in a preferential association, i.e. the *Xmn*I and *Mnl*I polymorphisms are in *linkage disequilibrium*, which means that results obtained with either are similar since one allele of the polymorphism is nearly always inherited with the same allele of the other. Therefore, not all five polymorphisms will necessarily be informative. This is a particular problem with Chinese and Asian Indian populations and the factor IX gene locus. Non-paternity and its effect on DNA polymorphisms is not an issue if male offspring are studied because the father does not contribute his X chromosome to males. However, the source of the paternal X chromosome is important if a female is being assessed for carrier status. Direct detection of mutations would overcome many of the above problems.

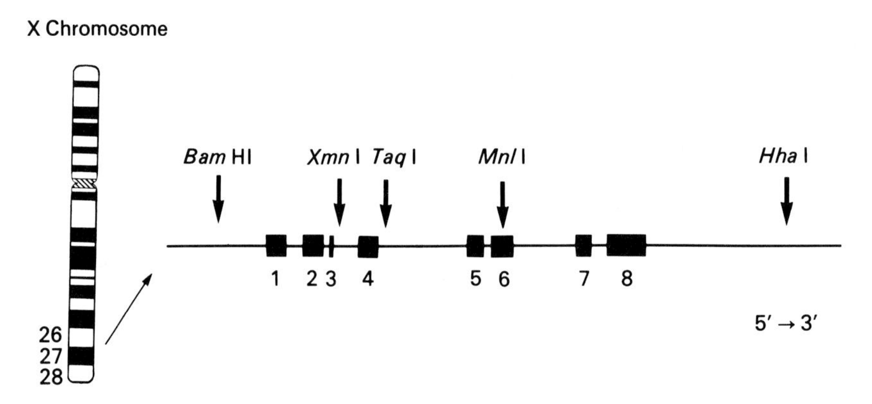

Fig 3.14 The structure of the factor IX gene.
Five polymorphic restriction enzyme sites giving restriction fragment length polymorphisms (RFLPs) indicated by ↓. Three occur within the gene (intragenic) and two (*Bam*HI and *Hha*I) are extragenic. Some of these polymorphisms are inherited in a preferential association known as linkage disequilibrium, e.g. *Xmn*I and *Mnl*I, and are therefore less useful in DNA testing.

Direct detection of mutations

The large size of the haemophilia genes (especially the factor VIII gene) and the heterogeneity of molecular defects (there were over 200 unique mutations reported in a recent haemophilia B database) meant that linkage analysis has been utilised for diagnosis. A novel approach has now been described in which haemophilia-specific mutations are sought in mRNA from peripheral blood lymphocytes using the polymerase chain reaction. This gets over the size problem since mRNA is considerably smaller than genomic DNA (Box 3.8).

Functional cloning

Factor VIII and factor IX had been isolated and characterised in human and other species by the late 1970s and early 1980s. These factors form part of the middle phase of the intrinsic clotting cascade and are serially activated and involved, in association with calcium and phospholipid, in the activation of clotting factor X. Since the protein structure was known it was possible to utilise functional cloning to look for the relevant genes. Oligonucleotide probes were synthesised from protein sequences of human and porcine factor VIII and bovine factor IX. Thus, each individual amino acid in the protein sequence was able to be reproduced in the form of a triplet codon in the oligonucleotide probe. The problem of a degenerate DNA code (i.e. there can be more than one codon for most of the amino acids – Table 2.1) was overcome by synthesising the codon which is more commonly used or alternatively making a mixture of 'degenerate' oligonucleotides. These comprised a cocktail of the possible combinations for codons. DNA libraries were screened with these oligonucleotides and the genes isolated. Not surprisingly, from its complex protein structure, the factor VIII gene is large (discussed further in Ch. 7). DNA probes from within and around the two genes have been isolated and these are now used for diagnostic purposes.

Knowledge of the structure and function of factor VIII and factor IX has been enhanced

Box 3.8 Detection of mutations using mRNA analysis

When it comes to diagnosis of genetic disease, a disadvantage of using mRNA compared to DNA is the former's tissue-specificity, i.e. mRNA is only found in tissues where a particular gene is functional. However, recently, it has been shown that mRNA for some genes can be isolated from peripheral blood lymphocytes despite the fact that these cells are not specifically involved in expression of those genes. This type of mRNA production has been called *'illegitimate transcription'* or 'leaky RNA' or 'ectopic RNA'. The amount of mRNA produced in this way is very small but detectable with the polymerase chain reaction. An additional step needs to be incorporated to convert mRNA to cDNA and then the cDNA can be amplified. 'Illegitimate transcription' in peripheral blood lymphocytes has been found for a number of genes associated with genetic disorders, e.g. haemophilia, Duchenne muscular dystrophy and familial hypertrophic cardiomyopathy. The advantages of this approach are: cells from the blood are suitable for assay; mRNA is considerably smaller than genomic DNA and so simplifies mutation analysis of larger genes. The major disadvantage of this technique occurs because mRNA only codes for exons. Therefore, mutations in introns or potential regulatory regions of the gene are less likely to be found.

Table 3.5 The major functional domains of factor IX. Structure/function of the factor IX coagulant protein as determined from its DNA sequence and gene organisation

Site	Activity
Exon 1	Hydrophobic signal peptide: allows secretion from the hepatocyte into the blood stream
Exons 2, 3	Contains a propeptide and undergoes post-translational modification (required for correct folding and calcium binding)
Exon 4	Epidermal growth factor-like domain; also binds additional calcium
Exon 5	Second epidermal growth factor-like domain
Exon 6	Activation domain for factor IX
Exons 7, 8	Serine protease (catalytic) domain; important for proteolysis of factor X to its active form

following cloning of the relevant genes and characterisation of their DNA sequences. For example, the eight exons of the factor IX gene encode for six major functional domains in the 415 amino acids of the glycoprotein. The domains are summarised in Table 3.5. Information gained from structure/function comparisons has been invaluable in our further understanding of the protein's biology. It has proven useful in the production of a recombinant DNA-derived product (see Ch. 7).

FRAGILE X MENTAL RETARDATION

Clinical and genetic features

The fragile X syndrome is the most common inherited form of mental retardation, affecting approximately 1 in 2000 children. The syndrome derives its name from the finding that there is a fragile site in the X chromosome at band Xq27.3. This is observed (although inconsistently) if cells for cytogenetic analysis are cultured under special conditions. The clinical picture in this syndrome can be quite variable. Affected males are moderately to severely retarded. Mild dysmorphic features can be difficult to recognise. They include facies with a high forehead, prominent lower jaw, large ears and subtle connective tissue abnormalities. Macroorchidism (large testes) is seen in adults. Behavioural problems are common and many affected children are misdiagnosed as autistic. Although X-linked, female carriers may be affected. Approximately one third of females with the fragile X site have mental retardation but usually the deficit is milder than that seen in males. Dysmorphic changes are infrequent in females.

There are a number of very unusual genetic features associated with the fragile X syndrome.

1. *Normal transmitting males.* This is unique amongst the X-linked disorders and describes males who are cytogenetically and intellectually normal but transmit the fragile X defect. The obligatory carrier status of these males has to be inferred from their position in a pedigree.

2. *Variable phenotypes in females.* Daughters of transmitting males, with very few exceptions, have normal phenotypes although *their* children can be affected. The risk of a fragile X carrier female having retarded children depends on her own phenotype, i.e. children who inherit the fragile X chromosome from a phenotypically normal female have a lower risk of getting the disease than if their mother has the phenotypic features of the syndrome.

3. *The fragile X syndrome shows anticipation.* There is increasing severity or earlier age at onset in successive generations. Anticipation has also been observed with an autosomal disorder – myotonic dystrophy.

Molecular defects in the fragile X syndrome

By positional cloning, a candidate gene has been isolated. The gene, called FMR-1, is expressed in many tissues including brain. FMR-1 is highly conserved across species and its mRNA has not

Box 3.9 Activity of genes and methylation

Over 60% of vertebrate DNA is methylated at the 5′ position in cytosine where it occurs in association with guanine (this is abbreviated to CpG). A small amount of DNA is hypomethylated and clustered into approximately 30 000 regions of DNA 1–2 kb in size and called CpG islands. CpG islands are frequently found at the 5′ end of genes and so are useful to identify in a positional cloning strategy since they indicate the potential location of genes. The methylation status of DNA correlates with its functional activity, i.e. inactive genes are methylated and actively transcribing genes are hypomethylated. One mechanism proposed by which methylation can modulate a gene's activity is through interference with protein-DNA binding which is essential for transcription to occur. Methylation plays a role in the maintenance of X-inactivation. Methylation may also be involved in imprinting. Whilst there is an association between the methylation status and a gene's activity, the dilemma arises as to which came first. For example, methylation may be an epiphenomenon which occurs once a gene has been inactivated by some other mechanism.

been found in most patients with the fragile X syndrome. DNA at the 5′ end of the FMR-1 gene has two interesting features. First, there is a CpG island which is methylated in patients with fragile X but hypomethylated in normal controls. This observation may have functional significance since it is frequently found that actively transcribing genes are hypomethylated whilst inactive genes are methylated (Box 3.9).

The second interesting feature of the FMR-1 gene is a CCG trinucleotide repeat. $(CCG)_n$ forms a DNA microsatellite (see Chapter 4 for further discussion of microsatellite DNA), where the number (n) can vary from 6 to 58 copies in the normal X chromosome. The $(CCG)_n$ microsatellite appears to play an important role in pathogenesis of the fragile X syndrome. The mechanism proposed involves instability of this region (and so the gene FMR-1) following amplification of the repeat. In normal families the $(CCG)_n$ sequences are stable and transmitted in the same way as is seen for other DNA polymorphisms. In families with fragile X, the number of repeats can be increased giving about 60–230 copies in DNA from asymptomatic carriers. Males and about half the females with the fragile X syndrome have a greater number of copies (>230) in their DNA. The number of copies can also change during transmission of this disorder to offspring. Males usually transmit about the same or fewer number of repeats whilst females can transmit an increased number. Once the amplified segment reaches what appears to be a critical size, around 200 copies, the CpG island described above becomes methylated. At high copy number the $(CCG)_n$ microsatellites also demonstrate *somatic* instability as evidenced by tissue mosaicism (cell lines of different genetic content in one individual). Therefore, individuals within the one family will inherit fragile X but can manifest different phenotypes because of the superimposed differences in copy number in their somatic cells.

How the fragile X phenotype is produced by these changes remains to be defined and will require characterisation of the FMR-1 gene and its protein product as well as additional family testing. Rare exceptions have been described in which an increase in the amplification unit is not seen in fragile X. However, in these circumstances, there is a deletion involving the FMR-1 gene. Thus, a novel molecular mechanism consisting of *triplet repeat amplification* has been proposed to explain the unusual inheritance patterns in the fragile X syndrome. The molecular characterisation of the fragile X site has also enabled DNA testing to be utilised in place of the unreliable cytogenetic analysis (see Ch. 4). The mechanism for mutation described above is not limited to the X chromosome, since unstable trinucleotide sequences have now been described in myotonic dystrophy and Kennedy syndrome (Box 3.10). More recently, the Huntington disease phenotype has been associated with amplification of an (AGC)n repeat.

Box 3.10 Triplet repeat amplification in myotonic dystrophy

Myotonic dystrophy (DM) is an autosomal dominant disorder that can affect a number of different organ systems. These include skeletal and smooth muscles, bone, eye and the neurological system. DM demonstrates variable expressivity and anticipation. The DM gene on chromosome 19q13.3 has now been isolated by positional cloning. An unstable trinucleotide repeat $(AGC)_n$* has been located 3′ to the DM gene. The severity of this disorder is paralleled by the number of $(AGC)_n$ repeats. The number of repeats is increased with successive generations and there is also somatic instability. Another disorder called Kennedy syndrome (spinal and bulbar muscular atrophy) has a similar triplet repeat to DM. Amplification occurs in Kennedy syndrome but not anticipation.

***There has been confusion as to nomenclature of the minisatellite repeats. The Human Gene Mapping Workshop 10.5 recommended that the repeats should be identified by their motifs in alphabetical order, e.g. the above AGCAGCAGCAGC etc. repeat could also be written GCAGCAGCAGCA etc. or CAGCAGCAGCAG etc. depending on where you started in the repeat sequence. To avoid using different repeat designations for what is the same sequence, the first nucleotide should be in alphabetical order, i.e. $(AGC)_n$ in the case of the above.**

DUCHENNE AND BECKER MUSCULAR DYSTROPHIES

Clinical features

Duchenne muscular dystrophy and Becker muscular dystrophy affect 1 in 3000 newborn males. Duchenne muscular dystrophy is a progressive muscle wasting disorder. Death occurs in the 2nd decade from the complications of muscle degeneration. Becker muscular dystrophy is a milder disease of late onset and slower progression but is otherwise identical to Duchenne muscular dystrophy. There are also intermediate cases with overlapping features. Both these muscular dystrophies are allelic i.e. they involve the same gene. An elevation in muscle enzymes such as creatine phosphokinase is found in all affected patients. However, because of random X-inactivation, female carriers are difficult to test by this method since only 70–80% show increases in creatine phosphokinase and these are modest. Therefore, DNA testing would be a better option for carrier detection.

The gene responsible for these dystrophies was the first to be isolated by positional cloning (Box 3.11). The large size of the gene may explain in part its high spontaneous mutation rate since one third of cases are considered to be new defects. Mutations in unrelated families are usually different abnormalities. In approximately 65% of patients with Duchenne or Becker muscular dystrophy, the molecular abnormality is a deletion or duplication involving exons in the gene. Present methods for DNA detection utilise DNA amplification to look for mutations in DNA and more recently mRNA via 'illegitimate transcription' (Box 3.8).

Box 3.11 Identification of the Duchenne muscular dystrophy gene

The location of the Duchenne muscular dystrophy gene to chromosome Xp21 was made in 1977 on the basis of a translocation involving this region which produced the muscle disorder in a female. In 1982, DNA polymorphisms linking Duchenne muscular dystrophy to the short arm of the X chromosome were described. Carrier detection and prenatal diagnosis using a number of DNA polymorphisms was possible in 1986. By 1987, the gene was cloned and characterised. It was, in fact, the finding of a male patient with a number of X-linked genetic defects apart from Duchenne muscular dystrophy which provided the means to isolate the gene. This patient had an extensive deletion involving Xp21 which enabled DNA probes specific to the deleted region to be obtained. One of the probes identified the Duchenne muscular dystrophy gene. The gene is very large (approximately 1% of the X chromosome itself) extending over 2300 kb and comprising 79 exons. The protein encoded by this gene is 427 kDa in size and has been named *'dystrophin'*. With the use of polyclonal antibodies, dystrophin has been localised to the inner surface of the sarcolemma in normal skeletal muscle. It is considered to be involved in the contractile apparatus of striated and cardiac muscles. Dystrophin is not found in patients with Duchenne muscular dystrophy or in animal models of this disorder such as the mdx mouse and the xmd dog.

Disease severity – the dystrophin protein

Polyclonal antibodies raised to the dystrophin protein have shown that Becker muscular dystrophy patients have reduced amounts of dystrophin or dystrophin of abnormal size. On the other hand, Duchenne muscular dystrophy patients have little or no dystrophin present in muscle biopsies. Thus, there is the potential to use a test based on muscle biopsy to distinguish the two. This would have a practical application given their similar phenotypes but different clinical outcomes.

Comparisons of the dystrophin-associated gene defects in the two dystrophies have shown an interesting feature which may explain the disease spectrum seen. This is called the 'frameshift hypothesis' and proposes that a mutation which results in a frameshift, or alteration of the normal codon reading frame, will produce an abnormal protein. This is responsible for producing the severe phenotype of Duchenne muscular dystrophy. On the other hand, maintenance of a reading frame, which might occur if there has been an interstitial deletion or duplication, does not grossly affect the overall structure of the dystrophin and a milder (Becker) phenotype is seen. Whilst attractive, the frameshift hypothesis does not explain all cases. Further studies comparing phenotypes and genotypes will clarify the pathogenesis and the clinical differences between these two dystrophies.

NON-TRADITIONAL INHERITANCE

MITOCHONDRIAL INHERITANCE

The nucleus is not the only organelle in eukaryote cells that contains DNA. Mitochondria have their own genetic material in the form of a 16.6 kb double-stranded circular DNA molecule. Mitochondrial DNA is characterised by a high mutation rate (5–10 times that of nuclear DNA), few non-coding (intron) sequences, a slightly different genetic code and maternal inheritance. The last occurs since spermatozoa make a negligable contribution to the conceptus in terms of mitochondrial DNA. Although most mitochondrial proteins are encoded by nuclear DNA, a few are encoded only by mitochondrial DNA. These include 13 proteins which are involved in the respiratory chain required for oxidative phosphorylation. This pathway allows mitochondria to play a crucial role in the cell's energy requirements.

Mitochondrial DNA and disease

Genetic disease

It is only recently that some genetic disorders, particularly those affecting organs with high energy requirements, e.g, brain, skeletal and heart muscles, have been connected to defects in mitochondrial DNA. Features which would suggest a mitochondrial DNA origin for an underlying disease are: (1) maternal inheritance, i.e. both males and females can be affected but the disorder is only transmitted by females, (2) the pathophysiology involves defects in mitochondrial oxidative phosphorylation, i.e. energy production, (3) there can be heterogeneity in affected individuals. This reflects *heteroplasmy* – the finding of a mixture of mutant and wild-type mitochondrial DNAs in the same cell – and (4) tissues will be affected differentially on the basis of their energy requirements. Thus, the central nervous system, skeletal and cardiac muscle fibres are at highest risk. Examples of some genetic disorders which arise from mitochondrial DNA defects are given in Table 3.6.

It is likely that the list of mitochondrial DNA-associated defects will grow considerably as DNA

Table 3.6 Mitochondrial DNA-associated genetic diseases. Since mitochondrial proteins can also be encoded by nuclear DNA it is possible to have what appears to be autosomal inheritance for a mitochondrial disorder. This has been reported in the Kearns–Sayre syndrome. As yet, mutations in nuclear DNA which produce mitochondrial defects have not been characterised.

Disease	Clinical phenotype	Mitochondrial DNA mutation(s)
Familial mitochondrial encephalomyopathy	Myoclonic epilepsy, myopathy (occasionally deafness, dementia, ataxia, hypoventilation, cardiomyopathy)	Heteroplasmic[1], point mutation affecting tRNA for lysine
Leber hereditary optic neuropathy	Late onset optic neuropathy giving rise to blindness in young adults and/or cardiac arrythmias	Homoplasmic, missense mutations at nucleotide positions 11778 or 3460
Kearns–Sayre syndrome	Progressive neuromuscular disorder: visual impairment, ophthalmoplegia, retinal degeneration, ataxia, muscle weakness and deafness	Deletions 4–8 kb; heteroplasmic 8 kb duplication; heteroplasmic

[1] Heteroplasmy is the finding of a mixture of mutant and wild-type mitochondrial DNAs in the same cell

amplification allows the mitochondrial genome to be studied with more ease. To date, the heterogeneity of the clinical phenotypes and the difficulty in using conventional biochemical approaches for study of mitochondria has meant that there has been little characterisation of these disorders. DNA technology prior to the availability of the polymerase chain reaction was demanding since large quantities of a tissue rich in mitochondria, e.g, the placenta, were required to enable sufficient DNA to be isolated. This is not a limitation to the polymerase chain reaction and strategies can be developed which allow the 16.6 kb genome to be amplified in segments using peripheral blood as a source of mitochondrial DNA. These are then screened for mutations by methods such as denaturing gradient gel electrophoresis. Segments which are shown to contain differences in nucleotide bases are confirmed to be abnormal by DNA sequencing (see Ch. 2).

Ageing

The high mutation rate in mitochondrial DNA reflects a combination of suboptimal DNA repair mechanisms, a high mutagenic environment secondary to the free radicals produced during respiration and a rapid turnover rate. The observations of a decline in mitochondrial respiratory activity with senescence and a concomitant accumulation of DNA mutations has led to the hypothesis that mitochondria play a role in the *ageing* process particularly in cells such as neurons which have a limited capacity for cell division. The relationship between mitochondrial DNA and ageing is presently the focus of much research. Deletions in mitochondrial DNA have also been observed in the basal ganglia of patients with Parkinson disease. The significance of this awaits further molecular characterisation.

Mitochondrial DNA as a population marker

The high mutation rate of mitochondrial DNA makes it more useful than nuclear DNA for evolutionary studies. In addition, the strictly maternal inheritance removes confounding effects such as recombination between the maternal or paternal alleles in these comparisons. Thus, extensive studies of the mitochondrial genome by population geneticists and molecular anthropologists have been reported. For example, a nine base pair deletion in one of the few noncoding mitochondrial DNA regions is a polymorphic marker for individuals of east Asian origin. One controversy surrounding the origin of the Polynesians in the south Pacific has been whether they derived from east Asia or South America. A study of mitochondrial DNA for the nine base pair polymorphism has shown that over 90% of Polynesians have the deletion. This is consistent with an east Asian origin (Fig. 3.15). In medical terms, the ethnic origin of populations may explain predisposition to certain diseases. For example, there is a high hepatitis B virus carrier rate in individuals of Asian origin. In view of the mitochondrial DNA studies in Polynesians it is interesting to note that they are also predisposed to infection with this virus.

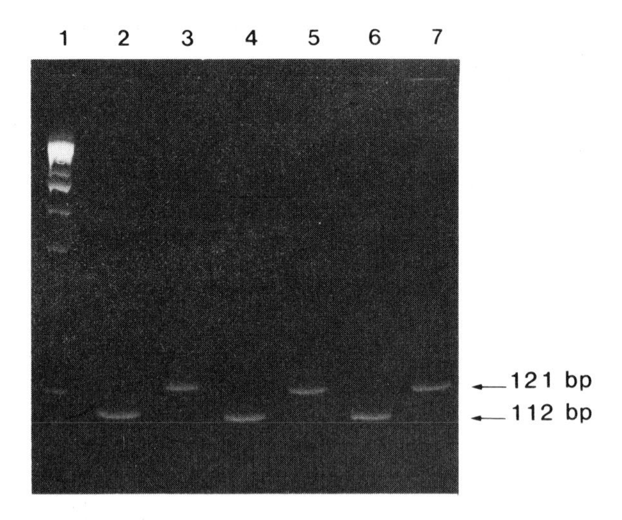

Fig. 3.15 Mitochondrial DNA as a population marker. The high mutation rate of mitochondrial DNA and the strictly maternal inheritance (avoiding recombination events) make it more useful than nuclear DNA for evolutionary studies. Mitochondrial DNA polymorphism involving a nine base pair deletion in noncoding region V. The picture shows an ethidium bromide stained polyacrylamide gel in which amplified mitochrondrial DNA has been electrophoresed. Track 1 = DNA size marker; tracks 3,5,7 show the normal 121 base pair fragment. The remaining tracks are examples of the nine base pair deletion (112 bp). (Reproduced with permission of the publishers of the American Journal of Human Genetics.)

MOSAICISM

Mosaicism refers to the presence of two or more cell lines which differ in genotype or chromosomal constitution but have been derived from a single zygote or fertilised ovum. Mosaicism is the result of a mutation which occurs during embryonic, fetal or extrauterine development. The time at which the defect arises will determine the number and types of cells (somatic and/or germ cells) which are affected. It is likely that mosaicism will be found in all large multicellular organisms to some degree. Mosaicism is now able to be studied in a greater number of circumstances and in more depth because DNA techniques allow an accurate genotypic assessment of multiple tissues. In this way the identity of individual cells can be established.

Chromosomal mosaicism

Females are examples of chromosomal mosaicism since there will be *random* inactivation of one of the two X chromosomes in all tissues. Both Turner syndrome (45,X) and Down syndrome (trisomy for chromosome 21) have had mosaicism demonstrated by cytogenetic analysis of cultured lymphocytes. The presence of a normal cell line in the mosaic Turner syndrome makes the clinical phenotype less severe.

With early fetal sampling made possible by chorion villus biopsy, it has become apparent that chromosomal mosaicism affecting the placenta occurs more frequently than previously considered (approximately 2% of samples). Chromosomal mosaicism confined to the placenta can produce false diagnostic results particularly in karyotypes obtained by chorion villus sampling (see Ch. 4). Retarded intrauterine growth in fetuses with normal karyotypes may result from aneuploidy (the addition or subtraction of single chromosomes) confined to the placenta. Chromosomal mosaicism also explains why some aneuploid fetuses can survive to term since mosaicism allows a normal cell line to be present in the placenta.

Single gene mosaicism

Somatic cell mosaicism is proposed as a mechanism which can produce phenotypic variation in single gene disorders. Clues to the presence of mosaicism may come from the finding in sporadic genetic disorders of marked tissue dysplasia which is patchy in distribution. Alternatively, mild phenotypic manifestations in a person with an apparent spontaneous single gene mutation or a mild phenotype in an individual whose offspring is severely affected may represent examples of mosaicism.

Somatic mosaicism may also be a mechanism for neoplastic change. A number of tumour suppressor genes (for example, the retinoblastoma gene) are involved in the development of malignancy once they are deleted or inactivated by a somatic event (see Ch. 6). It is hypothesised that focal areas of mutation (i.e. mosaicism) may lead to homozygosity for a mutant allele and so produce neoplastic change locally.

Germline mosaicism

From animal studies it has been estimated that the proportion of mosaicism in germ cells can vary from a few percent to 50%. Germline mosaicism is one explanation why parents, who are apparently normal on genetic testing, can have more than one affected offspring with an X-linked or dominant genetic disorder (e.g. X-linked: Duchenne muscular dystrophy, haemophilia A or B and autosomal dominant: osteogenesis imperfecta, tuberous sclerosis, achondroplasia). Therefore, the suspicion of germ cell mosaicism means that recurrence of a genetic disorder needs to be considered when individuals are being counselled.

Germline mosaicism affecting sperm has been sought using DNA amplification with the polymerase chain reaction. Normal DNA patterns obtained from somatic cells, such as peripheral blood, are compared with sperm DNA patterns. The latter would show both normal and mutant DNA forms if there is germline mosaicism. From the frequency of the mutant form, a theoretical recurrence risk can be estimated.

IMPRINTING

Contrary to Mendel's original theory that genes from either parent have equal effect, it is now clear that expression of some genes is dependent on the *parent of origin*. The mechanism involved is called *genomic imprinting*, which refers to the differential effects of maternally and paternally derived chromosomes or segments of chromosomes. The term imprinting has been used to signify that during a critical time in development, some genetic information can be marked temporarily so that its two alleles undergo differential expression. The critical period is considered to be the time of germline formation. Imprinting can be erased or reestablished in the germ cells of the next generation. The actual mechanism(s) involved in imprinting remain to be defined. It is not clear whether imprinting has a positive effect thereby activating a gene which would normally be silent or imprinting exerts a negative influence on a gene's function. An imprinted locus will be inherited along Mendelian lines whilst the expression of that locus will be dependent on the parent of origin (Fig. 3.16). Imprinting occurs in certain parts of the genome and has been shown to play a fundamental role both in normal development and in some pathological states such as genetic disease and cancer. Evidence for imprinting in placental mammals comes from a number of observations which are summarised in Table 3.7.

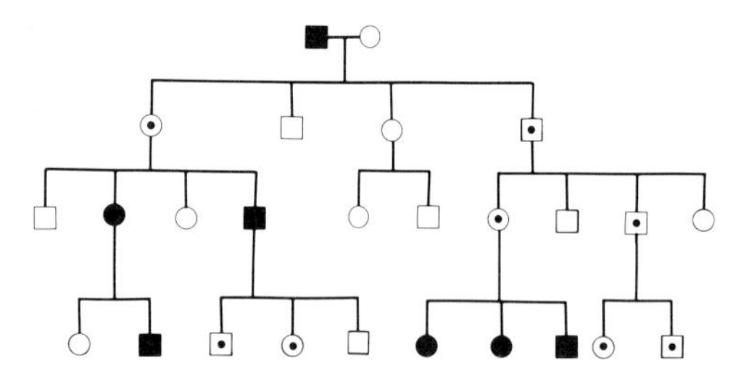

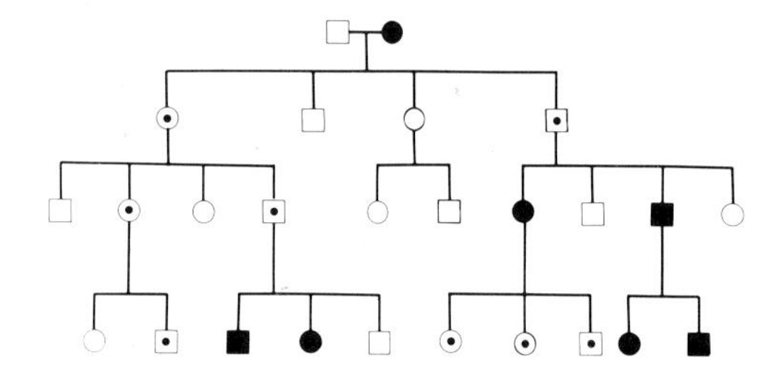

Fig 3.16 Hypothetical pedigrees which illustrate imprinting (parent of origin effects).
An imprinted locus is inherited in a normal manner but the expression of the two alleles will depend upon the parent of origin. (**a**) The paternal allele is inactive. There will be no expression of the mutant allele when transmitted by the father. For the mutant gene to express its phenotype it must pass through the maternal line. (**b**) In this case it is the maternal allele which is inactive and the disease phenotype only becomes apparent after paternal transmission of the mutant allele. In both cases there are carriers (indicated with a dot in the circle or square) who have normal phenotypes but can transmit the trait depending on their sex. There are equal numbers of affected and unaffected males and females in each generation.

Imprinting is more accurately detected and its implications have become better understood with the utilisation of molecular technology. This has enabled accurate assessment of the parental origin for chromosomal anomalies such as deletions, aneuploidies or uniparental disomies (see the next section for further description of uniparental disomy). At the DNA level, the parental source for a deletion can be defined. Genetic disorders and cancers whose inheritance is unclear are now being reassessed to look for parent of origin effects which may help to explain their irregular inheritance patterns.

Prader–Willi syndrome, Angelman syndrome

Parent of origin effects

Two syndromes with completely different phenotypes have been localised by cytogenetic analysis to chromosome 15q11-q13. The two are the Prader–Willi syndrome and the Angelman syndrome. Characteristic features of these disorders are summarised in Table 3.8. The *Prader–Willi syndrome* occurs sporadically with a frequency of approximately 1 in 25000 live births. The aetiology of this disorder remains unknown. Its inheritance pattern was confusing until cytogenetic data showed *paternal* loss of the 15q11-q13 locus. This was subsequently confirmed by molecular studies using DNA polymorphisms. Molecular analysis has now been demonstrated to be more sensitive and accurate than cytogenetics for diagnosis. It is evident that a paternal deletion of the Prader–Willi chromosome region is present in over half of these patients. A few examples of paternal allele loss in the Prader–Willi syndrome are associated with chromosomal rearrangements. The remaining cases remained a puzzle since cytogenetic analysis revealed two chromosome 15s in these patients. However, DNA characterisation has shown that both chromosomes have in fact originated from the one parent (i.e. there is uniparental disomy). In Prader–Willi syndrome, the two chromosome 15s are maternal in origin.

Table 3.7 Evidence for genomic imprinting in placental mammals. Chromosomes of maternal or paternal origin can have different expression of certain genes as a result of imprinting that probably occurs at germline formation

Observation	Parent of origin effect
Pronuclear transplantation gives zygotes with both sets of haploid chromosomes either maternal or paternal in origin	**Androgenetic**: (paternally-derived nuclear material): embryos have relatively normal development of membranes and placentas but very poor development of embryonic structures **Gynogenetic**: (maternally-derived nuclear material): the opposite occurs in development
Human chromosomal triploids	**Android**: (two paternal and one maternal haploid components): large cystic placentas **Gynoid**: (two maternal and one paternal haploid components): small underdeveloped placentas
Uniparental chromosomal disomies	**Mice**: some loci on chromosomes 2,6,7,11,17 display different phenotypes depending on whether maternal or paternal uniparental disomy **Human**: uniparental disomy chromosome 7 (cystic fibrosis), chromosome 15 (either Prader–Willi syndrome or Angelman syndrome)
Chromosome deletions	**Genetic disease**: deletion chromosome 15q11-q13: Prader–Willi syndrome (paternal deletion) and Angelman syndrome (maternal deletion) **Cancer**: maternal deletions of chromosome 13q (sporadic osteosarcoma) or chromosome 11p (Wilms' tumour)
Transgenic expression	There can be differences in the expression of a foreign gene by transgenic mice over a number of generations. Function or non-function of the transgene can be dependent on the sex of the transmitting parent. Methylation/demethylation of the transgene in association with the above has also been observed
Expression of specific genes	Parent of origin effect on phenotype, age of onset or severity, e.g. uncommon severe, rigid, juvenile form of Huntington disease (paternal transmission); severe, congenital form of myotonic dystrophy (maternal transmission)

Table 3.8 Clinical, cytogenetic and DNA features of the Prader–Willi and Angelman syndromes.
Although the genes involved have been localised to the same chromosome area, the phenotypes are completely different. Both disorders show imprinting, in the Prader–Willi syndrome the paternal allele is lost, in Angelman syndrome there is maternal loss

Parameter	Prader–Willi syndrome	Angelman syndrome
Clinical features	Obesity, waddling gait	Thin, ataxic gait with jerky voluntary movements
	Mental retardation (mild–moderate)	Mental retardation (severe)
	Hypogonadism	Microcephaly
	Behavioural problems	Happy, sociable mood, paroxysms of laughter
	Characteristic facies (narrow bifrontal diameter, almond shaped eyes, triangular mouth)	Characteristic facies (prominent lower jaw with tongue protrusion)
	Small hands, feet and stature	Epilepsy with characteristic EEG
	Floppy with feeding problems in the newborn period	Can be floppy at birth
	Hypopigmentation (variable feature)	Hypopigmentation (variable feature)
Cytogenetic findings on chromosome 15q11-q13	Interstitial deletion 50% Duplications or translocations 5% Normal 45%	Approximately 50% have a cytogenetic deletion
DNA findings on chromosome 15q11-q13	Paternal deletion 77% Uniparental disomy 20% Other 3%	Maternal deletion 74% Uniparental disomy 4% Other 22%

Uniparental disomy

Uniparental disomy is when two copies of a chromosome are inherited from one parent. There are two types of uniparental disomy: *isodisomy* in which both chromosomes from the one parent are identical copies and *heterodisomy* where the two chromosomes represent different

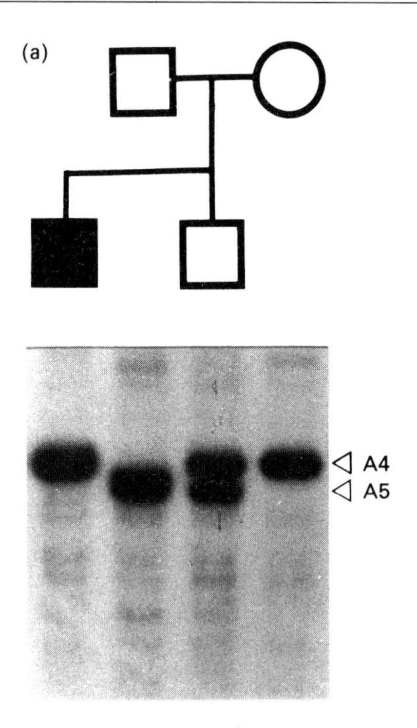

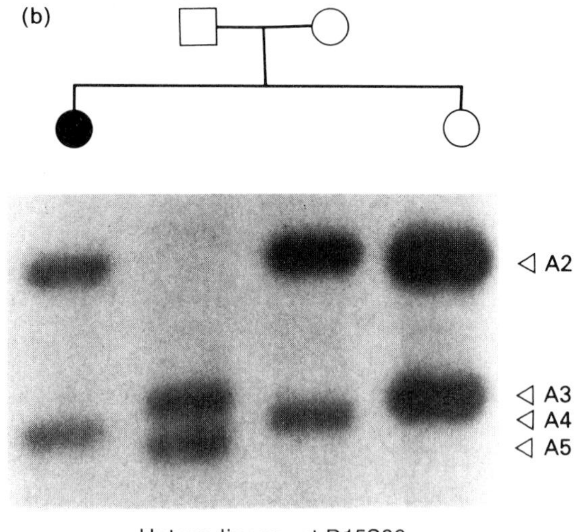

Fig. 3.17 Uniparental disomy in Prader–Willi syndrome. Southern blots illustrating paternal allele loss accompanied by maternal disomy in the Prader–Willi syndrome at DNA locus D15S86. (**a**) The child with Prader–Willi is indicated as (■) and his normal sibling as (□). DNA polymorphic markers for the paternal and maternal Prader–Willi chromosome regions are distinguishable as lower (A5) and upper (A4) bands respectively. The normal child has inherited both from his parents. The child with Prader–Willi does not have the paternal-specific band (A5). He has only his maternal band. On densitometry assessment of DNA in this individual there is no evidence that a deletion has occurred at this locus. Therefore, the maternal contribution involves both alleles, i.e. disomy. The polymorphic markers do not distinguish the two maternal alleles and so it cannot be ascertained if this is an example of isodisomy (identical copies) or heterodisomy (different copies). (**b**) Heterodisomy is found in another family in which the affected child is indicated (●) and her normal sibling (○). DNA patterns confirm both maternal specific alleles and neither of the paternal alleles have been inherited by the child with the Prader–Willi syndrome.

copies of the same chromosome. Both isodisomy and heterodisomy have been found in the Prader–Willi syndrome (Fig. 3.17). There are two explanations for uniparental disomy. Either there is fertilisation between disomic (diploid content) and nullisomic (no chromosomal content) gametes or a trisomic conceptus (formed from a normal gamete and a disomic gamete) loses one of its chromosomes (Fig. 3.18). The latter would seem more likely since chromosome 15 is one of the more common trisomies associated with spontaneous miscarriages. If trisomy were to follow from non-disjunction (uneven division of chromosomes during meiosis), a viable disomic conceptus is only possible if one of the three chromosomes is lost. If this were to occur, one third of the concepti will have the two remaining chromosomes originating from the same parent. Thus, one mechanism for the Prader–Willi syndrome is maternal non-disjunction giving a trisomic conceptus which is 'rescued' when the paternal chromosome is lost.

Uniparental disomy is not unique to chromosome 15. Cystic fibrosis occurring in the case of a carrier mother but normal father has been explained by uniparental disomy. In this situation, non-paternity was excluded and it was shown that affected children had inherited two copies of the mutant chromosome from their carrier mothers, i.e. isodisomy had occurred. It is noteworthy that the cystic fibrosis phenotype was also associated with developmental abnormalities, e.g. moderate to severe intrauterine and postnatal growth retardation. Thus, it is possible that paternally-derived gene(s) located on chromosome 7 are required for completely normal development. It has been estimated that perhaps 1 in 500 cases of cystic fibrosis are due to uniparental disomy.

The Angelman syndrome has a different clinical phenotype to that seen in Prader–Willi syndrome (Table 3.8). Nevertheless, the Angelman syndrome

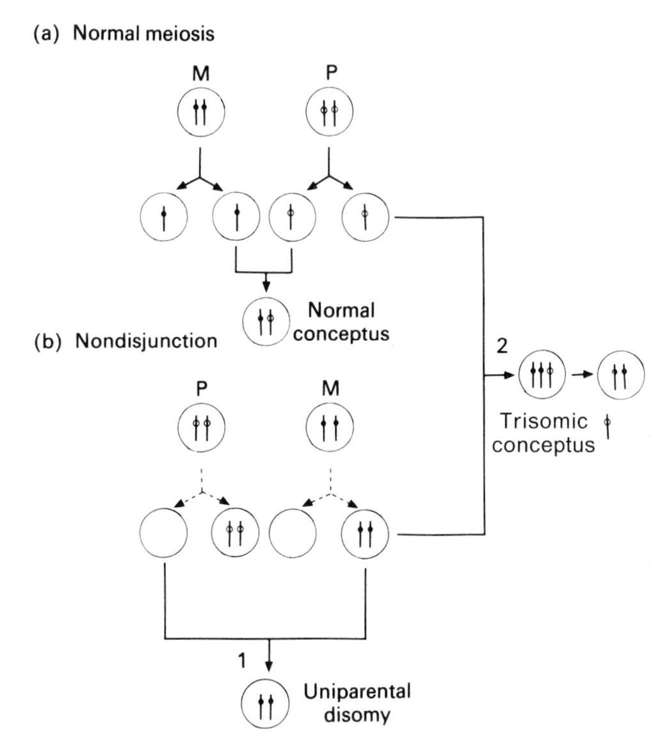

Fig. 3.18 Two ways in which uniparental disomy can occur. (**a**) During normal meiosis the haploid number of chromosomes is established in the germ cells (M or —●— = maternal; P or —○— = paternal). At fertilisation, the diploid chromosome content is produced with equal maternal and paternal contributions. (**b**) Illustrates nondisjunction during meiosis (indicated by -----). This will produce one gamete without any parental chromosomal contribution and the second will have the diploid content. Uniparental disomy can occur from fertilisation between nullisomic and disomic gametes (1) or more likely between a normal gamete and a disomic gamete to produce a trisomic conceptus (2). The latter is nonviable unless one of the trisomic chromosomes is lost. The end-result in both (1) and (2) is a normal chromosome content but loss of the paternal chromosome producing maternal uniparental disomy.

also demonstrates imprinting but unlike the Prader–Willi syndrome it is the *maternal* allele which is essential. It is also interesting that there have been fewer cases of uniparental disomy described in the Angelman syndrome. This may reflect the rarer occurrence of uneven distribution of chromosomes during meiosis in paternal germ cells.

Applications of molecular analysis in imprinting situations

Diagnostic applications

The ability to identify individual parental contributions by DNA testing will be important in studying imprinting. In a practical sense, diagnosis of the Prader–Willi syndrome or the Angelman syndrome is difficult until late childhood when the full clinical phenotypes become apparent. However, demonstration that there is differential paternal or maternal imbalance through a deletion or uniparental disomy is a critical diagnostic finding in these situations.

Although both Prader–Willi and the Angelman syndromes are found on a similar locus on chromosome 15, there is good evidence to indicate that the two involve different sets of genes some of which may overlap. Thus, they fall into a group of disorders called the *contiguous gene syndromes*. Although these disorders have some common features, e.g. mental retardation or growth abnormalities, the range of clinical symptoms may reflect the involvement of a number of physically related but otherwise distinct genes. Some of the features of these syndromes are summarised in Table 3.9. Compared to cystic fibrosis and Huntington disease, positional cloning

Table 3.9 Characteristic features of the contiguous gene syndromes.
These involve different sets of genes that may overlap but show some distinctive features

Parameter	Observations
Clinical features	Dysmorphic appearance and defective organ involvement Mental and growth retardation Heterogenous phenotypes since several functioning and unrelated genes are involved to variable degrees The severity of the disorder reflects the extent of the underlying mutation(s)
Genetic findings	Chromosomal deletions Normal chromosomal patterns shown on DNA testing to be microdeletions/microduplications or uniparental disomy if the involved region is imprinted Inheritance is usually sporadic although recurrences are possible if there is a mutation which can be transmitted from the parents and epigenetic factors, such as imprinting, can be satisfied. Affected individuals are unlikely to reproduce
Known disorders	Prader–Willi syndrome, Angelman syndrome, Beckwith–Wiedemann syndrome, retinoblastoma, Wilms' tumour, Miller–Dieker syndrome, Langer–Giedion syndrome, diGeorge syndrome, HbH-mental retardation syndrome

will be more difficult in the above two syndromes because genetic maps are of little help since the disorders are usually sporadic. Nevertheless, physical maps are being constructed for the chromosome 15q11-q13 region. Candidate genes will next be identified and the pathogenesis of these two syndromes determined. Recent work would suggest that one gene in this locus, that involved in hypopigmentation, which can be found in both the Prader–Willi syndrome and the Angelman syndrome, has now been cloned.

Significance of imprinting

A number of fundamental questions about imprinting remain to be answered. For example, how and when does imprinting occur? The mouse model, including the ability to produce transgenic mice, has provided important information in our understanding of imprinting. For example, methylation has been implicated as an important factor modifying the phenotype in the imprinting process (Box 3.9). Whether this is a cause or consequence of imprinting remains to be determined.

Another key issue is why imprinting occurs at all. To date, imprinting has only been observed in placental mammals and some flowering plants but not amphibians, reptiles, birds or marsupials. Answers to the above are being sought through molecular characterisation of loci shown to be imprinted in mice. Breeding experiments, which are easily undertaken in small animals, are impractical in the human situation. In the latter circumstance, the most useful strategy to follow involves an analysis of natural models of imprinting, i.e. human genetic disorders in which there is a parent of origin effect on the individual's development or phenotype, e.g. the Prader–Willi and Angelman syndromes. A further discussion of imprinting, with its potential effects in carcinogenesis, is given in Chapter 6.

MULTIFACTORIAL DISORDERS

Multifactorial disorders comprise a very significant proportion of the genetic diseases (Table 3.1). Study of the single-gene disorders, as illustrated above, has provided significant insight into their pathogenesis. The multifactorial disorders are now starting to be investigated. They are considerably more complex but will have far wider implications because of their association with many *common* diseases. Some of the common diseases where multifactorial inheritance has been implicated are:

- Coronary artery disease
- Hypertension
- Psychiatric illness
- Dementia
- Diabetes (insulin dependent)
- Cancer
- Mental retardation
- Congenital malformations e.g. cleft lip/palate; congenital dislocation of the hip, pyloric stenosis.

SCHIZOPHRENIA

Genetics

Family, twin and adoption studies have implicated a genetic component in schizophrenia. Risks of 10% and 3% in first and second degree relatives of individuals affected with schizophrenia are higher than the general population figure of approximately 1%. Concordance rate in monozygotic twins is approximately 50% compared to 10–15% in dizygotic twin pairs. There is a higher frequency of schizophrenia in the separated blood relatives of affected individuals compared to control adoptees whose parents have no known psychiatric history. At present the mode of inheritance is unknown. Autosomal dominant with reduced penetrance, autosomal recessive, multifactorial and a number of other combinations have all been proposed. In this description the schizophrenias are considered as multifactorial disorders since their mode of inheritance appears to be complex and may involve a number of genes and/or interactions between genes and the environment. Since clinical and biochemical studies had failed to explain the genetics or pathogenesis of schizophrenia, the next obvious step was to try positional cloning.

Linkage analysis

There was a clue where to start looking in the genome since it had been observed in one family that two schizophrenic males were partially trisomic for chromosome 5q11.2-q13.3. DNA polymorphic markers for this region were then used in linkage analysis. In 1988, lod scores between 3–6 were obtained in two British and five Icelandic families. This is a statistical analysis of the odds favouring linkage: a lod score of +3 means the odds in favour of linkage are 1000:1. Thus, DNA studies were consistent with the cytogenetic observation and pointed to a schizophrenia gene in association with chromosome 5q11-q13. Subsequently, multiple pedigrees have been investigated and these have failed to confirm the chromosome 5 findings. One initial explanation was that schizophrenia is caused by a number of mutations which involve other loci. However, reassessment of the chromosome 5 linkage studies with more families and additional DNA probes has now confirmed that the original results were probably related to statistical problems.

Two other loci have been implicated in schizophrenia. They are chromosome 15, through an association with Marfan disease, and the dopamine D_2 receptor on chromosome 11. To date neither locus has been confirmed by linkage studies. *Exclusion maps* of the human genome are also being constructed, i.e. by linkage analysis certain loci can be shown *not* to be linked to schizophrenia. This is a laborious strategy and less likely to be successful, particularly if heterogeneity is a significant factor in schizophrenia.

Positional cloning

Schizophrenia illustrates the difficulties which are inherent in recombinant DNA techniques when these are used to study complex disorders with an apparent genetic component in their aetiology. It is possible that there are multiple genetic loci involved in schizophrenia and chromosomal localisations described above are simply one of many options. Alternatively, the equivocal data may reflect the use of linkage analysis programs which are more appropriate for single-gene disorders that follow traditional Mendelian-type inheritance. Some reasons for discrepant results when applying linkage analysis studies to multifactorial disorders:

- Genetic heterogeneity
- Variable penetrance
- Variable expressivity
- Late age of onset
- Existence of non-genetic cases
- Assortive mating (more than one disease gene in a family).

More sophisticated computer programs are presently being developed to deal specifically with the multifactorial disorders.

The example of schizophrenia also illustrates the importance of accurately defining phenotypes in linkage analysis studies. Clinical criteria have been proposed to assist diagnosis of schizophrenia. However, the criteria are broad and there is overlap with other psychiatric disorders. The confounding effects of drug or alcohol abuse and the potential for some neurological disorders to produce schizophrenic-like features must also be considered when criteria for inclusion and exclusion are being determined in linkage studies. Even more basic is the unresolved question whether schizophrenia and another group of psychiatric disorders called the affective psychoses are in fact two distinct entities or a spectrum of the same disorder. Is schizophrenia that has an obvious genetic component the same disease as the more frequently occurring sporadic forms?

Despite the shortcomings described above it is still appropriate to utilise molecular strategies since there is a reasonable chance that success will follow. In this respect, the experience from sporadic cancers is encouraging where DNA changes initially observed in the less common familial cancer syndromes are now being found to be similar in the genetically more complex sporadic cases (discussed further in Ch. 6).

HYPERTENSION

Multifactorial inheritance

Approximately 20% of the population has essential hypertension. This has long-term complications affecting the brain (stroke), heart (ischaemic heart disease, cardiac failure) and the kidney (renal failure). Family and twin studies have shown that hypertension is inherited as a multifactorial trait. It is highly likely that several genes are involved in the regulation of blood pressure. Therefore, identifying the genetic factors involved in hypertension in the human is considerably more difficult than the procedure described previously for the single gene disorders. For example, two problems with a human model for hypertension are: (1) the definition of the phenotype, i.e. what is hypertension and what is the normal variation in blood pressure measurements and (2) the ascertainment of informative families for studies and the likely complexity brought about by genetic heterogeneity including the environmental effects which may modulate a gene's phenotypic effects.

Approaches to multifactorial genetics include the use of very large pedigrees in which there are multiple affected individuals. Highly polymorphic DNA markers which span (ideally) the whole genome are then tested in the linkage analysis strategy for co-segregation of the hypertension phenotype with a particular DNA marker. Since this is not involving a single gene disorder, the conventional statistical methods to analyse the data are no longer valid and more sophisticated mathematical calculations are required to enable multiple contributing loci to be identified.

An animal model

Genetic maps in humans require extensive pedigrees. These may be difficult to obtain. There are inherent problems when dealing with humans such as non-paternity, unavailable individuals and so on. Genetic maps are established on the basis of phenotype segregation, chromosomal analysis and, more frequently these days, from DNA polymorphic markers. Genetic maps in laboratory animals are much easier to identify because of the breeding options available. Thus, an alternative approach in multifactorial disorders avoids the human situation altogether and looks at comparable animal models. In the present context, one such example is the hypertensive rat. Searching the rat genome for genes associated with hypertension is more likely to be successful. Once found, candidate genes in the rat are then sought in the homologous (syntenic) region in the human.

In one study, two different rat breeds were used: the stroke-prone-hypertensive rat which shows increases in blood pressure in response to dietary sodium and Wistar–Kyoto rats with normal blood pressure in which sodium loading has no effect. Crosses to produce F2 hybrids were undertaken with both Wistar-Kyoto males and stroke-prone-hypertensive-rat females as progenitors and vice versa. Over 100 DNA markers were sought in the F2 population to test for linkage. Linkage analysis revealed two loci cosegregrating with hypertension. The first locus was assigned to the rat's X chromosome. This was considered to be one factor contributing to hypertension in female rats.

The second locus was of more interest. It was associated with the rat growth hormone promotor region and the rat fast nerve growth factor receptor. These genes are on rat chromosome 10. The comparable region in the human is on chromosome 17q which contains amongst other loci the gene for the angiotensin I converting enzyme (ACE). This would be a good candidate gene for essential hypertension since it plays a key role in the production of angiotensin II, a vasoactive peptide which can elevate the blood pressure. Inhibitors to the angiotensin I converting enzyme are very effective agents for the control of hypertension. To date, analysis in the human has failed to show positive linkage between the angiotensin I converting enzyme gene and hypertension. However, genes which are located nearby and might have a direct effect on blood pressure or even modify the expression of the angiotensin I converting enzyme have not been excluded. Other candidate genes, such as renin, are also being investigated by use of genetic and physical mapping in humans and animal models of hypertension.

DNA BANK

Three aspects of recombinant DNA technology have led to the concept of a DNA bank being developed. These are: (1) the rapid advances in gene identification which have meant that what is a genetic defect of unknown aetiology today is very likely to have an associated DNA marker tomorrow, (2) the necessity in linkage analysis to have key family members available for testing and (3) the long-term potential for storage of DNA.

Purpose of a DNA bank

Genes or DNA polymorphisms associated with single gene disorders are being identified at an exponentially increasing rate. The information generated from the Human Genome Project (see Ch. 10) will accelerate this process. The multifactoral disorders are starting to provide DNA data so that the genetic component for common diseases will be identifiable in the near future. Interesting or esoteric genetic diseases can be the focus of a research study but as the molecular defect becomes characterised the research emphasis shifts to a service (diagnostic) component. Therefore, individuals or families in which there is a genetic component to disease (be it single gene or multifactorial) should be informed that future technological developments may allow further definitive study of the disorder even though nothing tangible can be offered at present. In these circumstances, it is essential that health care professionals are aware of the potentials of recombinant DNA technology so that counselling given to such individuals or families is accurate and permits them access to future developments. The end-result is DNA (in the form of a tissue or DNA itself or immortalised cell lines) stored in a professional DNA banking facility which will make it available if required at some future date. A list of single gene disorders for which DNA banking is considered appropriate is found in Table 3.10.

Table 3.10 Single gene disorders for which DNA banking should be considered

Cystic fibrosis
Huntington disease
Myotonic dystrophy
Neurofibromatosis 1
Tuberous sclerosis
Osteogenesis imperfecta
Adult polycystic kidney disease
Familial adenomatous polyposis coli
Thalassaemia
Fragile X syndrome
Duchenne/Becker muscular dystrophy
X-linked immunodeficiency
Haemophilia A,B
Familial hypertrophic cardiomyopathy

A number of genetics societies have proposed guidelines for a DNA bank. These cover: actual physical facilities; the relationship between depositors, their families and various health professionals; confidentiality of information; safety precautions and quality assurance measures. The word *depositor* rather than donor is used because the individual giving the sample maintains ownership and is not acting as a donor in the broadest sense. Depositors need to have clear statements of the length of banking, the potential problems and their rights in respect of the banked DNA.

The strategy of linkage analysis for identification of genes requires a family study in which there is one or more key individuals which allow the phase of the polymorphisms to be determined. The types of key individuals required in the various genetic disorders have been described above. Similarly, parents are important to help identify inheritance patterns for the DNA polymorphisms. The unavailability of either of the above through death, separation or loss to contact does not prevent a linkage study provided there are appropriate family members available which will allow DNA markers for that parent to be derived. However, the amount of work is now considerably increased and the more meioses which have occurred the greater the risk for recombination (and non-paternity). Thus, it is essential to store blood or DNA from key family members if it is likely to be useful in the future or if there is the potential for that individual to become unavailable for whatever reason.

Long-term storage of DNA

DNA can be stored in the form of whole blood or as DNA for a few years at −20°C or −70°C. Transformation of lymphocytes with the Epstein–Barr virus produces an immortalised cell line which can be cryopreserved over many years in liquid nitrogen. Aliquots can be thawed and propagated as required. The advantage of immortalising lymphocytes is the availability of an unlimited amount of DNA which is also suitable for techniques such as pulsed field gel electrophoresis and amplification of illegitimately transcribed mRNA. One disadvantage of immortalised lymphocyte cell lines is the technical demands required to prepare and then maintain the lines. The utility of the polymerase chain reaction which can identify DNA mutations in the smallest amount of tissue has also meant that it may not be necessary in future to transform DNA. Material which is suitable for DNA banking includes blood, hair follicles, liver/spleen and other tissues from a deceased individual, abortus specimens, buccal cells from mouth washes and dried blood spots (Guthrie spots). The last are collected as part of a neonatal metabolic screening service (see Ch. 4).

Potential problems

A DNA bank is not simply a diagnostic laboratory which keeps a number of DNA specimens in the refrigerator for 'future purposes'. It is a planned activity with very strict operating guidelines. There are legal requirements, defined above, that the depositors will need to understand. What can be done with the DNA, particularly in terms of research, brings out both legal and ethical issues which need definition. For example, can the banked DNA be used for research and where does research end and clinical service (diagnosis) begin? Can the courts direct that a banked DNA specimen be tested for legal purposes? Confidentiality becomes an important consideration particularly when other family members may need access to information derived from the banked DNA. These issues will be discussed further in Chapter 9. Guidelines for DNA banking have been published by a number of genetics societies (American Journal of Human Genetics 1988; 42: 781–783; Journal of Medical Genetics 1989; 26: 245–250).

FURTHER READING

Baird P A, Anderson T W, Newcombe H B, Lowry R B 1988 Genetic disorders in children and young adults: a population study. American Journal of Human Genetics 42: 677–693

Bell J 1992 ACE (or PNMT?) in the hole. Human Molecular Genetics 1: 147–148

Butler M G 1990 Prader–Willi syndrome: current understanding of cause and diagnosis. American Journal of Medical Genetics 35: 319–332

Collins F S 1992 Cystic fibrosis: molecular biology and therapeutic implications. Science 256: 774–779

Giannelli F, Green P M, High K A, et al 1991 Haemophilia B: database of point mutations and short additions and deletions – second edition. Nucleic Acids Research 19(suppl): 2193–2219
Graham J B, Kunkel G R, Egilmez N K, Wallmark A, Fowlkes D M, Lord S T 1991 The varying frequencies of five DNA polymorphisms of X-linked coagulant factor IX in eight ethnic groups. American Journal of Human Genetics 49: 537–544
Hall J G 1988 Somatic mosaicism: observations related to clinical genetics. American Journal of Human Genetics 43: 355–363
Hall J G 1990 Genomic imprinting: review and relevance to human diseases. American Journal of Human Genetics 46: 857–873
Harding A E 1989 The mitochondrial genome – breaking the magic circle. New England Journal of Medicine 320: 1341–1343
Hertzberg M, Mickleson K N P, Serjeantson S W, Prior J F, Trent R J 1989 An asian-specific 9 base pair deletion of mitochondrial DNA is frequently found in Polynesians. American Journal of Human Genetics 44: 504–510
Hilbert P, Lindpaintner K, Beckmann J S et al 1991 Chromosomal mapping of two genetic loci associated with blood-pressure regulation in hereditary hypertensive rats. Nature 353: 521–529
Pritchard C, Cox D R, Myers R M 1991 The end in sight for Huntington disease? American Journal of Human Genetics 49: 1–6
Richards R I, Sutherland G R 1992 Heritable unstable DNA sequences. Nature Genetics 1: 7–9
The Huntington's disease collaborative research group 1993 A novel gene containing a trinucleotide repeat that is expanded and unstable on Huntington's disease chromosomes. Cell 72: 971–983
Tsui L-C, Markiewicz D, Zielenski J, Corey M, Durie P 1992 Mutation analysis in cystic fibrosis. In: Dodge J D, Brock D J H, Widdicombe J W (eds) Current topics in cystic fibrosis. Wiley, New York, vol 1, in press
Wallace D C 1989 Mitochondrial DNA mutations and neuromuscular disease. Trends in Genetics 5: 9–13
Wicking C, Williamson B 1991 From linked marker to gene. Trends in Genetics 7: 288–293

4

FETAL MEDICINE

INTRODUCTION

The 1980s have been described as a new era in reproductive technology. Developments have included the availability of in vitro fertilisation and the donation of gametes or embryos. The fetus is able to be visualised with greater clarity using modern ultrasound equipment. Biopsy of tissue from the fetus has been simplified by the procedure of chorion villus sampling (CVS). Recombinant DNA techniques enable fetal-specific DNA to be characterised. The last two developments have made it possible to undertake many types of prenatal diagnoses. The long-term diagnostic potentials have become endless with the availability of the polymerase chain reaction. Fetal therapy has also started. Blood transfusions and surgical corrections of some defects are possible in the fetus. Attempts at in utero gene therapy are conceivable options for the future. Following on from the many changes described above has come the development of *fetal medicine units* which are primarily responsible for the care of the fetus in utero.

PRENATAL DIAGNOSIS

Sources of fetal tissues

Detection of genetic disorders by DNA testing in adults has been simplified since a blood sample serves as a universal source of DNA. In contrast, access to the fetus because of size and location is more difficult. Sources of fetal tissues for prenatal diagnosis include: amniocytes, fetal blood, chorionic villus and specific biopsy material, e.g. liver.

Amniocentesis

Amniocytes are shed fetal cells present in amniotic fluid. Amniocentesis is usually performed at the 15th week in pregnancy although there are presently trials under way to determine if earlier amniocentesis (approximately 13 weeks) is possible. Amniocytes are the sources of fetal tissue for enzyme analysis (prenatal diagnosis of metabolic disorders), DNA characterisation (prenatal diagnosis of genetic disease) and cytogenetic studies (chromosomal disorders in pregnancy). Amniocytes can be cultured, if insufficient numbers are obtained from the amniocentesis sample. The use of the polymerase chain reaction with amniocyte-derived DNA is not an ideal combination since there will be the worry that contaminating maternal cells might be present in the amniotic fluid. Although safer and simpler to perform than either fetal blood sampling or chorion villus sampling, amniocentesis has the major disadvantage of a late prenatal diagnosis during the 2nd trimester of pregnancy.

Fetal blood sampling

Fetal blood sampling is also undertaken in the 2nd trimester of pregnancy since at this time access to fetal blood vessels becomes possible. In the 1970s and early 1980s fetoscopy (fetal blood sampling from the chorionic plate or umbilical cord controlled by direct vision via a fetoscope) or ultrasound guided puncture of the chorionic plate (placenta) were the usual means to obtain fetal blood. The latter was not always pure but mixed to a variable degree with maternal blood.

Today, an experienced operator can get consistently pure fetal blood samples by ultrasound guided umbilical vein puncture (called cordocentesis). This has been possible because of the improved imaging obtainable with ultrasonography. Fetal blood can be tested in a number of ways including the isolation of DNA for prenatal diagnosis of genetic disorders or detection of infectious agents.

One disadvantage of fetal blood sampling for prenatal diagnosis is the delay inherent in this technique. Thus, termination of pregnancy for a

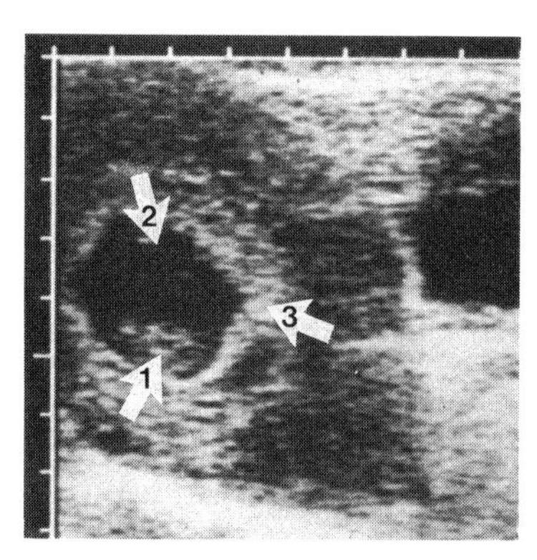

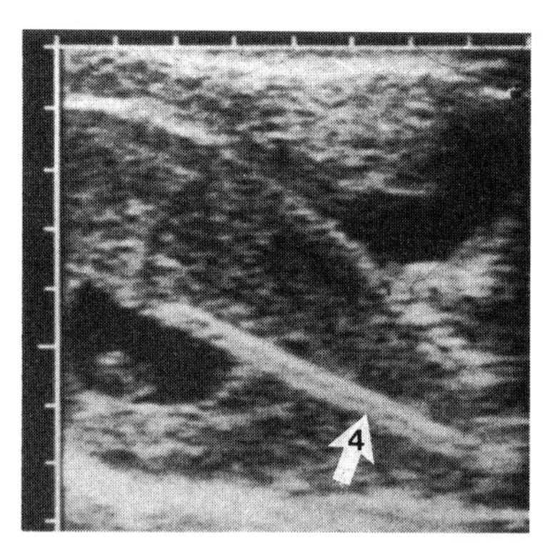

Fig. 4.1 Ultrasound pictures of a chorion villus sampling. The (→1) indicates the fetus in the amniotic sac (→2). An echo-dense region around the fetus (→3) defines the chorionic tissue. Under ultrasound guidance, a sampling probe (→4) has been placed into the chorion. The probe can be inserted through the mother's abdomen or per vaginam.

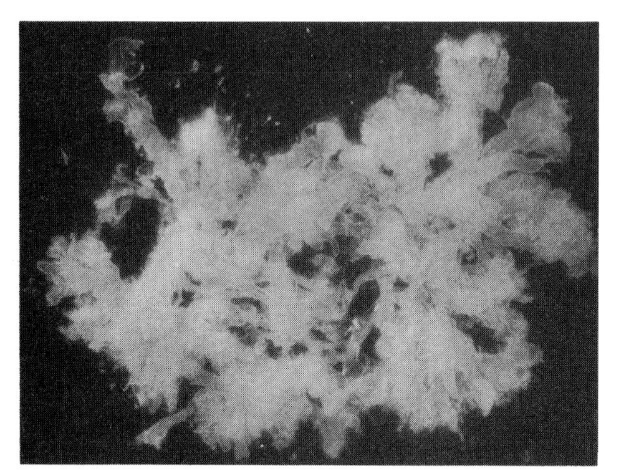

Fig. 4.2 The appearance of chorionic villus tissue under a dissecting microscope.

fetal abnormality is only possible during the 2nd trimester. The first prenatal diagnosis for thalassaemia on the basis of fetal blood sampling was reported in 1974. Fetal blood sampling for the haematological disorders (thalassaemia, haemophilia) is now only used if earlier diagnostic approaches such as chorion villus sampling (discussed below) are unable to be performed.

Chorion villus sampling

Tissue surrounding the developing embryo and called the chorion frondosum is fetal in origin and will eventually become the placental site. It can be biopsied by the technique of chorion villus sampling which was described in the late 1960s and became a routine procedure in China during the 1970s. Chorion villus sampling has now been used in many centres throughout the world and, despite some concern about possible side-effects including miscarriage and damage to the fetus, it has proven to be a reliable and safe procedure in the hands of the experienced operator.

A number of trials have looked at the safety issues and results suggest that women undergoing chorion villus sampling have a slightly less chance of a successful pregnancy outcome than those who have had 2nd trimester amniocentesis. However, actual figures for the miscarriage rate following chorion villus sampling are difficult to obtain because of the spontaneous miscarriages which normally occur during the 1st trimester of pregnancy. Some cases of limb deformity following chorion villus sampling have been reported. At this stage, this does not appear to be a significant problem although chorion villus sampling at 8 weeks or earlier may be more risky and consequently is discouraged. The usual time for chorion villus sampling is the 1st trimester of pregnancy at approximately 10 weeks (Figs 4.1, 4.2).

A chorion villus sample is an excellent source of DNA. Provided the operator is experienced and the sample is carefully dissected under a microscope, there should be no contaminating maternal tissue. This is relevant since DNA amplification by the polymerase chain reaction is frequently used.

In utero DNA testing

Genetic disorders

The indications for prenatal diagnosis can vary in

Table 4.1 Some genetic disorders for which prenatal diagnosis is available

Thalassaemia α, β
Haemophilia A, B
Cystic fibrosis
Huntington disease
Adult polycystic kidney disease
Fragile X mental retardation
Duchenne muscular dystrophy and a number of other muscular dystrophies
Retinoblastoma
Phenylketonuria
Ornithine transcarbamylase deficiency
Other less common disorders

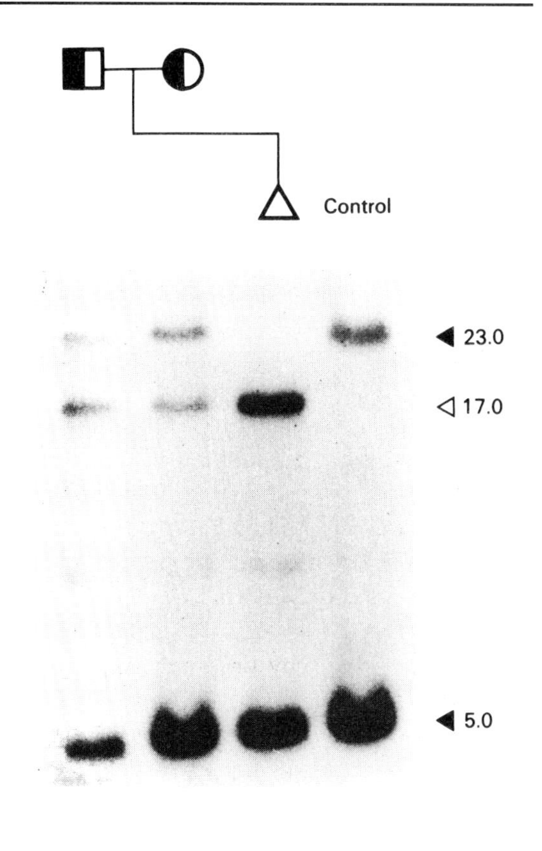

Fig. 4.3 Prenatal diagnosis of α thalassaemia.
An example of DNA mapping for a deletion disorder. The two parents in this example have heterozygous α^0 thalassaemia of the South East Asian type. On Southern blotting the mutation-specific DNA restriction fragment is 17 kb in size (◁). The corresponding normal restriction fragment is 23 kb. Fetal DNA (Δ) has only the 17 kb band. There is no 23 kb fragment. Therefore, the fetus is a homozygote for the α thalassaemia mutation and will develop haemoglobin Bart's hydrops fetalis, a fatal anaemia. A normal control is included and shows no thalassaemia-specific 17 kb restriction fragment.

different communities. In general they exclude genetic defects which are not sufficiently life-threatening (e.g. adult polycystic kidney disease) or for which effective forms of treatment are available (e.g. phenylketonuria). Nevertheless, circumstances arise when prenatal diagnosis (even if termination of pregnancy is not being considered as an option) is useful to allay anxiety in a couple. A list of some genetic disorders which can be detected prenatally is given in Table 4.1.

DNA strategies for genetic diagnosis have been described in Chapters 2 and 3. They fall into two groups:

1. *DNA mapping* for deletional disorders (e.g. α thalassaemia) or DNA linkage using polymorphisms for disorders in which the mutant gene has not been found or the defect is non-deletional (e.g. Huntington disease) (Figs 4.3, 4.4).

2. *DNA amplification* by the polymerase chain reaction to detect small deletions (e.g. cystic fibrosis) or point mutations (e.g. hereditary fruc-

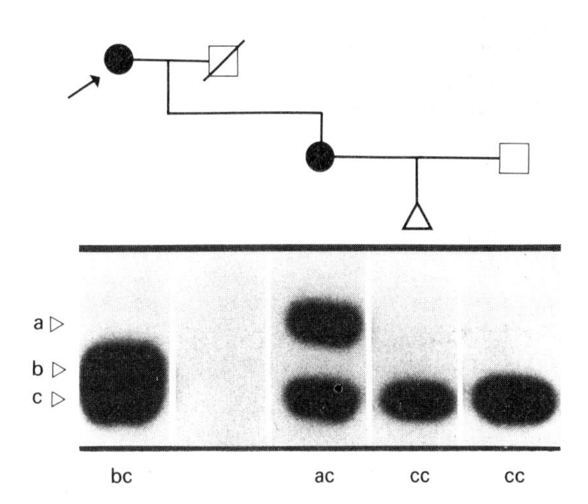

Fig. 4.4 Prenatal diagnosis for Huntington disease.
The use of DNA polymorphisms for a disorder in which the mutant gene has not been found or the defect is non-deletional. The proband (→) has Huntington disease. Prenatal diagnosis for Huntington disease has been requested by her daughter who is also affected. The consultand's unaffected father is deceased but DNA polymorphisms associated with the Huntington disease locus in this family are nevertheless still informative. DNA polymorphisms are: consultand = ac; her spouse = cc; proband = bc. Since a in the consultand could have only been inherited from her deceased father, the c marker in the consultand signifies Huntington disease. DNA from a chorion villus biopsy has the polymorphisms cc. Therefore, the fetus has Huntington disease. A potential error of 3–4% in this diagnosis reflects the recombination risk with the DNA polymorphism used.

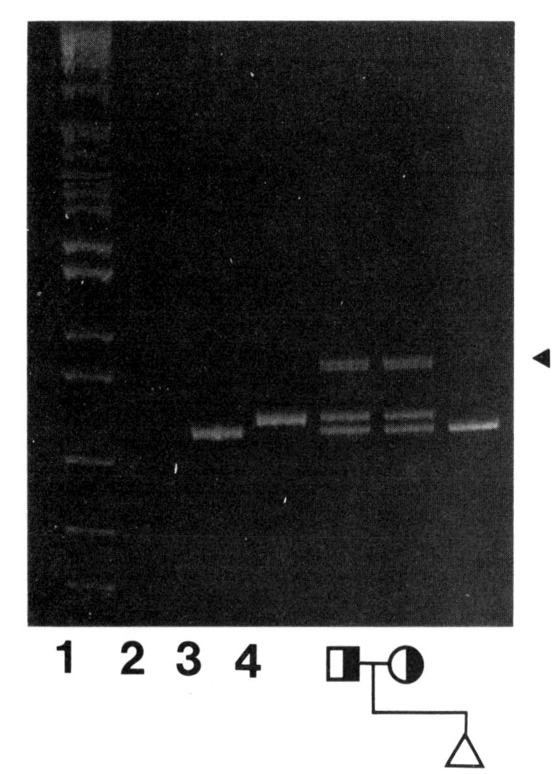

Fig. 4.5 Prenatal diagnosis of cystic fibrosis.
The use of DNA amplification by the polymerase chain reaction to detect small deletions. The figure shows an ethidium bromide stained gel on which are amplified DNA products. The oligonucleotide primers were constructed so that amplification occurred at the locus affected by the ΔF508 mutation. Track 1, DNA size markers; 2, no DNA control; 3. homozygote ΔF508 control; 4, normal control. The two parents heterozygous for the ΔF508 mutation have both the upper (normal) and the lower (ΔF508-specific) band. The difference between these two bands is 3 nucleotide bases. The chorion villus sample from the fetus (△) has only the lower band and so is homozygous for the ΔF508 mutation. DNA from the heterozygous parents shows a larger band (indicated by ◀). This is an artefact called a heteroduplex band which results from annealing of normal and mutant amplified products.

tose intolerance) (Figs 4.5, 4.6). The polymerase chain reaction technique is also useful as a rapid means of detecting DNA polymorphisms and in many cases is replacing conventional DNA mapping in this respect (Fig. 4.7).

Fetal sexing

There are two additional applications of DNA technology in prenatal diagnosis. It is possible to use Y-specific DNA probes for gene mapping or Y-specific oligonucleotide primers for DNA amplification. Thus, fetal sex can be determined from the DNA in a chorion villus sample at the 10th week in pregnancy. This has proven to be very useful in the X-linked genetic disorders such as haemophilia and Duchenne muscular dystrophy since identification of the fetus as a female excludes a severely affected individual unless X-inactivation has not been random (see Ch. 3).

Prenatal exclusion

Another application of DNA testing is in the unusual situation where one parent is at 50% risk for an autosomal dominant disorder, e.g. Huntington disease, but he/she does not want to know his/her genetic status or it cannot be reliably determined because of uninformative DNA polymorphic patterns or a key family member cannot be studied. In this case, the fetus starts off with a 25% risk of inheriting the genetic disorder. However, the risk can be defined more closely by prenatal exclusion testing, i.e. DNA markers from

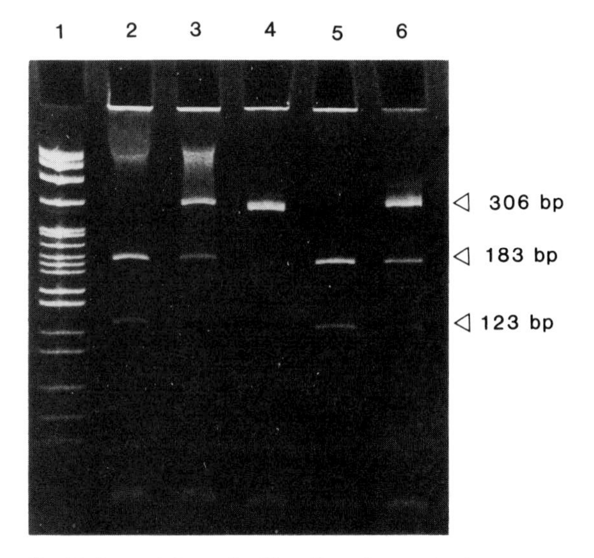

Fig. 4.6 Prenatal diagnosis of hereditary fructose intolerance.
Use of the polymerase chain reaction to identify point mutations in DNA. Hereditary fructose intolerance is a rare autosomal recessive disorder associated with metabolic defects following weaning in infancy. Subsequently, intolerance to fructose and related sugars can lead to episodic hypoglycaemia, liver disease, renal tubular acidosis and growth retardation. The most common mutation in this disease involves exon 5 of the aldolase B gene. The point mutation fortuitously creates a recognition sequence for the restriction endonuclease *Aha*II. Wild-type (normal) DNA when amplified for the exon 5 locus produces a fragment 306 bp in size (see also Fig. 2.12). Digestion of the amplified fragment with *Aha*II gives two bands of 183 bp and 123 bp in the presence of the mutation. Track 1 of the ethidium bromide stained gel is the DNA size marker; tracks 2 and 3 are homozygous and heterozygous affected controls; track 4, normal control; track 5 shows an individual who is homozygous for the above mutation and track 6 a heterozygous-affected individual.

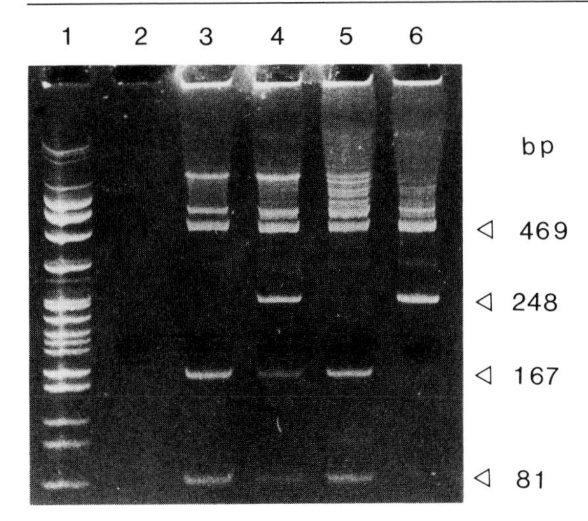

Fig. 4.7 Factor VIII (haemophilia A) related intragenic DNA polymorphism (RFLP) detectable by DNA amplification. Lanes 1, 2 = size marker, no DNA control respectively; 3–6 = amplified DNA digested with *Hin*dIII. This produces a constant 469 bp fragment and polymorphic fragments (RFLPs) of 248 bp or 167 bp + 81 bp. Lane 3 = a male with (167 + 81) RFLP. Lane 4 = female proband, a haemophilia A carrier with (248/167 + 81) RFLPs. Lane 5 = proband's spouse (167 + 81 RFLP). Lane 6 = proband's brother with haemophilia A (248 RFLP). Therefore, the marker for haemophilia A in this family is (248 RFLP). Genetic status for sons of the proband can now be predicted, i.e. (248 RFLP) will mean haemophilia A; (167 + 81 RFLP) will mean normal. Carrier status of daughters is also predictable (see Fig. 3.13).

the at-risk grandparent are sought in the fetus. If the fetus has not inherited the grandparent's chromosome 4 markers (in the case of Huntington disease), then the a priori risk is reduced to 1–5% (which represents the recombination error for the DNA probes). On the other hand, if fetal DNA demonstrates the affected grandparent's polymorphisms, the risk for Huntington disease in the fetus now increases to that of his/her at-risk parent, i.e. 50%. Irrespective of the prenatal diagnosis outcome for the fetus, the parent's risk remains unchanged (Box 4.1).

Congenital infections

The screening of pregnant women for infections which can produce fetal abnormalities is a complex issue. A number of factors need to be considered. These include: the prevalence of the infection in each population, the risk to the fetus, the availability of treatment for the infection, the sensitivity and specificity of screening tests, access to resources to provide the screening tests and the counselling which will be required.

Box 4.1 Prenatal exclusion testing to define fetal risk

The consultand, a 35-year-old woman, is at 50% risk for Huntington disease. The proband (→) is her affected father who is deceased although his DNA had been banked prior to his death. The woman is 8 weeks pregnant and wants to know if the a priori risk of 25% for her fetus can be better defined by DNA testing. She specifically requests that her carrier status remains unknown. A chromosome 4 DNA polymorphism which is estimated to be 1–5 cM (i.e. 1–5% recombination distance) from the Huntington disease locus is used to test DNA from the consultand, a chorion villus sample (Δ), her spouse and parents. Results are given below.

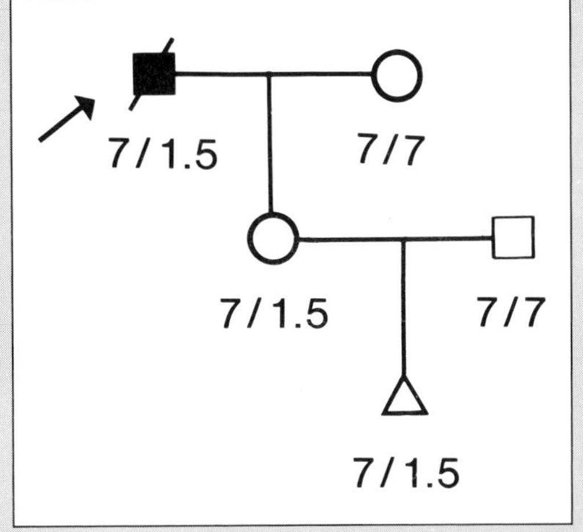

The consultand has inherited the 1.5 kb DNA marker from the proband. However, from this limited family study it is not known if this or the paternal 7 kb marker co-segregates with Huntington disease. Therefore, the consultand's risk is not altered in the study. However, the fetus has also inherited the 1.5 kb marker from the proband (since the '7' polymorphic marker in the fetus must have come from his/her father who is '7/7') and so the a priori 25% risk for the fetus now increases to 50% just like his/her mother. If the fetal DNA polymorphisms had in fact been 7/7 then the DNA marker from the proband's spouse rather than the proband himself would have been present in the fetus. In this case, the risk for Huntington disease would have been reduced as indicated in the text.

Some infections which can be transmitted vertically from the mother to her fetus and cause damage to the latter are: rubella, syphilis, hepatitis B virus, cytomegalovirus, *Toxoplasma gondii*, human immunodeficiency virus, herpes simplex virus, group B streptococcus, *Chlamydia trachomatis* and *Neisseria gonorrhoeae*. In many communities the first three are routinely screened for during pregnancy. Whether additional ones are indicated for screening will depend on local requirements.

During pregnancy, acute infections in the mother which might involve any of the above pathogens require careful investigation. Conventional microbiological detection systems, described in Chapter 5, may or may not lead to a definitive diagnosis. The fetus introduces an additional complexity, i.e. infection in the mother, even if proven definitively, does not necessarily mean that the fetus will also be infected. Some examples which illustrate the potential use of recombinant DNA tests in the above circumstances follow.

Rubella

Fetal abnormalities associated with rubella infection, particularly early in pregnancy, are numerous (Table 4.2). Although these complications have been reduced by vaccination programmes they still occur. In these circumstances pregnancy is frequently terminated unless accurate diagnostic tests are available to exclude fetal infection. Diagnosis of intrauterine infection depends on culture of the virus or measurement of a specific immunoglobulin M (IgM) response in fetal blood. An important disadvantage of the latter test is the need to wait until the 2nd trimester since the fetus does not fully develop his/her IgM responses until about 22 weeks of gestation. Viral infections per se can inhibit the fetal immunological response and so false negative results are possible in these circumstances.

Rubella-specific RNA sequences can be sought in the fetus (through blood sampling or chorion villus biopsy) by Southern blotting or the more sensitive DNA amplification by the polymerase chain reaction. Trials are presently under way to assess the utility of nucleic acid-based tests in rubella as well as a number of other intrauterine infections such as toxoplasmosis, cytomegalovirus and the human immunodeficiency virus.

Table 4.2 Fetal complications associated with rubella infection in utero

System	Complications
Cardiac	Patent ductus arteriosus, ventricular septal defect, pulmonary artery stenosis, aortic arch anomalies
Neurological	Mental retardation, seizures, deafness
Eye	Cataracts, chorioretinitis, micropthalmia, glaucoma
Immunological	Humoral and/or cellular immunodeficiency
Miscellaneous	Hepatosplenomegaly, intrauterine growth retardation, thrombocytopenic purpura, interstitial pneumonia, metaphyseal bone lesions

Toxoplasmosis

Congenital toxoplasmosis results from transplacental infection with *Toxoplasma gondii*. The affected fetus can develop serious complications such as chorioretinitis, cerebral calcifications, hydrocephalus and neurological damage. The fetus is only at-risk if maternal infection is acquired during pregnancy. Screening pregnant women for toxoplasmosis is a debatable issue. There are still no clear data to indicate the percentage of women who are at-risk and the accuracy of the serum toxoplasma-specific IgM test. Transmission rate from mother to fetus is also considered to be low so the finding of a high (or rising) IgM titre in maternal serum does not necessarily indicate infection in the fetus. To confirm the latter requires culture of amniotic fluid or fetal blood in fibroblast cell cultures. This may provide a diagnosis within a week but only half the cases will be positive. Inoculation of mice is more sensitive although it involves a longer culture time (3–6 weeks). The necessary expertise for the culture systems described would be available in few laboratories. Testing for toxoplasma-specific IgM in fetal blood has the same difficulties as described above for rubella.

To overcome these problems a number of DNA oligonucleotide primers directed at toxoplasma-specific DNA sequences have been designed.

Results obtained to date are as sensitive as the mouse culture technique with the added advantage that the time required for diagnosis is reduced to 1 or 2 days. This would have practical implications if treatment of the affected fetus is being considered. Alternatively, termination of pregnancy would be an option if congenital infection could be confirmed.

A pregnant woman who is shown to have a recent toxoplasma infection is placed in a dilemma, particularly late in pregnancy when there is no guarantee that her fetus has been infected but there is little time to wait for culture results. In these circumstances, termination of pregnancy might be avoided if a rapid detection system, such as is possible with the polymerase chain reaction, could be used to exclude infection in the fetus.

Cytomegalovirus

Similar dilemmas to those described above are found with congenital cytomegalovirus infection which can produce mental retardation and deafness. In some communities over 50% of healthy adults are seropositive. Confirmation of active infection requires either detection of viral antigens or isolation of the virus by tissue culture. The latter is time consuming. Therefore, diagnosis of intrauterine infection is difficult. A further complexity arises since our knowledge of the effects of cytomegalovirus infection remain incomplete. For example, the majority of congenitally affected children show no long-term sequelae. This presents ethical problems in relation to termination of pregnancy even if cytomegalovirus infection can be detected during pregnancy. DNA-based technology will offer additional approaches both in terms of diagnosis and research into the effects of the cytomegalovirus on the fetus.

Future directions

Fragile X syndrome

Chromosomal analysis for important genetic disorders such as Down syndrome and fragile X mental retardation are, or will soon be, replaced by DNA-based technologies. To date, prenatal testing for fragile X has relied on cytogenetic detection of the fragile X site in cultured amniocytes, chorion villus sample or fetal blood. Cytogenetic studies give an overall error rate of 5% which includes both false positive and false negative results. DNA polymorphisms have more recently become available but risk calculation from the DNA patterns is difficult. The observation that the fragile X phenotype is associated with an unstable DNA sequence has enabled a more accurate and direct DNA test for this disorder (see Ch. 3) (Fig. 4.8).

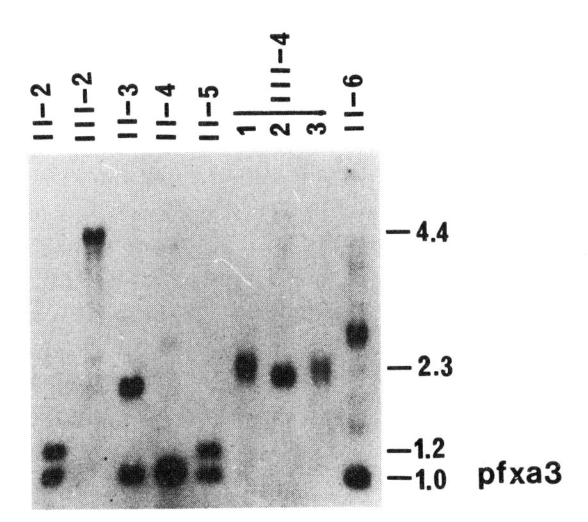

Fig. 4.8 DNA analysis for fragile X syndrome.
A Southern blot analysis which illustrates direct detection of the fragile X DNA mutation by identification of an unstable DNA sequence resulting from (CCG) triplet repeat amplification (see Ch. 3). The DNA probe is pfxa3 which hybridises to a normal X-chromosome specific fragment of 1.0 kb when DNA is digested with the restriction endonuclease *PstI*. III-4 = DNA from prenatal diagnosis of a male fetus (1 and 3 = chorion villus samples; 2 = muscle biopsy from the same fetus). II-2 and II-5 are female relatives who did not have the fragile X site detectable cytogenetically although DNA studies indicate that they are carriers, i.e. they have the normal 1.0 kb band and in addition a larger (amplified) fragment of 1.2 kb which represents the fragile X mutation. II-4 is a normal female relative with only the 1.0 kb band. III-2 is a male relative with the fragile X syndrome. He has only the amplified band which measures 4.4 kb. As expected the latter was symptomatic since *normal* transmitting males have smaller amplified fragments, i.e. between 1.1–1.6 kb. Therefore, the fetus in this case will become symptomatic with the fragile X syndrome since he has an amplified fragment of 2.3 kb. Two other female relatives, II-3 and II-6, have additional (amplified) bands in the vicinity of 2.3 kb. Both had fragile X sites also detectable cytogenetically. (Southern blot courtesy of Professor G Sutherland, Adelaide Children's Hospital, Adelaide and was reproduced with permission of the publishers of the New England Journal of Medicine.)

Maternal blood sampling

Non-invasive sources of fetal tissue will become available if it is possible to identify and then isolate the occasional fetal cells thought to be present in the maternal circulation during pregnancy. These cells are: syncytiotrophoblasts, lymphocytes and erythroblasts (nucleated red blood cells). A number of monoclonal antibodies have been described which are reported to be specific for fetal syncytiotrophoblast (the outer layer of the chorion frondosum). These cells can be isolated by flow cytometry or by using antibody-coated beads. The beads are mixed with 10–25 ml of maternal blood and are then separated from the blood on the basis of their centrifugation or magnetic properties. After washing, the mixture of beads plus fetal cells becomes the substrate for DNA analysis. Very few cells are isolated in this way but the polymerase chain reaction can be utilised to overcome the volume problem.

Lymphocytes are the second source of fetal cells. They are also few in number, e.g. one estimate would suggest that there is approximately 0.05 ml of fetal blood in the maternal circulation. However, on the basis of different maternal and paternal HLA antigen types, it would be possible to utilise antibodies to select fetal–specific lymphocytes, i.e. cells which carry a paternal HLA antigen type not found in the mother. These cells could then be tested as described above. A third and potentially very promising source of fetal tissue is the erythroblast. At present, a major problem is the lack of an antibody which can identify unique determinants on this cell.

A variant of the maternal blood sampling approach relies on the detection of paternal-specific DNA sequences in the fetus. This avoids the requirement for an antibody to isolate fetal cells. For example, Y-specific DNA sequences have been sought in maternal blood with the polymerase chain reaction. Total maternal DNA can be tested without the necessity to fractionate fetal cells or DNA since the mother does not have a Y chromosome. A number of studies have reported promising results with accurate prediction of male fetuses whose sex was confirmed following delivery. In another approach, couples of Mediterranean origin have had prenatal diagnosis by looking for an abnormal haemoglobin (haemoglobin Lepore) in the mother's blood. In these circumstances, the mothers had β thalassaemia and their spouses the haemoglobin Lepore defect. The latter is a variant haemoglobin which results from a hybrid β and δ globin gene. The hybrid can be detected by DNA amplification. Inheritance of both the β thalassaemia and haemoglobin Lepore defects (a risk to the fetus of 25% in the above circumstances) leads to a severe form of thalassaemia. By looking specifically for the haemoglobin Lepore defect in the maternal blood, it would be possible to *exclude* a severely affected fetus if the haemoglobin Lepore defect were not found. Prenatal diagnosis will only be possible 50% of the times since the identification of the haemoglobin Lepore defect does not indicate whether the fetus has inherited the β thalassaemia defect as well.

A number of questions need to be resolved before maternal blood sampling by the methods described becomes an acceptable approach for prenatal diagnosis. How specific are the antibodies for fetal cells? What is the likelihood of maternal contamination being present in the sample? This is relevant since there is little fetal tissue available and so the polymerase chain reaction becomes mandatory. In this circumstance, amplification of maternal DNA could easily occur. The approaches which avoid the use of antibodies to select for fetal cells are limited by the necessity to identify a paternal-specific HLA or DNA component which is not present in the mother. The DNA amplification procedure is technically more demanding since it is essential to include appropriate controls to ensure that fetal cells were in fact present *and* the DNA amplification had worked, otherwise false negative results will occur.

Genetic diagnosis in the preembryo

Two other approaches to prenatal diagnosis have been described in the circumstances where termination of pregnancy is unacceptable or the couple have had difficulty conceiving. Couples in the latter situation are now seeking in vitro fertili-

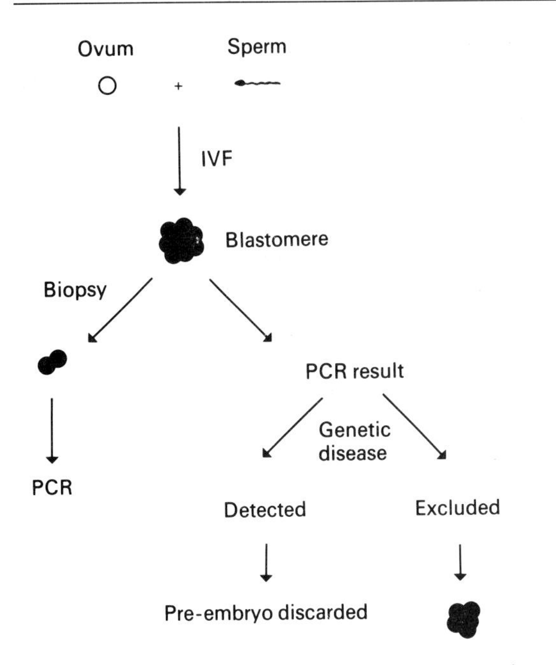

Fig. 4.9 Preimplantation diagnosis of genetic disease. Preimplantation biopsy of an in vitro fertilised oocyte isolates a few of the blastomeres of the preembryo before differentiation. Amplification of the DNA allows screening for possible genetic disease without the additional risks involved in prenatal screening methods. IVF = in vitro fertilisation; PCR = DNA amplification by the polymerase chain reaction.

sation. However, an additional complexity arises if there is a family history of genetic disease. Prenatal diagnosis following in vitro fertilisation might be an unacceptable risk after the effort the couple have had to go through to conceive.

Preimplantation biopsy of the fertilised oocyte as a source of tissue to detect genetic defects is possible since blastomeres which make up the early preembryo remain undifferentiated. Thus, isolation of a few blastomeres by microdissection will provide DNA for amplification by the polymerase chain reaction. If a genetic defect is excluded from the biopsied cells, the remaining blastomeres can be implanted into the mother and they will develop into a normal embryo (Fig. 4.9). Preliminary results have indicated that preimplantation diagnosis can be used for sexing to identify male preembryos in situations where the mother is a carrier for an X-linked disorder. Preimplantation diagnosis in a couple whose offspring was at risk for cystic fibrosis has been undertaken and the result confirmed to be correct following delivery.

Another strategy involves biopsy of the first polar body in the *preconception oocyte*. The first polar body following meiosis I divides away from the unfertilised oocyte but remains under the zona pellucida. Using microdissection techniques similar to those required for blastomere biopsy, it is possible, in some cases, to isolate the first polar body after removing the zona pellucida. Finding that the first polar body contains the mutant gene would indicate that the oocyte itself has the normal allele and vice versa. At the laboratory level, the preconception genetic diagnosis approach has been shown to be feasible in diseases such as cystic fibrosis and the haemoglobinopathies.

Further studies, particularly using animal models, are required to determine the sensitivity and specificity of the polymerase chain reaction in preembryo diagnosis. In particular, the potential for contaminating material to cause error is increased if only few cells form the template for the DNA amplification reaction.

NEWBORN SCREENING

Current strategies

Newborn screening programmes are now available in many communities. They are directed at diseases which are difficult to diagnose because they are asymptomatic in the newborn but are treatable particularly if detected early. The first example of this type of screening occurred in the 1950s and involved the testing of newborn's urine for phenylketonuria. In the 1960s, the impetus and scope for newborn screening increased with the availability of *blood spots*. These were obtained from heel pricks with the blood being spotted onto filter papers. Newborn blood spots

Table 4.3 Metabolic and genetic defects which can be screened for in the newborn (from Filkins & Russo 1990)

Screening possible on a mass population basis	Phenylketonuria Congenital hypothyroidism Galactosaemia Cystic fibrosis Maple syrup urine disease Sickle cell disease Homocystinuria Congenital adrenal hyperplasia Biotinidase deficiency
Screening possible on pilot study basis	Glucose-6-phosphate dehydrogenase deficiency Duchenne muscular dystrophy γ-Glutamyl cycle disorders (pyroglutamic aciduria) Urea cycle disorders

were initially tested for phenylketonuria using the Guthrie bacterial inhibition assay. The blood spots or 'Guthrie spots' as they are now frequently called can be utilised for a number of diseases. Screening will depend on what at-risk conditions are present in particular communities. For example, a common combination of tests involves the 'Guthrie spot' being screened for phenylketonuria, hypothyroidism, galactosaemia and cystic fibrosis (see Table 4.3 for a list of potential candidates for screening in the newborn).

Future directions

Genetic disorders

The efficacy of the present cystic fibrosis newborn screening programme involving assaying 'Guthrie spots' for immunoreactive trypsin has been proven. However, a problem with the immuno-reactive trypsin screening assay is the large number of false positives which result. In these circumstances infants have to be recalled, retested and if again positive the diagnosis is confirmed with a sweat test. This is a time consuming process. Apart from economic considerations, recall causes considerable anxiety to the involved families.

Preliminary data are now emerging to suggest that the problems relating to false positive results can be reduced by looking for the ΔF508 mutation in the *original* blood spot. The finding of a homozygote for ΔF508 will confirm the diagnosis of cystic fibrosis without the necessity for further tests. A heterozygote for the ΔF508 mutation will either be a carrier of cystic fibrosis or have a second (unknown) mutation associated with the disease. The two will be differentiated by a sweat test. The exclusion of ΔF508 (which will be the usual outcome of DNA testing) will make it more likely that the immunoreactive trypsin result was a false positive, particularly in those communities where the ΔF508 mutation accounts for over 70% of the cystic fibrosis defects. Using this strategy, some newborns will be missed since both their cystic fibrosis mutations will be non-ΔF508 but these will be few. The utility of the scheme described above needs to be assessed in prospective studies which compare it with the conventional protocol (Fig. 4.10).

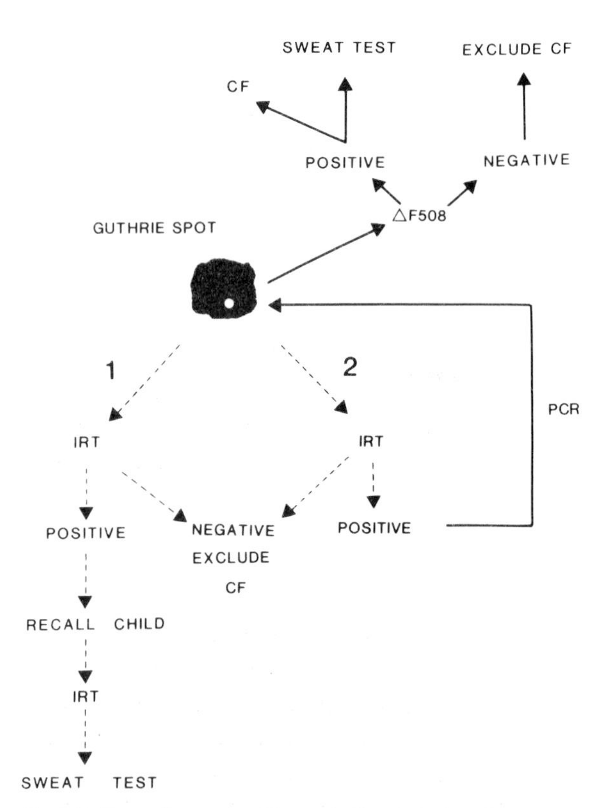

Fig. 4.10 Alternative protocol for cystic fibrosis (CF) newborn screening.
This method combines the immunoreactive trypsin (IRT) test and DNA amplification to detect the ΔF508 mutation. **1**, the conventional screening approach; **2**, the addition of DNA testing.

The applications of the polymerase chain reaction on newborn blood spots are in many ways unlimited. Therefore, decisions which involve the testing of DNA from blood spots need to be taken with care and foresight (see Ch. 9). Factors to be considered when determining the types of screening tests undertaken in newborns are: cost-effectiveness, community knowledge, the availability of counselling facilities and whether there is DNA sequence information to synthesise oligonucleotide probes for DNA amplification. Of the four, the last is the least limiting at present and will become even less so as more genes are sequenced (see the Human Genome Project discussion in Ch. 10). Therefore, conflicts are bound to arise. These reflect an increasing pool of DNA information but limited (or diminishing) resources required to ensure this knowledge is utilised appropriately (see Chs 9, 10).

Infectious disorders

The potentials of the DNA amplification procedure in the identification of intrauterine infections has already been discussed. However, controversy still exists about the significance of some infections in different communities. For example, in France and Austria there are pregnancy screening programmes to detect primary maternal infection with toxoplasmosis. In other countries, this infection is not considered a significant factor in causing fetal abnormalities. However, the lack of epidemiological data makes decisions about screening programmes difficult. The ability of the polymerase chain reaction to be automated and thus screen many samples simultaneously will be of considerable help in constructing suitable studies to resolve some of the above and related controversies.

Cytomegalovirus is an important congenital infection in humans. A rapid and inexpensive test for cytomegalovirus infection may prove useful in a newborn screening programme. DNA amplification has already been shown to detect accurately cytomegalovirus-specific sequences in urine from newborns. Infected newborns would benefit from early diagnosis which would enable special developmental follow-up and prompt diagnosis of hearing impairment. Before complications become established therapeutic interventions might offer some assistance in the future.

FETAL THERAPY

Current options

In utero fetal therapy is available in a limited number of situations. Haemolytic disease of the newborn, usually secondary to rhesus immunisation (Rh disease), can now be treated by intrauterine blood transfusions through cordocentesis until the fetus is considered to have reached an age where the risk of further transfusion is greater than the complications associated with prematurity. The ability to obtain pure fetal blood samples by cordocentesis means that the clinical progress of an affected fetus can be monitored more accurately through serial haemoglobin estimations rather than the less precise bilirubin levels in amniotic fluid.

At the molecular level, cloning and characterisation of the genes which code for the different blood groups is well under way. In particular, the isolation of genes for the Rh antigens will mean that the Rh status of the fetus will soon be obtainable by DNA typing in utero. This will be useful in the situations where the mother has already developed antibodies to the Rh antigens and the confirmation of an Rh-negative fetus will be very reassuring to the parents and their medical attendants. Conversely, the finding of an Rh-positive fetus in this circumstance will ensure closer monitoring of the pregnancy.

In utero surgery to correct abnormalities such as obstructive uropathy, abdominal wall defects, congenital hydrocephalus and a number of other conditions is available in a few centres. Fetal surgery is still in its early days, so it remains an experimental form of therapy. Nevertheless, a start has been made which enables the fetus in utero to undergo surgical manipulation.

Future directions

In utero transplantation, particularly that involving bone marrow, would be useful to correct genetic defects by somatic gene therapy where the transfer of the relevant genetic material is made into non-germline cells (see Ch. 7). Prenatal correction might be necessary to minimise end-organ damage which would develop once the fetus was born. Fetal tissues for transplantation might also provide a better source of stem cells into which can be inserted normal genes. The underdeveloped immunological system in the fetus would be useful in situations where transplantation from a genetically dissimilar donor must occur because tissue from the same person was unavailable. Implicit in the scenarios described above is an early detection system (i.e. DNA analysis) for the underlying genetic defect.

Germ cell therapy in the human is prohibited (see Ch. 7). Nevertheless, there is now a method available to biopsy blastomeres or the first polar body to determine genetic status in the preembryo or unfertilised oocyte. If the problems inherent in germ line therapy can be overcome, it is possible in the future that genetic defects identified in the preembryo could be corrected prior to implantation in the mother.

FURTHER READING

Adinolfi M 1992 Breaking the blood barrier. Nature Genetics 1:316–318 (a report on a recent conference on early prenatal diagnosis)

Anonymous (editorial) 1991 Chorion villus sampling: valuable addition or dangerous alternative? Lancet i: 1513–1515

Camaschella C, Alfarano A, Gottardi E, Travi M, Primignani P, Cappio F C, Saglio G 1990 Prenatal diagnosis of fetal hemoglobin Lepore-Boston disease on maternal peripheral blood. Blood 75: 2102–2106

D'Alton M E, DeCherney A H 1993 Prenatal diagnosis. New England Journal of Medicine 328: 114–120

Filkins K, Russo J F 1990 Human prenatal diagnosis, 2nd edn. Marcel Dekker, New York

Hammond K B, Abman S H, Sokol R J, Accurso F J 1991 Efficacy of statewide neonatal screening for cystic fibrosis by assay of trypsinogen concentrations. New England Journal of Medicine 325: 769–774

Ho-Terry L, Terry G M, Londesborough P 1990 Diagnosis of foetal rubella virus infection by the polymerase chain reaction. Journal of General Virology 71: 1607–1611

Sutherland G R, Gedeon A, Kornman L, Donnelly A, Byard R W, Mulley J C, Kremer E, Lynch M, Pritchard M, Yu S, Richards R I 1991 Prenatal diagnosis of fragile X syndrome by direct detection of the unstable DNA sequence. New England Journal of Medicine 325: 1720–1722

5

MEDICAL MICROBIOLOGY

INTRODUCTION

Traditional methods for the detection of pathogens include visual identification under the microscope, culture and growth characteristics of an organism, recognition of antigenic determinants related to the organism and the host's specific immune responses, i.e. production of antibodies. However, direct visualisation or culture is not always possible. These techniques can also be time-consuming and technically difficult. Host immune responses can be delayed or conversely may remain persistent even after resolution of a previous infection. Cross-reacting antibodies acquired from natural infections or vaccination can produce false positive results. Phenotypic variation can occur during a pathogen's life-cycle, e.g. eggs, larvae and adult forms of a species may alter depending on the stage of development, the associated host/vector and whether the organism is free-living. Thus, antibodies (polyclonal or monoclonal) or isoenzyme techniques for detection may become limiting and dependent on certain stages in the life cycle.

In the long term, many of the conventional diagnostic approaches will be complemented or even replaced by detection of DNA/RNA-specific sequences through nucleic acid hybridisation techniques or DNA amplification by the polymerase chain reaction (Box 5.1). An understanding of a pathogen's life cycle and the host's responses to the infectious agent will be enhanced by characterisation of the former's genome using molecular technology. The spread of epidemics or hospital-acquired (nosocomial) infections will be followed and characterised with more accuracy by the identification of unique DNA fingerprints for individual pathogens.

Box 5.1 Diagnostic approaches in infectious diseases

Conventional approaches

- **Direct visualisation (light, electron microscopy)**
- **Culture (isolation, characteristics, biochemistry)**
- **Antibody/antigen reaction (enzyme-linked, fluorescence, agglutination)**

Nucleic acid detection

- **Solution hybridisation (Southern blots, dot blots, pulsed field gel electrophoresis)**
- **In situ hybridisation**
- **DNA amplification by the polymerase chain reaction**

LABORATORY DETECTION OF PATHOGENS

BACTERIA

Intestinal infections

Intestinal infections occur involving a range of pathogens in an environment containing normally non-pathogenic bacterial species.

Advantages of the rDNA tests

An important consideration in intestinal organisms, for example, *Escherichia coli*, is the differentiation of pathogenic from non-pathogenic strains. There are also fastidious-growing organisms, for example, *Campylobacter* species, which will give a delayed diagnosis if detection is sought through traditional means, e.g. culture.

Recombinant DNA methods are helpful in the above circumstances. DNA probes specific for toxins associated with enterotoxigenic *Esch. coli* or enteroinvasive organisms such as *Esch. coli* or *Shigella* species are available. *Staphylococcus aureus* produces a widespread spectrum of clinical infections including food poisoning and the toxic shock syndrome. The latter is found in

Table 5.1 Gastrointestinal infections in which diagnosis by the DNA approach is/will be useful

Organism	Utility of DNA test
Esch. coli	Distinguishes the pathogenic strains by looking for virulence factors and invasiveness. Resistance to antibiotics sought by detecting the relevant plasmids
Shigella spp.	Invasiveness plasmid can be identified
Cl. difficile	Time-consuming isolation and toxin identification can be avoided
Campylobacter spp.	Fastidious growing; DNA probes useful for strain classification
Y. enterocolitica	Special culture conditions avoided; serological changes delayed
H. pylori	DNA tests compare favourably with urease tests
Cl. perfringens	DNA strain classification useful in food-poisoning outbreaks
E. histolytica	Time-consuming microscopy avoided; pathogenic/non-pathogenic can be distinguished by DNA testing
G. lamblia	Still in development stage
Rotaviruses groups A,B,C	Polymerase chain reaction more sensitive than enzyme immunoassay or electronmicroscopy; applicable to all three groups
T. solium/saginata	Distinguishable at the DNA level

women who use certain types of hyperabsorbent tampons. Clinical features of the toxic shock syndrome include fever, confusion, oedema, rash and hypotension. Genes for the various *Staph. aureus*-related toxins can be identified and distinguished by DNA testing. Detection, as well as subclassification of *Campylobacter* strains, identification of virulent *Yersinia enterocolitica, Helicobacter pylori* and *Clostridium perfringens* may now be undertaken through recombinant DNA approaches (Table 5.1).

Rotaviruses have become recognised as important pathogens in humans and animals. Three of the six identified groups (rotavirus A, B, C) infect humans. Group A is the most common cause of diarrhoea in infants and young children and accounts for a large proportion of hospital admissions for severe diarrhoea in this age group. The rotavirus itself is the most common cause of hospital-acquired gastroenteritis. Detection of rotaviruses relies on solid phase enzyme immunoassay for group A and electron microscopy or RNA characterisation for groups B and C. Recently, tests based on DNA amplification (polymerase chain reaction) have been described for each of the three groups. DNA amplification results have also shown that shedding of rotavirus group A in the paediatric hospital population occurs over a greater time period than previously considered. The significance of the latter observation in terms of nosocomial infections will need to be assessed.

Isolation of DNA probes

To obtain the appropriate DNA probe it is necessary to clone and sequence the organism's DNA to identify species-specific regions. From these, DNA probes can be isolated for nucleic acid hybridisation or primers can be synthesised for DNA amplification by the polymerase chain reaction. A useful target for a probe is *repetitive* DNA. An example of this is ribosomal RNA (rRNA). rRNA as well as the rRNA genes in chromosomal DNA provide naturally-derived amplified products which enhance their hybridisation potential. In these circumstances, non-^{32}P labelled DNA probes become feasible since the signal to noise ratio is increased because of the amplified target sequence (Fig. 5.1 and Table 5.2). rRNA probes have another useful property, i.e. their ability to distinguish different levels of relatedness between organisms. For example, whether organisms are

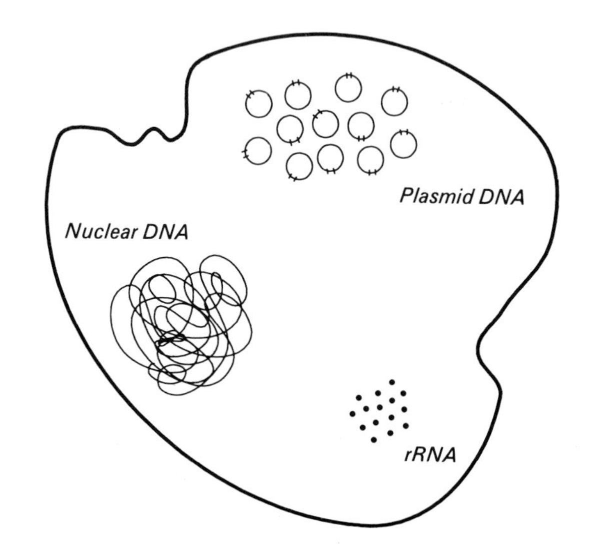

Fig. 5.1 Potential targets useful for DNA probes in a microorganism.
Successful DNA probes need to be species specific although certain rRNA probes are genus specific which can be useful clinically.

Table 5.2 Comparison of nucleic acids as potential probes

Useful properties of a DNA probe	Nuclear DNA	Ribosomal RNA and DNA	Plasmid DNA
Ubiquitous and stable	Yes	Yes	Not ubiquitous or stable
Species specific ± repetitive regions identifiable	Yes	Yes	Identification of plasmid not necessarily species-specific
Abundance	may need amplification to detect	abundant, ~10^5 copies/cell	Multiple copies may be present

bacteria or more closely related to another kingdom, e.g. fungi, can be determined by their rRNA hybridisation patterns. A genus-specific probe against rRNA will hybridise to many species of the organism in the one reaction. On the other hand, a species-specific probe will identify only a single species.

Neurological infections

Meningitis

Bacterial meningitis is a treatable disorder but delays or ineffective treatment can lead to serious consequences. Infections by the bacterium *Neisseria meningitidis* (meningococcus) can develop rapidly into a septicaemia or meningitis leading to disseminated intravascular coagulation, shock and death. Rapid diagnosis and treatment is essential in this infection. Detection of the meningococcus by Gram stain of cerebrospinal fluid and/or culture is simple and able to be undertaken in most laboratories. However, culture of cerebrospinal fluid will not give a positive result for 24 hours. Additional delays in making a diagnosis can arise if antibiotics had been used earlier on in the illness or the meningococcus does not survive transportation of the specimen (e.g. cerebrospinal fluid) from the referring centre. Tests directed at identification of the meningococcal antigen are available but these have the disadvantages of low sensitivity and false positives. DNA oligonucleotide primers specific for the meningococcus have now been described. They are likely to provide a rapid DNA amplification-based detection system for this microorganism.

Serotyping and serogrouping based on *protein antigens* are useful in epidemiological studies but do not differentiate between outbreak strains. Meningococci from the nasopharynx are often difficult to type or group. Therefore, the alternative approach utilising DNA typing by Southern blots to identify strain similarities is now being attempted (see Nosocomial infections below for further discussion on this use of recombinant DNA technology in tracing the sources of infections).

Encephalitis

Encephalitis following infection with herpes simplex virus type 1 is associated with significant mortality unless treatment is prompt. However, this type of viral encephalitis is difficult to diagnose clinically. Non-invasive investigations such as CAT scans (CAT is computer axial tomography) and EEGs (electroencephalograms) are non-specific and may be normal in the acute phase of the disease. Antibodies to herpes simplex virus in peripheral blood or cerebrospinal fluid are unreliable indicators of infection because of a delayed response and cross-reactivity with type 2 herpes simplex virus. Searching for herpes simplex virus-specific antigen in cerebrospinal fluid may enable an earlier diagnosis to be made although this approach is relatively insensitive. To obtain a specimen for direct detection and culture of herpes simplex virus requires brain biopsy, a procedure which has morbidity of its own.

The above difficulties have meant that empirical treatment with an antiviral agent such as acyclovir is usually started before a definitive diagnosis of herpes simplex virus encephalitis is made. The diagnostic dilemma will be resolved with the increasing availability of DNA amplification tests for herpes simplex virus. For example, in one retrospective study which looked at cerebrospinal fluid from patients with proven herpes simplex virus encephalitis, over 95% of cases tested were positive with the polymerase chain reaction.

Varicella zoster virus

Varicella zoster virus is a human herpes virus which produces varicella (chickenpox) or herpes zoster (shingles). The virus is highly contagious and can be transmitted by direct contact with infected skin lesions and perhaps through respiratory secretions. Reactivation of dormant varicella zoster virus, particularly in the immunosuppressed patient, is associated with debilitating shingles or potentially life-threatening neurological complications such as encephalitis, nerve palsies and meningoencephalitis. If available, electron microscopy will allow direct detection of varicella zoster virus. However, a better understanding of the virus has been hindered by inadequate systems for laboratory detection. This partly reflects the virus's slow replication cycle and the difficulties in isolating it from clinical samples such as the pharynx or cerebrospinal fluid.

Preliminary studies utilising the polymerase chain reaction to amplify DNA specific for the herpes zoster virus have shown promising results. For example, virus was identified in over 90% of throat swabs from a group of patients with clinical varicella. This is a major improvement to the more usual 5% detection rate obtainable by conventional culture techniques. DNA from the varicella zoster virus has also been found in cerebrospinal fluids from patients with herpes zoster. Results for the polymerase chain reaction were obtained within days of the appearance of vesicles and prior to serological confirmation becoming possible. Recombinant DNA-based tests for a range of neurological infections have been described. A summary is given in Table 5.3. In the near future, DNA tests will be used more frequently and the range of organisms detectable will expand considerably.

Table 5.3 Neurological infections in which diagnosis by the DNA approach is/will be useful

Infection	Microorganisms detectable
Meningitis	*N. meningitidis*, *M. tuberculosis*, enteroviruses, *B. burgdorferi*
Encephalitis	Human immunodeficiency virus, herpes simplex virus, herpes zoster virus, *T. gondii*, enteroviruses, JC virus
Encephalopathy	*P. falciparum*
Slow-virus infection	Measles (subacute sclerosing panencephalitis), Creutzfeldt–Jakob disease

Mycobacterial infections

Mycobacterium tuberculosis

M. tuberculosis illustrates some of the difficulties which can be encountered in the diagnosis of bacterial infections. Although searching for mycobacteria by microscopy (on the basis of their acid-fast staining characteristics) is rapid, it does not allow *M. tuberculosis* to be distinguished from the atypical mycobacteria (e.g. *M. avium*, *M. intracellulare*). Large numbers of organisms ($>10^4$/ml) must also be present to be detected reliably. The specificity of the acid-fast bacilli approach is close to 100% in experienced hands but the sensitivity is variable with a wide range from 10–60% being described. Further discussion on the implications of *specificity* and *sensitivity* in diagnostic assays follows below (human immunodeficiency viruses p. 101).

Culturing of mycobacteria to determine antibiotic sensitivity or to distinguish *M. tuberculosis* from atypical forms is an additional problem since it will take a number of weeks to get sufficient organisms to grow. This has practical significance since the atypical mycobacterias are unresponsive to conventional therapy. The increasing emergence of *M. tuberculosis*, as well as the atypical mycobacteria in immunocompromised hosts, particularly the acquired immunodeficiency syndrome (AIDS), has encouraged exploration for alternative and more rapid means of detection.

DNA probes distinguishing the various mycobacteria through hybridisation procedures are now commercially available. Commercial kits utilise either radioisotopically labelled DNA probes or, increasingly popular, are the kits with non-radioactive labelling methods. The latter avoid the potential hazards of radioactivity, the probe has a longer shelf-life and there is no radioactive disposal problem. As indicated previously, DNA probes are often designed to hybridise to repetitive sequences (e.g. rRNA) in order to provide sufficient target DNA/RNA to identify microorganisms *directly* in clinical specimens. This is possible with some organisms, e.g. *Neisseria gonorrhoeae* or *Chlamydia* spp. However, in the case of the mycobacteria, even DNA probes to rRNA are not sensitive enough if clinical specimens are used for direct detection. An additional step, which involves a short-term culture of the sample, is required to overcome this sensitivity problem.

Special liquid media are now available which allow rapid culture and from this an aliquot is taken for testing with the DNA probes. In practical terms, hybridisation with two sets of DNA probes (one for *M. tuberculosis* and the second for the atypical mycobacteria) provides a sensitive and relatively rapid method to diagnose mycobacterial infection as well as distinguish *M. tuberculosis* from the atypical mycobacteria. In a number of laboratories, the DNA approach described above is now being preferred over the time-consuming and slower conventional methods.

An alternative DNA strategy, i.e. amplification by the polymerase chain reaction, has been successful in limited trials in which DNA from clinical samples (sputum, gastric aspirate, biopsy, blood and abscess material) was amplified with primers derived from different mycobacterial antigens including repetitive regions and rRNA. The amplified fragments were hybridised to species-specific oligonucleotides for *M. tuberculosis, M. avium* and *M. intracellulare*. Results were obtainable within 24–36 hours. This approach showed greater sensitivity than direct examination and was more rapid than culture. In some patients with AIDS and proven mycobacterial infections it was possible to identify mycobacterial-specific DNA in their peripheral blood. The DNA amplification strategy would be particularly valuable in the case of immunocompromised patients in whom diagnosis needs to be both rapid and sufficiently specific to distinguish the atypical forms.

Tuberculous meningitis

This type of meningitis occurs in approximately 10% of patients with tuberculosis in developing countries. Despite the availability of effective chemotherapeutic drugs against the mycobacterium, mortality and morbidity remain high which partly reflects the delay in diagnosis. In one clinical study comparing three diagnostic approaches – culture, polymerase chain reaction and enzyme immunoassay for cerebrospinal fluid antibodies – it was shown that the overall sensitivities of the above three methods were 12%, 65% and 44% respectively. Although false positives were found with the enzyme immunoassay test (6%), the polymerase chain reaction technique was associated with an even greater number of false positives (12%). However, the latter represented cross-contamination of specimens since repeat analyses on uncontaminated cerebrospinal fluid samples were shown to be negative. This study confirmed both the advantages and potential problems of the polymerase chain reaction for diagnosis of tuberculous meningitis. At present, this approach remains useful within the confines of a specialised laboratory. As the technology is simplified, DNA testing would become suitable for a wider range of diagnostic laboratories. Further trials to determine sensitivity and specificity and to ensure that the potential for cross-contamination can be avoided are awaited.

Mycobacterium leprae

M. leprae cannot be cultured. Furthermore, the demonstration of an acid-fast bacillus from tissue scrapings does not necessarily indicate *M. leprae*. Serological assays and skin tests are unsatisfactory since they lack both sensitivity and specificity. Even if positive serology is obtained, it does not define current bacteriological status but may simply reflect a past infection. DNA probes specific for *M. leprae* are now available and these allow detection of the organism in biopsy material from patients with lepromatous leprosy. If there are few bacilli, such as might be found in tuberculoid leprosy or in a subclinical infection, the DNA hybridisation approaches are less sensitive. In these circumstances, a DNA amplification strategy is more appropriate. Oligonucleotide primers specific for *M. leprae* have been described and they enable the detection of few organisms. The utility of the DNA approaches for diagnosis will take time to assess since there is no satisfactory gold standard with which to compare results from DNA tests. Nevertheless, the inadequate options which the laboratory presently has at its disposal make it essential to find an alternative approach to diagnosis of the mycobacteria.

Chlamydial infections

Detecting chlamydia by conventional tests

Although chlamydia are bacteria they are difficult to culture. Thus, only a limited number of laboratories are willing to provide a diagnostic service for chlamydial infections which are very common in some regions (Box 5.2). Transport of specimens requires care since the organism is fragile and this increases costs for diagnosis as well as being a source of false negative results if organisms do not survive transportation.

Commercial kits are now available for detecting chlamydia. These are based on direct immunofluorescence or enzyme immunoassay using monoclonal antibodies. Antibodies can be directed against the outer membrane protein which is species-specific (e.g. *C. trachomatis*) or the genus-specific lipopolysaccharide (*C. trachomatis, C. pneumoniae, C. psittaci*). Direct immunofluorescence has the added advantage that a rapid diagnosis is possible and a small number of specimens can be managed (the situation which would be found in non-reference laboratories). One drawback of this technique is the potential for false positives which partly reflects the subjectivity of the operator. Enzyme immunoassay can overcome this problem through automation. Nevertheless, antibody-based assays have a lower sensitivity than culture and there is the possibility of cross-reactivity to other bacterial antigens, particularly if rectal specimens are used.

Box 5.2 Clinical manifestations of chlamydial infections

The chlamydiae are a group of pathogens causing disease in humans and a wide variety of animals. Due to extreme specialisation, they lead an obligatory intracellular existence.

Ocular effects
- Neonatal ophthalmia, inclusion conjunctivitis, trachoma (blindness), ophthalmia conjunctivitis

Genitourinary effects
- Urethritis, cystitis, epididymitis, prostatitis, proctitis, cervicitis, lymphogranuloma venereum

Respiratory effects
- Pneumonia, adult and neonatal respiratory infections, psittacosis

Reiter's syndrome
- Urethritis, arthritis, conjunctivitis, colitis

Detecting chlamydia by DNA tests

Commercial diagnostic kits based on DNA hybridisation have been produced. One such kit from the Gen-Probe Corporation (San Diego, California) is proving, like the mycobacterial kit from this company, to be reliable in the setting of a routine diagnostic laboratory. The chlamydia kit has the additional advantage that *C. trachomatis* can be detected directly in clinical specimens. DNA amplification protocols for the polymerase chain reaction are available for *C. trachomatis*. In one clinical study utilising urethral swabs from males with acute non-gonococcal urethritis, the polymerase chain reaction compared favourably with direct immunofluorescence testing. However, the frequently occurring problem of contamination made DNA amplification no more attractive than direct immunofluorescence.

Lyme borreliosis

Clinical and laboratory detection

Lyme borreliosis was first recognised as a distinct clinical entity in the USA in 1975. Since then it has been reported in a number of European countries, Canada, Asia and Australia. Lyme borreliosis is a tick-transmitted disorder shown only in 1984 to be due to the spirochaete *Borrelia burgdorferi*. Although a rare condition on a worldwide basis, it is associated with considerable morbidity and in the USA it has spread rapidly with more than 6000 human infections reported annually. Accurate diagnosis and a knowledge of the organism's life cycle is essential to initiate treatment early and to attempt preventative measures when it first appears.

There are a number of problems associated with current diagnostic approaches to Lyme borreliosis. A history of tick bite may not be obtained and a characteristic skin rash need not be present. Symptoms associated with the early stages of the disease, e.g. malaise and fever, are non-specific. Potentially debilitating sequelae of Lyme borreliosis such as arthritis, neurological and cardiac problems are not diagnostic. Direct detection of the spirochaete by microscopy rarely gives a positive result since there are few organisms present. Thus, the mainstay of diagnosis is serology using immunofluorescence or enzyme immunoassay methods to detect IgG or IgM antibodies. However, serological testing is suboptimal because antibody responses are slow to develop and there is cross-reactivity with other spirochaetes, e.g. *Treponema pallidum*, *B. hermsii* (another Borrelia but not a cause of Lyme borreliosis).

Recombinant DNA technology

Characterisation of DNA from *B. burgdorferi* has shown that there are two distinct types (1) low molecular weight, multiple copy, plasmid DNA and (2) a high molecular weight chromosomal DNA. In one study, a sequence of chromosomal DNA was identified which was specific to *B. burgdorferi* but not the closely related *B. hermsii*. Chromosomal, rather than plasmid, DNA was selected since the latter can be unstable. From the DNA sequence, oligonucleotide primers for DNA amplification by the polymerase chain reaction were constructed. A diagnostic test based on DNA amplification could now be used for *B. burgdorferi*. The test is highly sensitive and capable of detecting few organisms, although trials to date have been based on *culture-derived* organisms. Clinical specimens will be more difficult to handle because of inhibitory factors which can be present and the greater potential for contamination.

Epidemiological features

The specificity of the polymerase chain reaction in Lyme borreliosis is high although a European-derived isolate could not be detected in one study. In this circumstance a reassessment of the DNA sequence from various Borrelias enabled alternative DNA amplification primers to be constructed. These were then able to detect both European and US isolates. As further knowledge is gained about the DNA structure of *B. burgdorferi* it will be possible to design even better primers. This is important since clinical heterogeneity is geographically related. For example, late clinical features in Europe are predominantly neurological and dermatological, whilst arthritis is the most frequent complication in the USA.

The few species of *Ixodes* ticks implicated in transmission of the disease in the USA and Europe have not been found in relatively isolated areas such as Australia. Identifying the insect vector in this region will be facilitated with a test such as the polymerase chain reaction. This will be an important preliminary step in planning effective preventative measures. The value of DNA amplification is further illustrated by a study of museum tick *Ixodes* specimens for *B. burgdorferi* specific DNA. From this it was possible to show that a number of ticks from different geographical locations had been infected with the spirochaete many years before Lyme borreliosis was described.

VIRUSES

Human immunodeficiency viruses 1,2

Emergence of the acquired immunodeficiency syndrome (AIDS)

Human immunodeficiency virus (HIV), the agent which causes AIDS, is considered to have evolved from Central African monkeys some time ago. There are two types of this virus (HIV-1 and HIV-2). Human immunodeficiency virus type 2 demonstrates similarity to SIV-SM (SIV–simian immunodeficiency virus) which can cause an AIDS-like illness in some primates. There is less clear similarity between human immunodeficiency virus type 1 and SIVs except for the virus found in a chimpanzee (SIV-CPZ). Here it remains uncertain in which direction the virus has spread, i.e. chimpanzee to human or vice versa. The emergence of the human immunodeficiency virus has only become apparent in the past 2 decades. Regions affected include Africa, Haiti, the USA, South America, Europe, Australia and more recently there has been rapid spread in South East Asia. Factors associated with the worldwide pandemic include the independence of former African colonies, the greater availability of travel, changes in sexual behaviour and the increasing frequency of intravenous drug addicts. An additional factor has been the improved access to blood or blood-derived products.

In the west, AIDS has affected mostly homosexual men and intravenous drug users. Today, particularly in the developing countries, intravenous drug addicts, heterosexual spread (particularly in terms of prostitution) and mother to infant transmission are concerns in an epidemiological sense since the numbers infected in these groups is increasing. Although the risk from blood products has diminished with comprehensive screening programmes under way in developed countries (p. 101), blood and its products remain a significant source of infection in the developing countries. It is estimated that by the turn of the century there will be more than one million new cases of AIDS each year and it will be predominantly a heterosexual disease, as is already found in Africa and the developing countries.

The first human immunodeficiency virus (HIV-1) was isolated in 1984. A year later, a similar virus (HIV-2) was found in Paris from West African patients with AIDS but seronegative for human immunodeficiency virus-1. Fortunately, HIV-2 has not spread as rapidly although it has been reported in the USA, Europe and more recently in Australia. This has required changes in blood screening protocols (see p. 103). The urgency and potential implications of the AIDS pandemic has led to many research programmes to produce better diagnostic tests, more effective therapeutic regimens and a greater understanding of the virus's biology. In all these areas, recombinant DNA technology is playing a major role. In the context of the present chapter, molecular medicine is particularly relevant to AIDS since opportunistic infections are the chief cause of death. All who have AIDS will develop infections at some time in their illness. The pathogens involved in-

Box 5.3 Screening tests for HIV

There are essentially two types of screening tests available for HIV-1. The first identifies viral specific antibodies in the serum, the second looks for viral antigens. Antibody-based tests are the most frequently used, with the enzyme immunoassay (EIA) preferred because it can be automated. With this test, components of the viral antigen(s), the quality of which has improved dramatically with the availability of recombinant DNA-derived products, are coated onto a solid surface (e.g. beads) and allowed to react with the patient's serum. HIV-specific antibodies in the serum will bind to the viral antigen(s) on the beads. The beads are then exposed to a second antibody which attaches to human antibody. The second antibody is conjugated to an enzyme. The latter will produce a colour change which indirectly indicates the presence of HIV antibody. Samples positive by EIA are then confirmed by Western blotting, an assay which has higher specificity. In this test various combinations of the HIV antigens have been transferred by electroblotting onto a strip of nitrocellulose (this is the Western blot). Sera are reacted with these strips. Just as for EIA, antibody binding is detected with a second antibody which is labelled with a colour producing enzyme. In one commercial kit, results are grouped into positive, negative and indeterminate. A positive requires reactivity to at least one viral-specific antigen in each of the three major HIV groups, i.e. Gag, Env and Pol. Negatives have no reactivity with any of the HIV-specific bands and indeterminates are somewhere in between. Indeterminates require follow-up and retesting since they can represent early infection, individuals who cannot mount an adequate immune response, infected infants and normals.

clude both those that normally do not produce overt disease, e.g. *Candida albicans*, and the more exotic organisms that will not usually be seen in clinical practice, e.g. JC virus which gives rise to a demyelinating neurological disorder called progressive multifocal leukoencephalopathy.

Screening tests available

There are many commercial enzyme immunoassay kits which detect human immunodeficiency virus-1 and 2 antibodies. These are used to screen those who are at-risk and blood donors (Box 5.3). The assays are well standardised with good sensitivities and specificities. A high sensitivity (sensitivity refers to the ability to detect antibodies in sera that contain antibodies) will mean that few false negatives should occur. Specificity refers to the ability to identify as negative those sera which do not have antibodies, i.e. the number of false positives will be reduced if the specificity is high. Screening tests must have a high sensitivity so that positives will not be missed. False positives are then excluded by using a second assay which has a high specificity. In the case of human immunodeficiency virus infection, positives from the initial enzyme immunoassay are confirmed by a second test, e.g. a Western blot (Box 5.3).

The present serological assays have proven to be very effective in identifying those who are infected. However, there remain a number of unanswered questions:

1. What is the prognosis for those who are seronegative for human immunodeficiency virus but positive by DNA amplification? The latter is extremely sensitive and can detect as few as 10–20 copies of viral genome. The biological significance of this finding remains to be established. It will require longer follow-up studies.
2. What is the significance of isolated and persistent reactivity with some of the human immunodeficiency virus-1 core proteins (e.g. anti-p24 specificity on Western blots) in otherwise low risk individuals who are blood donors? To date recipients of these transfusions have not become human immunodeficiency virus-1 seropositive. To confirm that the isolated core activity detected serologically does not represent potentially infectious blood products, sera from such a group and their sexual partners were tested in one study. DNA amplification failed to detect human immunodeficiency virus-1 specific viral sequences, thereby confirming that the above are unlikely to be infected or infectious. Additional long term follow-up is also required in this circumstance.
3. Are potentially infectious individuals not being identified because of a delay in the development of a detectable antibody titre? The latent period before a positive antibody response can be detected has been estimated to be as long as 35 months. Therefore, in terms of blood donor screening, this window period as well as the possibility of a donor acquiring human immunodeficiency virus infection around the time of donation must be considered.
4. What is the human immunodeficiency virus-1 status of infants born to seropositive mothers? This will become an increasingly more important issue following the results of a recent European collaborative study which showed that the rate of vertical transmission from an infected mother to her unborn child was approximately 14%. However, serological testing will not identify the human immunodeficiency virus status of the newborn during the first 6 months of life because of the presence of maternal antibodies in the newborn's sera and because the relatively immature newborn immune system means that neonatal IgG is either not yet present or at very low levels. In this circumstance, an alternative diagnostic test is required.

DNA detection of human immunodeficiency virus-1

Human immunodeficiency virus-1 testing by DNA analysis can be directed towards the *gag*, *env* or *pol* genes of the virus (Fig. 5.2). Detection protocols utilising various combinations of these sequences have been constructed in order to minimise false positive or false negative results. In one multicentred trial comparing DNA amplification results in 105 human immunodeficiency virus-1 seronegative and 99 human immunodefi-

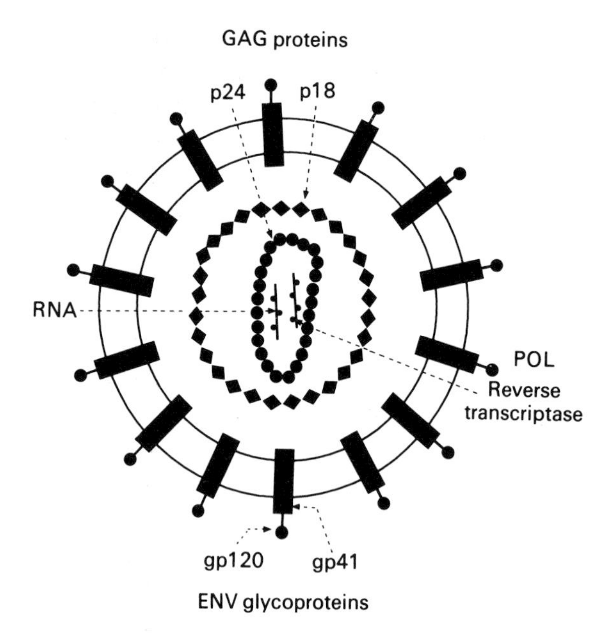

Fig. 5.2 DNA probes for detecting the human immunodeficiency virus (HIV).
Probes can be directed towards a number of genes which make up this virus. ENV (envelope) codes for a glycoprotein precursor gp160 which is cleaved to give envelope structural proteins gp120 and gp41. GAG (**g**roup specific **a**nti**g**ens) codes for a protein p55 which is cleaved to give core structural proteins including p18 and p24. POL codes for reverse transcriptase and an endonuclease.

ciency virus-1 seropositive/culture positive specimens it was shown that the overall sensitivity rate for the polymerase chain reaction was 99.0%, specificity 94.7% and there was a 3.2% (32 out of 1005 determinations) misclassification rate. The last included 1.8% false positives, 0.8% false negatives and 1.9% in which a diagnosis could not be made. The above trial illustrates the potential of the polymerase chain reaction in human immunodeficiency virus-1 testing. However, serological testing for human immunodeficiency virus-1 and 2 will remain the preferred screening method until there is better standardisation of DNA sample preparation and more stringent and uniform polymerase chain reaction protocols become available.

Cytomegalovirus

Clinical features

Infection with cytomegalovirus is significant in the following clinical situations:

- Immunosuppressed patients in whom disseminated infection can be fatal
- Transplanted organs, tissue or bone marrow where rejection or disseminated infections may occur secondary to cytomegalovirus being transmitted with the transplant or blood products given during transplantation
- Intrauterine infections, which can produce mental retardation and deafness (discussed further in Ch. 4).

Except in nurseries, cytomegalovirus is unlikely to represent a nosocomial (hospital-acquired) infection and so the viral status of immunosuppressed recipients, donors and blood products is important. With improved processing and preservation of human tissues, there has been an increase in the range and number of transplants involving solid organs (e.g. kidney, heart, heart/lung, pancreas) and tissues (e.g. bone, semen, cornea, skin etc.). Tissues can be preserved for months whilst solid organs need transplantation within hours of collection. In the context of the above, it should be noted that conventional serological or enzyme immunoassay methods for cytomegalovirus detection are associated with false positive and false negative results. Active cytomegalovirus infection can be identified through viral cell culture but this takes time and is expensive.

DNA testing for cytomegalovirus

DNA amplification tests for cytomegalovirus are now available and in the future will be useful in screening donors, transplants, blood products and recipients for this virus. However, further refinements of the DNA-based tests for cytomegalovirus are required. For example, transplantation-related cytomegalovirus infections can be detected and the progression of infection and response to treatment followed by looking in circulating neutrophils for antigenaemia (using enzyme immunoassay), viraemia (by culture) and 'DNAaemia' (by the polymerase chain reaction). Cytomegalovirus DNA which has been amplified is detectable earlier than antigenaemia or viraemia although in this situation the patient is usually

Table 5.4 Infections associated with transplantation of tissues

Transplanted tissue/organ	Transmitted infections
Cornea	Rabies, Creutzfeldt–Jakob (prion)
Semen	*N. gonorrhoeae*, *T. vaginalis*, human immunodeficiency virus, hepatitis B virus
Human heart valves	*M. tuberculosis*, various bacteria, fungi, yeasts, *T. gondii*
Bone marrow	Human immunodeficiency virus, cytomegalovirus, bacterial infections
Skin	Human immunodeficiency virus

asymptomatic. Does the DNA finding indicate imminent infection or latent (non-replicating) virus? Quantitation of amplified DNA is difficult but would be helpful in this clinical setting. At the other end of the spectrum, it has been noticed that some patients, following treatment with the antiviral drug ganciclovir, remain persistently positive for cytomegalovirus by DNA amplification testing. The biological significance of this finding remains to be determined. It may indicate those who are at risk for recurrence of infection.

Cytomegalovirus is not the only infectious agent which can be transmitted during blood transfusion or transplantation. A wide range of bacteria, fungi and viruses has been described in these circumstances (Table 5.4). In many cases serological testing will identify active infection in the donor. In other situations this will not be possible or the necessity for transplantation to be undertaken quickly excludes many of the more conventional diagnostic approaches. DNA amplification to detect infectious agents in the donor or transplant will prove very useful in these circumstances.

Hepatitis viruses

Screening blood donors

Blood Transfusion Services screen blood products for hepatitis B virus, hepatitis C virus, human immunodeficiency virus-1 (and in some regions human immunodeficiency virus-2, human T-cell lymphotropic virus types I and II), syphilis and cytomegalovirus. The last is undertaken in selected cases (e.g. blood required for an immunosuppressed recipient, fetus or neonate). Enzyme immunoassay-based assays are available for hepatitis B and C viruses and human immunodeficiency viruses 1 and 2. Positive results may require additional investigations for confirmation or assessment. Serological testing is used for cytomegalovirus and syphilis.

Polymerase chain reaction technology is now starting to be included in blood transfusion screening protocols. This partly reflects the potential of DNA amplification to become a more efficient and sensitive test. Data are also accumulating to suggest that individuals who are negative by conventional screening, e.g. surface antigen for the hepatitis B virus, can still harbour the viral genome as demonstrated by the polymerase chain reaction. This is not surprising given the exquisite sensitivity of DNA amplification over both immunoassays and DNA hybridisation tests (Box 5.4).

Post-transfusion hepatitis is caused by hepatitis B virus in up to 10% of cases. Serological assays for hepatitis B virus antigens (surface or e antigens) and antibodies (directed towards the surface, e or core antigens) are well established and will continue to play a role in screening for some time. On the other hand, hepatitis C virus is implicated in 80–90% of cases of non-A non-B

Box 5.4 Sensitivity of different viral assay methods

Current HBsAg (hepatitis B surface antigen) enzyme immunoassays can detect approximately 3×10^7 hepatitis B virus (HBV) particles/ml. These tests will not detect individuals with low levels of circulating antigens/infectious virions. DNA hybridisation tests are more sensitive, measuring 3×10^4 to 3×10^5 HBV genomes/ml serum. Even better is DNA amplification by the polymerase chain reaction which can detect as little as 10 HBV DNA copies. The biological significance of such small numbers of virions awaits confirmation, although HBsAg-negative (serological) but DNA amplification-positive blood has been shown to produce acute hepatitis if transfused into chimpanzees (Jackson 1991).

post-transfusion hepatitis. This is defined as hepatitis in patients who do not develop antibodies to hepatitis A or B viruses, cytomegalovirus, Epstein–Barr virus and there is no clinical history for other potential causes of hepatitis. Since the hepatitis C virus's single-stranded 10 kb RNA sequence was reported in 1989, it has been possible to use an enzyme immunoassay to detect antibodies against a number of viral antigens in blood donor screening. However, the original assay gave rise to many false positives and there can be a delay of up to 1 year between exposure to hepatitis C virus and seroconversion. The false positivity problem has been overcome to a large extent by using an immunoblot assay which incorporates multiple recombinant DNA-derived specific antigens for this virus. These antigens have been prepared by using DNA expression systems such as *Esch. coli* or yeast (see Chs 2, 7).

DNA testing for hepatitis viruses

Preliminary data would suggest that blood donors found to be positive by enzyme immunoassay have been infected with the hepatitis C virus. These individuals may carry the virus and be capable of transmitting non-A non-B hepatitis. However, current enzyme immunoassays do not identify all infected individuals or predict whether an individual who is antibody positive will become a chronic viral carrier. For this reason, better detection systems have been sought. Assays based on DNA amplification have been tested against panels of blood donors, drug addicts and at-risk individuals such as haemophiliacs. Because hepatitis C virus is an RNA virus, it is necessary to include an additional step in the polymerase chain reaction. This utilises reverse transcriptase to produce cDNA (copy or complementary DNA) which can then be amplified (see Ch. 2). Preliminary results for the DNA-based detection system are reassuring with rapid diagnosis and a sensitivity level comparable to the better immunoassays.

In the clinical setting, the diagnosis of hepatitis C as the cause of *chronic* non-A non-B hepatitis is possible with the enzyme immunoassays. However, in *acute* non-A non-B hepatitis there is a much lower positivity rate with enzyme immunoassay which may reflect late seroconversion. This becomes even more of a problem if the enzyme immunoassay is predominantly detecting IgG antibodies. One further application of recombinant DNA lies in the area of hepatitis induced liver disease such as cirrhosis or liver cancer. An association between chronic liver disease and the hepatitis B and C viruses has been established. Hepatitis B virus is also implicated in primary liver cancer. In the above situations, the demonstration of viral sequence(s) in liver cells is important in the long-term understanding of epidemiology and pathogenesis. DNA methodologies, both in situ hybridisation and the polymerase chain reaction, have and will become increasingly more useful in this respect (see also Disease pathogenesis, p. 108 and Ch. 6).

Enteroviruses

Clinical spectrum

Enteroviruses are important human pathogens with the gut as their primary site of infection. Despite this, gastrointestinal symptoms are rare as enteroviruses spread to other sites particularly the central nervous system. One of the enteroviruses (poliovirus) was responsible for major epidemics during the 1950s until a vaccine was produced. Today, poliomyelitis remains a problem in developing countries where vaccines may not be available or they are ineffective. Non-polio enteroviruses include coxsackieviruses, echoviruses and other enteroviruses. These viruses are responsible for many symptomatic infections which usually affect children.

The most frequent clinical presentation for an enterovirus infection is a non-specific febrile illness, with or without a rash. Clinical syndromes associated with the enteroviruses are varied and affect a number of organs (Table 5.5). Apart from their primary pathological effect, another consideration in dealing with the enteroviruses is their ability to mimic other infections at the clinical level. For example, many children are hospitalised or unnecessarily treated with antibiotics or the

Table 5.5 Organ systems involved in enteroviral infections. Although the gut is the primary site of infection, the enteroviruses quickly spread to other sites. Clinical symptoms are varied and differential diagnosis difficult

System	Infections
Neurological	Aseptic meningitis, encephalitis, poliomyelitis
Respiratory	Upper respiratory infections
Cardiovascular	Myocarditis, pericarditis
Miscellaneous	Neonatal sepsis, non-specific exanthematous illness, febrile illnesses, herpangina, Bornholm disease, hand foot mouth disease in children

antiviral drug acyclovir because of the difficulty in distinguishing between a meningeal-like illness that is a benign (aseptic) meningitis secondary to the enteroviruses and the more serious bacterial or herpes simplex virus meningitis.

Diagnostic options

The gold standard for diagnosis of enterovirus infections is viral cell culture. However, apart from the labour intensive nature of the work, a number of enteroviruses do not grow well, if at all, in culture. Suckling mouse inoculation provides a better yield for enterovirus detection but is not a practical proposition in most laboratories. Delay, if growth does occur, is unacceptable since part of the differential diagnosis of meningitis is the potentially more serious bacterial or herpes simplex virus infections. Serological testing is only available for polio and some of the coxsackieviruses. There is also the problem of a delayed host response. Antigen detection is further limited by a lack of shared antigens amongst the enteroviruses.

Nucleic acid detection systems would be ideally suited for the enterovirus infections. Already it has been shown that there are a number of similar RNA sequences shared between a variety of enterovirus serotypes. Thus, appropriately designed oligonucleotide probes for DNA amplification would identify a number of these viruses. To date preliminary results are very promising. For example, detection of enterovirus-derived RNA in cerebrospinal fluid specimens will make it possible to diagnose aseptic meningitis accurately and rapidly.

FUNGI, PROTOZOANS, YEASTS, OTHERS

Pneumocystis carinii

Immunosuppressed host

Fatal pneumonia in immunosuppressed patients can be caused by *P. carinii*. Diagnosis of a pneumocystis infection is difficult since, for practical purposes, it does not grow in vitro. Examination of bronchoscopic lavage specimens or induced sputums for organisms which stain with silver, Giemsa or toluidine blue is unsatisfactory except perhaps in the context of AIDS when many organisms may be present. More recently, the development of monoclonal antibodies to *P. carinii* has improved both the sensitivity and specificity of direct detection.

DNA studies

DNA oligonucleotide primers which allow amplification of pneumocystis-specific sequences have been described. The polymerase chain reaction will offer an alternative and perhaps better approach to direct detection by immunofluorescence. Sensitivity has been increased in the polymerase chain reaction strategy by 'oligoblotting', i.e. the amplified product is transferred to a nylon membrane in a similar way to Southern blotting. The amplified product can then be visualised by hybridization to an internal oligonucleotide which is labelled with ^{32}P.

The natural history and epidemiology of *P. carinii* infections remain to be understood. Whether pneumocystis pneumonia in patients with AIDS, following transplantation or in association with the immunosuppression of malignant disease is the result of reactivation of latent *P. carinii* infection or an acquired infection is not entirely clear. The ability to identify small numbers of organisms which can then be subtyped is essential before answers to the above will be forthcoming.

Malaria

Diagnosis of malarial infection relies on direct microscopic detection of Giemsa-stained blood smears. With this approach, a limited number of slides can be surveyed per day and the microscopist's experience becomes a limiting factor in the overall accuracy. A number of alternative strategies to malaria diagnosis are presently being evaluated. In terms of recombinant DNA technology these follow the conventional steps, i.e. hybridisation to targets which contain repetitive DNA or rRNA genes and amplification of DNA by the polymerase chain reaction. Results obtained in several field studies have been very promising when compared to conventional microscopy. Development of a better DNA test is still awaited but when complete it will be possible to have an automated test which will improve standardisation as well as increase considerably the number of samples that can be analysed. The latter becomes an important consideration in regions which are endemic for malaria.

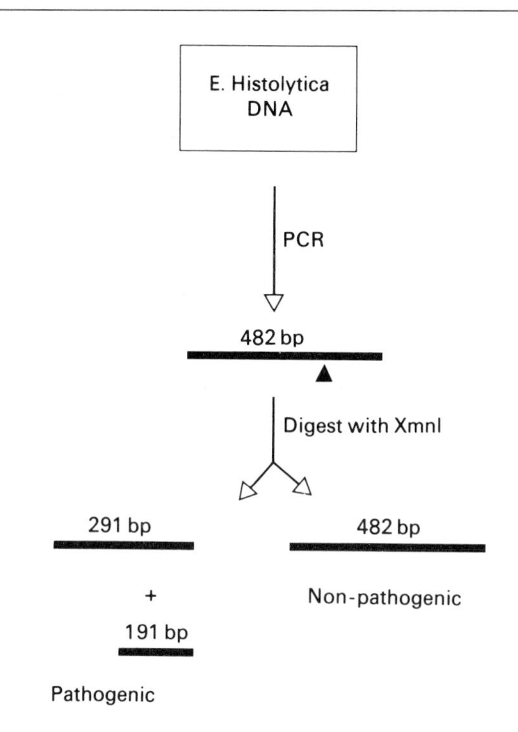

Fig. 5.3 Strategies for restriction enzyme polymorphism diagnosis in *Entamoeba histolytica*.
Infectivity rates are high and those infected can be either asymptomatic or develop complications, which is thought to reflect pathogenic variation in *E. histolytica*. Pathogenic and non-pathogenic isolates can be distinguished because in the former a 482 base pair amplified fragment has a unique *XmnI* restriction endonuclease recognition site (▲) (Tannich & Burchard 1991).

Amoebae

Entamoeba histolytica

Approximately 500 million people worldwide are infected with the amoeba *E. histolytica*. Those infected can be asymptomatic or develop clinical complications such as haemorrhagic colitis or extraintestinal abscesses. The difference between infectivity rates (which are high) and complications (relatively low) is thought to reflect the presence of non-pathogenic and pathogenic varieties of *E. histolytica*. These can be distinguished by their isoenzyme patterns or by monoclonal antibodies. DNA sequences are also different in the two forms. One approach, which is based on DNA amplification, can separate pathogenic from non-pathogenic species on the basis of a restriction endonuclease fragment polymorphism (Fig. 5.3).

Naegleria fowleri

There are six species of the amoeba *Naegleria*. The various species lack distinguishing morphological characteristics and there is serological cross-reactivity which makes differentiation difficult. One of the species (*N. fowleri*) is ubiquitous in many aquatic environments and is the agent responsible for primary amoebic meningoencephalitis, a human disease produced by aspirating warm water containing the amoebae. There is no known effective treatment for *N. fowleri* primary amoebic meningoencephalitis which is a rare disorder but usually fatal. The disease progresses rapidly and diagnosis is difficult. If there is to be any improvement in treatment, early and accurate identification of the amoeba is essential.

Oligonucleotide primers for the polymerase chain reaction have been designed to amplify repetitive DNA from *Naegleria*. By lowering the annealing temperature of the primers and target DNA it is possible to have a less-stringent DNA amplification reaction in which DNA from a number of *Naegleria* species is amplified. Increasing the annealing temperature results in more stringent amplification cycles so that only DNA from *N. fowleri* is targeted by the polymerase chain reaction. Thus, it has been proposed that environmental samples which might contain non-pathogenic species could first be tested with a screening-based DNA amplification and this product then used as the template for a subsequent and more stringent amplification to identify the pathogenic amoebae.

DISEASE PATHOGENESIS

CERVICAL CARCINOMA AND HUMAN PAPILLOMAVIRUS

Viruses can cause certain tumours in animals and these tumour viruses may contain DNA or RNA. The papillomaviruses are of medical interest because of their relation to tumours.

Conventional diagnosis

There are experimental and epidemiological data to implicate the sexually transmitted human papillomaviruses and, to a much lesser extent, herpes simplex virus type 2 in the pathogenesis of cervical cancer and its precursor, cervical intra-epithelial neoplasia. Human papillomavirus can induce cellular transformation and may also interact with other viruses or oncogenes to bring about neoplastic changes. At present, there are 60 types of human papillomaviruses which are known to infect skin or mucosal surfaces.

Detection of the human papillomavirus by conventional techniques is difficult since there are no suitable serological responses to measure and these viruses cannot be cultured. By immunohistochemical and electron microscopic means it is possible to identify the appropriate antigens but these may only be present at certain stages of infection. Histological examination of tissue biopsies will detect the characteristic changes of human papillomavirus but the different types cannot be distinguished and changes may be patchy in distribution. In one comparison, it was shown that the sensitivity of cytological criteria for human papillomavirus detection was approximately 15% compared to DNA techniques. Thus, it is not surprising that epidemiological studies to date have been unable to confirm definitively the experimental evidence implicating human papillomavirus in the aetiology of cervical cancer. This dilemma may be resolved with the use of the more specific and sensitive DNA tests which will make it possible to detect, type and characterise the human papillomaviruses.

DNA studies

To date all standard DNA approaches have been tried, i.e. DNA Southern blotting, dot blotting, in situ hybridisation and the polymerase chain reaction (Fig. 5.4). Preliminary results would suggest that human papillomavirus types 16 and 18 are

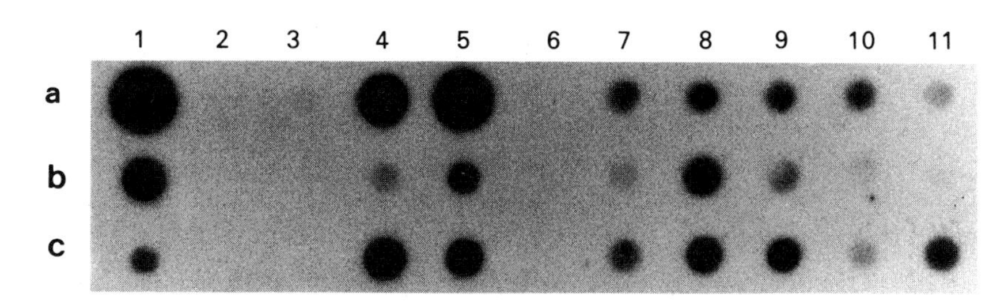

Fig. 5.4 DNA testing by dot blotting for the detection of the human papillomavirus type 11.
Detection of papillomavirus by conventional techniques is difficult which has hampered epidemiologial studies of its role in cervical carcinoma. Lane 1 is positive control DNA in three concentrations a,b,c which range from 125 pg, 12.5 pg, 1.25 pg. Lanes 2 and 3, 6 negative controls. Lanes 4–11 demonstrate 24 different clinical specimens tested for papillomavirus (dot blot courtesy of Professor Y Cossart, Department of Infectious Diseases, University of Sydney).

more likely to be associated with high grade squamous precancer (called cervical intraepithelial neoplasia III), cancer of the cervix, and endocervical adenocarcinomas. In contrast, human papillomavirus types 6, 11 are found in more benign lesions, e.g. condylomas or low grade dysplasia (cervical intraepithelial neoplasia I). Data obtained from the more objective and accurate DNA assessment of human papillomavirus infections in non-invasive and invasive carcinoma of the cervix have provided additional information on the frequency of the virus in various lesions.

Just as in the human immunodeficiency virus example (Box 5.4), the exquisite sensitivity of the polymerase chain reaction has enabled very few organisms to be detected and the biological significance of this remains to be determined. Thus, the implication of human papillomavirus type 16 in cervical cancer becomes less clear when studies utilising DNA amplification show that the virus is also detectable in populations of normal, sexually active women. Furthermore, the potential for false positive results secondary to contamination when using the polymerase chain reaction becomes even more significant when smaller numbers of microorganisms are involved. Technological improvements will be required to remove this potentially confounding factor.

Questions will remain for some time about the role of human papillomavirus in cervical cancer and why some cases of non-invasive neoplasias regress and other progress to an invasive form. The first step towards answering these questions is in place now that there are DNA methods for viral identification, typing and quantification. Assessment of viral aggressiveness and/or interaction with other genetic or environmental factors (e.g. cigarette smoking) may also be solved through DNA technology (see Ch. 6 for further discussion).

VIRULENCE FACTORS

Hepatitis B virus and liver disease

The hepatitis B virus is a DNA virus. The intact virus particle (virion) is also called a Dane particle. Three different antigens are associated with this virus: HBsAg (hepatitis B surface antigen); HBcAg (hepatitis B core antigen) and HBeAg (hepatitis B e antigen). Following infection with the hepatitis B virus, antibodies to the various antigens described are formed. These assist with recovery from infection and the maintenance of immunity for several years thereafter. Occasionally, antibodies are not formed but antigenaemia persists. In these circumstances, the infected individual becomes a chronic carrier. The finding of hepatitis B e antigen in acute cases or the chronic carrier state also correlates with infectivity.

The hepatitis B virus was first identified in 1963. Today, over 300 million people are chronically infected with this virus. A wide spectrum of liver disease is associated with infection. Clinical features include: (1) acute self-limited hepatitis, (2) chronic hepatitis which can progress to cirrhosis including liver failure or primary liver cancer, (3) an asymptomatic carrier state and (4) fulminant hepatitis. The last is a rare complication of hepatitis B infection but leads to high mortality. Liver damage in this instance is not thought to reflect a direct hepatotoxic effect of the virus but rather an exaggerated immune response. The role of the hepatitis B virus in primary liver cancer is well proven and will be discussed further in Chapter 6.

There are a number of controversial issues surrounding the hepatitis B virus. For example, what is the role played by hepatitis B virus in those who are hepatitis B surface antigen negative but have chronic liver disease? Studies based on serological and DNA hybridisation techniques have shown conflicting data. More recently, hepatitis B virus particles both circulating and in liver cells have been detected by the polymerase chain reaction in hepatitis B surface-antigen negative patients with liver disease. Whether they represent intact virions or free viral DNA shed from necrotic cells remains to be determined. If the former can be proven, then hepatitis B virus plays a much more significant role than hitherto suspected in chronic liver disease and perhaps primary liver cancer which is hepatitis B 'negative'. Individuals who are persistently positive for hepatitis B surface antigen may also be positive

Box 5.5 Mutation of hepatitis B virus and virulence

The hepatitis B virus e protein comprises both the hepatitis B core (capsid) protein and a pre-core component. The latter contains a signal peptide which enables secretion of the hepatitis B virus e protein into the serum through cellular secretion pathways. Hepatitis B e antigen-negative patients with anti-e antibodies and severe liver disease were studied by using the polymerase chain reaction to isolate hepatitis B virus DNA from their sera. Sequencing of DNA identified a mutation which produced a premature stop codon at the end of the pre-core region. Thus, synthesis of the hepatitis B e protein could not be completed. A similar type of mutation in the pre-core region has now been detected in a number of other studies including one where the mutation was associated with fatal fulminant hepatitis. How this mutation in viral DNA leads to a more severe form of liver disease is presently being investigated particularly in relation to possible immunological sequelae which might result from the body's failure to develop tolerance to the e antigen rather than from a direct cytopathic effect of the virus (Liang et al 1991).

for hepatitis B e antigen and/or develop antibodies to the e antigen. The significance of hepatitis B e antibodies has at times been unclear. One suggestion was that these antibodies indicated loss of viraemia. Recent molecular analysis of hepatitis B virus DNA has shown how serological assays can confuse the picture, e.g. it is possible to have persistent viral replication leading to very severe liver disease despite the development of hepatitis B e antibodies (Box 5.5).

Vaccine-acquired poliomyelitis

Although rare, poliomyelitis may occur following administration of the oral attenuated poliovirus (Sabin) vaccine particularly types 2 and 3. It is known that these two Sabin type viruses revert to the wild-type (neurovirulent) phenotype on passage through the human gastrointestinal tract. To understand further the change in virulence of the poliovirus, its RNA was sequenced and a point mutation uracil (vaccine) to cytosine (wild-type) was detected at position 472 in the non-coding region of the viral genome. This was a consistent finding and the significance was further reinforced when fecal samples were progressively examined following oral intake of the Sabin vaccine. In one study, at 24 hours after vaccination, polio virus in fecal samples demonstrated uracil at position 472. At 35 hours, a mixture of viruses with either uracil or cytosine at the 472 locus was present. From 48 hours, all viral isolates had cytosine at this position. Two questions remain unresolved. Why should there be selection in the gut for 472 cytosine-bearing polioviruses? What is the molecular basis for the increase in neurovirulence which alone does not produce the complete poliomyelitis phenotype? A clue to the latter may come from molecular analysis which suggests that the 5′ non-coding region at position 472 has functional significance since nucleic acid comparative studies show this area to be highly conserved between the various polio serotypes and a rhinovirus.

Toxigenic bacteria

As mentioned above in the section 'Intestinal infections' (p. 93), a number of bacterial toxins are identifiable through recombinant DNA testing and so their clinical effects can be predicted. Molecular technology will also enable the toxin's mode of action to be understood. For example, toxigenic strains of *Clostridium difficile* are responsible for pseudomembranous colitis which follows the use of broad-spectrum antibiotics. The organism produces two toxins: A and B. The latter requires previous tissue damage by toxin A before it is toxic. How the two toxins exert their effects is unclear. The *Cl. difficile* toxin A gene has now been cloned into *Esch. coli* and expressed. It has been shown that all known toxin A effects (haemagglutination, cytotoxicity, enterotoxicity and lethality) can be elicited by the cloned toxin A gene. The gene is now being subcloned into defined segments and the various toxin effects analysed in more detail, i.e. what is the mimimal size of DNA required? How is expression of the toxin regulated?

Antibiotic resistance

A DNA probe to detect the β lactamase gene has been isolated and will enable resistance to the antibiotic penicillin to be predicted. Similarly, resistance to the aminoglycoside antibiotics can be sought by identification in bacteria of the aminoglycoside 2-*O*-adenyltransferase gene.

IN SITU HYBRIDISATION

Pathogenesis of many infections, particularly viral ones, has been deduced from experimental strategies based on light and electron microscopy, cell culture and immunoassay. To these research tools can now be added nucleic acid (DNA, RNA) probes for in situ hybridisation. Advantages provided by this technique include the ability to detect latent (non-replicating) viruses and to localise their genomes to nuclear or cytoplasmic regions within cells. Tissue integrity remains preserved during in situ hybridisation and so histological evaluation can also be undertaken. Nucleic acid probes or the hybridisation conditions can be manipulated so that a broad spectrum of serotypes will be detectable. This is particularly valuable in those emerging infections where the involved serotypes are unknown.

The utility of in situ hybridisation to detect viral DNA/RNA sequences is illustrated by animal and human studies of enterovirus-induced cardiomyopathy. Enteroviruses of the group B coxsackievirus (types 1–5) are considered to be the most common causes of viral myocarditis. Infections by these viruses can produce life-threatening arrhythmias or acute dilated cardiomyopathy. Treatment of the latter may require cardiac transplantation.

Molecular cloning and characterisation of the single-stranded genomic RNA from the cardiotropic B3 coxsackievirus enabled cDNA to be made. With this as a probe, it was possible to show by in situ hybridisation in a mouse model for acute dilated cardiomyopathy, that B3 coxsackievirus could be detected in muscle cells (myocytes) and small interstitial cells thought to be fibroblasts in a multifocal and random distribution in heart muscle. Lymphocytes involved in the inflammatory process were not infected. Progression of the infection could also be seen from areas with myocardial fibrosis to as yet uninfected myocytes indicating possible cell-to-cell spread of the virus. In human patients with acute dilated cardiomyopathy, the incidence of enterovirus infection has been estimated to be approximately 30% using in situ hybridisation on cardiac biopsies.

More controversial is the role of viruses in the chronic dilated cardiomyopathies. Using in situ hybridisation, individuals with more chronic forms of dilated cardiomyopathies were also shown to have persistence of enterovirus-specific RNA sequences in their cardiac myocytes. Study of serial endomyocardial biopsies up to 12 months after the onset of dilated cardiomyopathy has identified enteroviral RNA. Therefore, cardiotropic viruses may be playing an even greater role than previously considered in the aetiology of the cardiomyopathies, although it remains to be excluded that in the more chronic conditions the enterovirus infection is secondary to heart muscle damage.

ANTIGENIC VARIATION

African trypanosomes, e.g. *Trypanosoma brucei*, are unicellular parasites that infect a variety of mammals. Infection is usually fatal if left untreated. The life cycle of the trypanosome can be divided into two main stages: blood stream forms in mammals and the procyclic form in the midgut of the vector, *Glossina* (tsetse fly). Each form is covered by a coat composed of characteristic surface proteins: the variant surface glycoprotein in blood stream forms and procyclin in procyclic forms. Variant surface glycoproteins are highly immunogenic and antibody-mediated destruction of trypanosomes is an important host response to infection. To evade the host's immune response, *T. brucei* undergo continuous switching of their variant surface glycoproteins.

At the molecular level, the genetic events

underlying variation in these glycoproteins is being defined. The gene for each variant surface glycoprotein is present in all trypanosomes of a stock regardless of whether the gene is being expressed. Thus, antigenic variation is not produced by recombination of coding segments as occurs with the immunoglobulin genes (see Ch. 6). From a repertoire of several hundred variant surface glycoprotein-specific sequences, only one is transcribed at any time. Antigenic variation occurs either through DNA rearrangements at the expressed site or through activation of a new expression site. Further work is required at the molecular level to unravel this parasite's interesting strategy to escape antibody responses generated by the host. Other aspects of the trypanosome's life-cycle may also be explained through a molecular approach, e.g. infection may persist for months in an animal and can show waves of serologically distinct parasites appearing at 7–10 day intervals. The mechanism for this remains to be determined.

EPIDEMIOLOGY OF INFECTIOUS DISORDERS

Conventional typing of pathogens based on their phenotypes (phage susceptibility; biochemical, antigen profiles; antibiotic resistance; immune response; fimbriation) is not always successful in epidemiological studies. A changing spectrum of infectious agents particularly in the immunocompromised host and in hospital outbreaks has meant that newer epidemiological approaches are required to complement or replace the more traditional methods. Four DNA strategies are possible in these circumstances:

- Characterisation of pathogen DNA by nucleic acid hybridisation
- Plasmid identification
- Chromosomal DNA banding patterns
- The polymerase chain reaction.

NOSOCOMIAL INFECTIONS

It is important in hospital-acquired infections to have methods to type pathogens to enable their source(s) and mode of spread to be identified. A composite of DNA restriction fragments is able to provide a unique DNA pattern for an individual. DNA fingerprinting, as it is popularly known, is becoming increasingly more established in forensic practice (see Ch. 8). The DNA/RNA profile of infectious agents is similarly being exploited for diagnostic or epidemiological purposes. One approach to fingerprinting microorganisms is called chromosomal fingerprinting or RFLP (restriction fragment length polymorphism) analysis. Digestion of genomic DNA with one or more restriction enzymes and then hybridisation to a DNA probe will produce a number of fragments. Probes can be selected from a conserved region of the microbial genome if less discrimination is required and so species or even subspecies within an organism will be detected. On the other hand, choosing a DNA probe from a highly variable (i.e. polymorphic) region of the genome will allow discrimination between closely related organisms.

The above can be illustrated by reference to the *Legionella* spp. These are ubiquitous bacteria which are present in domestic and industrial water systems, tanks and other sources of pooled or collected water. More than 30 different species of the genus *Legionella* have been described. The bacteria cause Legionnaire's disease a proportion of which is considered to be nosocomial in origin. Both sporadic cases and outbreaks of this respiratory disorder have been reported. The organism (e.g. *L. pneumophila*) is difficult to grow and it is frequently necessary to obtain bronchial aspirates or lung biopsies for direct detection. Serological testing is the mainstay of diagnosis although a number of DNA-derived tests are now available. These frequently involve DNA to DNA/RNA hybridisation with probes which are

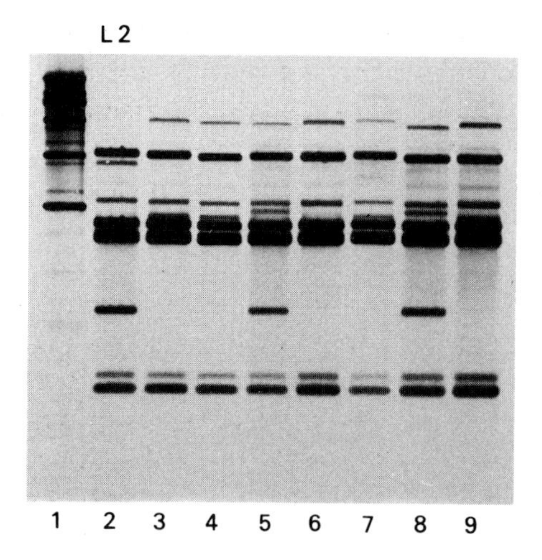

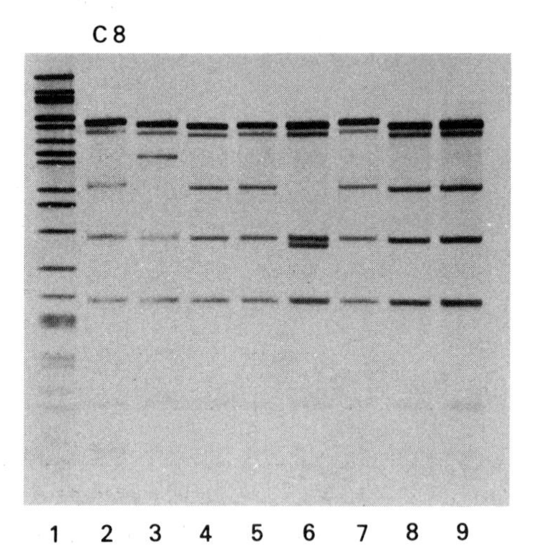

Fig 5.5 Typing of legionella for epidemiology.
The DNA/RNA profile of infectious agents can be used to 'fingerprint' them enabling their source and mode of spread to be identified. This is particularly important in hospital-acquired infections. Chromosomal fingerprints (RFLPs) of *Legionella longbeachae* serogroup 1 strains which are usually less common causes of human infection. Microbial DNA was obtained from various sources (human infections, environment – in this case commercial potting mixes) and digested with two restriction enzymes (*Hin*dIII and *Bam*HI). The two DNA probes used for hybridisation were obtained from random clones of the organism's DNA in a λ phage (L2) or a cosmid (C8). Lanes 1 are DNA size markers; the remaining lanes are the isolates in identical order for each gel. Considerable variability between the eight samples of the one serogroup can be seen (Southern blot courtesy of Dr J Lanser, Institute of Medical and Veterinary Science, Adelaide).

often rRNA-specific. Typing of legionellae for epidemiological purposes follows similar strategies (Fig. 5.5).

More recently, another recombinant DNA approach to fingerprinting bacterial genomes has become available. This is illustrated by the following study which confirmed that an outbreak of legionella in patients admitted to an intensive care ward was caused by *L. pneumophila* isolated from water taps in that ward. The DNA strategy involved pulsed field gel electrophoresis. The feature of this technique is that it allows *large segments of DNA* to be separated by making use of restriction enzymes such as *Not*I, *Sfi*I which digest DNA very infrequently. The upper limit for resolution in conventional DNA gel electrophoresis is approximately 30 kb whereas for pulsed field gel electrophoresis it is approaching 10 Mb (10×10^3 kb) (see Ch. 2 for further discussion). Pulsed field gel electrophoresis is particularly attractive in microbiological work because the genomes of infectious agents are relatively small. For example, total DNA from *Legionella* spp. can be cleaved into a limited number of fragments (5 to 10 with *Not*I depending on the strain) and these

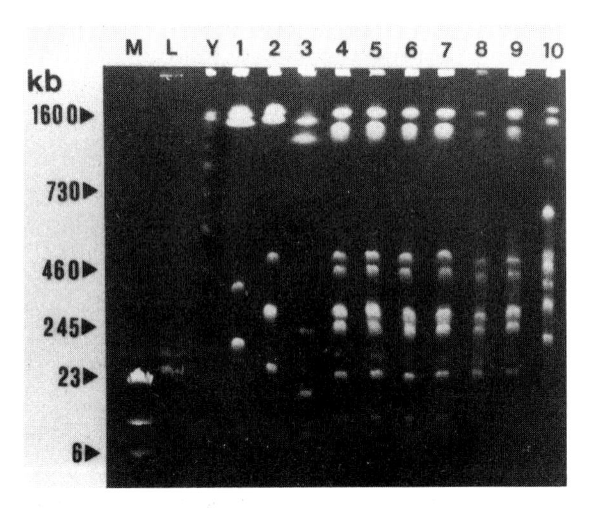

Fig. 5.6 Analysis of a legionella outbreak in an intensive care ward.
Pulsed field gel electrophoresis map of *Legionella pneumophila*. Bacterial-derived DNA has been digested with the rare cutting restriction enzyme *Not*I which produces large segments of DNA. Tracks M,L,Y are molecular weight size markers. Tracks 1–3 and 10 are control samples of *L. pneumophila* (track 10 is DNA from bacteria obtained from a patient who acquired legionella pneumonia outside a hospital environment). Tracks 4–6 are bacterial isolates taken from three individuals who developed legionella pneumonia whilst in the intensive care unit of a hospital. Tracks 7–9 are bacterial isolates taken from the water supply in the same intensive care unit. It can be seen that legionella-specific DNA in tracks 4–9 are identical (compared to the different DNA patterns seen in tracks 1–3 and 10). This identifies the source of the *L. pneumophila* in the three patients (pulsed field gel map courtesy of Dr M Ott, Wurzburg, Germany).

fragments or 'fingerprints' can be directly visualised from an ethidium bromide stained gel thereby avoiding the requirement for a DNA probe and a hybridisation step. Identical restriction fragment patterns from the *L. pneumophila* in the patients described above and the water tap-derived strains identified the common source. Legionella isolates from other geographical regions were distinguished by their different pulsed field gel electrophoresis patterns (Fig. 5.6).

Not surprisingly, a third DNA approach utilises the polymerase chain reaction to amplify DNA. With this technique the identity or relatedness of organisms is based on DNA sequence or variations in single or small segments of DNA. DNA sequencing is now possible by a number of methods and the technology will only improve as the Human Genome Project progresses. This is in fact a misnomer since components of this project involve the sequencing of model organisms such as yeast, *Esch. coli* and *Mycobacterium tuberculosis* (see Ch. 10). Once a segment of microbial DNA is amplified, it is also possible to use it for detection purposes or for comparison with amplified fragments from other organisms by using techniques such as denaturing gradient gel electrophoresis or the chemical cleavage method (see Ch. 2). The latter has been applied successfully in distinguishing RNA from different dengue virus isolates.

HOST–VECTOR IDENTIFICATION

Leishmania illustrates some of the difficulties facing epidemiologists in their study of infectious diseases. Conventional identification of leishmania involves biochemical isoenzyme characterisation or detection by monoclonal antibodies. The former approach is difficult since a minimum of 10^7 cells is required and the cells need to be grown in experimental animals or in culture. Monoclonal antibodies will identify leishmania but are not satisfactory in distinguishing the various species. To find the insect vectors for leishmania, it may be necessary to dissect thousands of sandflies looking for the organism. Even when found, it is difficult to be absolutely sure that the leishmania is human-specific and has not come from other animals such as birds and reptiles. In view of the above, it is understandable that prevalence figures, particularly in tropical and subtropical countries where leishmaniasis is an important public health problem, are inadequate.

Alternative diagnostic strategies based on DNA identification will help to overcome many of the above problems. Parasites from cultures or within insect vectors can be smeared onto microscope slides. DNA is gently denatured with alkali to make it single-stranded but at the same time retain the cell's morphology. In situ hybridisation with DNA probes will then allow this protozoan's detection within its potential vectors. Any leishmanias found can be discriminated to the subspecies level if necessary. The potential for a polymerase chain reaction-based DNA strategy to identify both hosts and vectors in bacterial infections has been discussed above in reference to Lyme borreliosis and would apply equally to the leishmania example.

DRUG RESISTANCE

Malaria

Two major classes of drugs used in treatment of malaria are the antifolates (sulphonamides or the inhibitors of dihydrofolate reductase such as pyrimethamine or proguanil) and the cinchona alkaloids (e.g. chloroquine and its derivatives). The spread of drug-resistant *Plasmodium falciparum* and to a lesser extent *Plasmodium vivax* has made the prophylaxis and treatment of malaria a major public health problem. Chloroquine-resistant *P. falciparum* is present in South America, Africa and South East Asia. Resistance is thought to have spread in the past 40 years from a limited number of foci in South East Asia and South America. The epidemiology is consistent with multigenic effects, i.e. rare events occurring initially in South East Asia or South America and then spreading slowly worldwide. In contrast, resistance to pyrimethamine and proguanil has arisen independently in many different regions where these drugs have been used which would suggest a single gene is involved. Identification of the molecular basis for resistance is now possible. Information derived from this will enable modification to the above drugs in terms of regimens used or possibly their chemical structure so that drug efficacy can be restored or maintained for a longer time period.

Resistance to the dihydrofolate reductase inhibitors pyrimethamine and proguanil occurs on the basis of point mutations in the parasite's dihydrofolate reductase inhibitor gene. These point mutations are thought to exert their effect by inhibiting binding of the antimalarial drugs to the enzyme's active site. The inhibition can interfere with binding by both the drugs or can be selective for one alone (Table 5.6). As the molecular defects become more comprehensively characterised it will be possible to attempt manipulation of the dihydrofolate reductase inhibitors to overcome the effects of the point-mutations on binding, e.g. data in Table 5.6 would suggest that combinations of pyrimethamine and proguanil are better than either drug given alone as the mutations which lead to resistance are different.

Less is known about chloroquine resistance although putative gene(s) for this effect have been mapped to chromosomes 5 and 7 in the parasite. Chloroquine-resistant *P. falciparum* parasites remove this drug 40–50 times faster than do drug-sensitive parasites. The actual basis for the increased efflux is unknown but it appears to be the same in all resistant parasites. Resistance has been reversed in vitro with the drug verapamil but how this interferes with rapid efflux is unknown. Some properties of resistance to chloroquine are similar to that found with cytotoxic drug-related resistance produced by stimulation of P-glycoprotein (see Ch. 7 for further discussion of P-glycoprotein). In the near future, candidate gene(s) will be identified in the chromosome loci targeted by DNA mapping studies to be implicated in drug resistance. From this it will be possible to understand why rapid efflux occurs, how verapamil works and whether the problem can be bypassed so that chloroquine, a safe and effective drug, can be reintroduced for malarial therapy.

Table 5.6 DNA changes in the dihydrofolate reductase (DHFR) gene leading to drug resistance in malaria (Wellems 1991).
Point mutations lead to single amino acid changes which are thought to exert their effect by inhibiting drug binding to the enzyme's active site

Drug	Mutation in DHFR	
	Residue No.	Amino acid change
Pyrimethamine	108	Serine → asparagine
Increased resistance	108	Serine → asparagine
to pyrimethamine	51	Asparagine → isoleucine
	59	Cystine → arginine
Proguanil	108	Serine → asparagine
	16	Alanine → valine
Pyrimethamine and	108	Serine → asparagine
proguanil	164	Isoleucine → leucine
	59	Cystine → arginine

Methicillin-resistant Staphylococcus aureus

Methicillin-resistant *Staphylococcus aureus* is an important cause of nosocomial infections in hospitals. Infection control procedures are able to identify outbreaks of this infection. The source(s) of the methicillin-resistant *S. aureus* must next be sought to determine if there has been a breakdown in infection control practices. For this to occur the different bacterial subtypes must be distinguished. Strategies for this include: antimicrobial susceptibility profiles, phage typing,

Box 5.6 Utility of DNA subtyping for infection control

A nosocomial methicillin-resistant *S. aureus* (MRSA) infection was detected in a hospital. DNA fingerprint profiles of MRSA-derived plasmid DNA identified 10 distinct subtypes from 24 patients infected or colonised with MRSA. Nine of 24 had a single subtype (A2). The nine came from the surgical service and eight had been hospitalised in the surgical intensive care unit. None of the health care workers in that unit were colonised by MRSA. On the basis of this, it was concluded that the MRSA outbreak represented a breakdown in infection control procedures within that unit so that nosocomial transmission of a single MRSA subtype (A2) had occurred from patient to patient (Pfaller et al 1991).

immunoblot patterns and DNA plasmid analysis. The last approach can be very useful in deriving a fingerprint for the various methicillin-resistant subtypes since it is possible to use a number of restriction enzymes to produce different restriction fragment patterns (Box 5.6). More recently, profiles of bacterial DNA obtained by pulsed field gel electrophoresis have been used to monitor *S. aureus* and other infections in the hospital setting (Fig. 5.7).

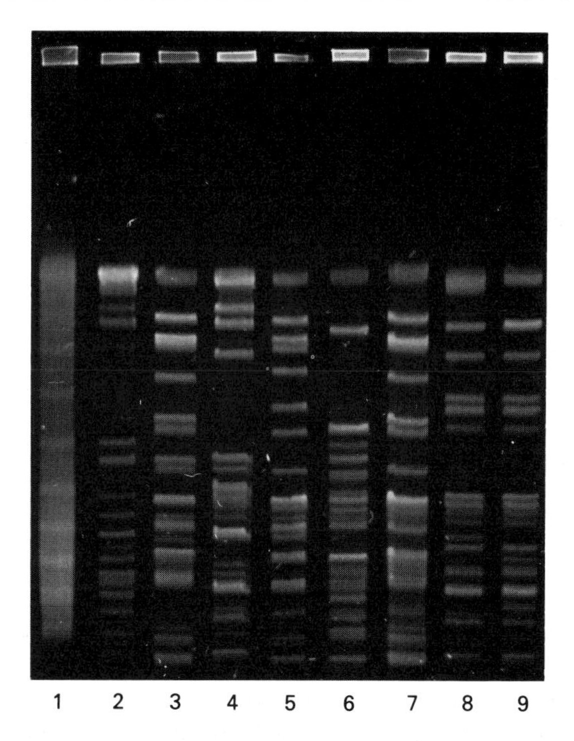

Fig. 5.7 Monitoring hospital infections by DNA analysis. A pulsed field gel electrophoretic (PFGE) analysis of chromosomal DNAs from *Pseudomonas aeruginosa* strains that were digested with the restriction enzyme *Spe*I. These strains were isolated from children in a neonatal intensive care unit. Isolates were recovered from patients in the unit over a period of several years. From left to right: lane 1 is λ DNA concatamers used as molecular weight standards; lanes 2–9 are *P. aeruginosa* isolates from neonates. Isolates in lanes 2–8 show highly distinct patterns and are probably not related to one another, while isolates in lanes 8 and 9 are very similar but not identical. These strains are most likely related to one another (PFGE courtesy of Ms Zaiga Johnson and Professor M L Vasil, University of Colorado Health Sciences Center, Denver).

EVOLUTION AND TAXONOMY

Characterisation of new microorganisms

In the past, the emergence of infectious agents reflected changed patterns of human movements which disrupted traditional geographical boundaries. For example, yellow fever is thought to have emerged in the New World as a result of the African slave trade which brought the mosquito *Aedes aegypti* in ships' water containers. More recently, *Aedes albopictus*, a potential vector for dengue virus, has become established in the USA following its conveyance from South East Asia in old car tyres. With this, the threat of dengue in the North American continent has become real. Most emergent viruses are zoonotic, i.e. they are acquired from animals which are reservoirs for infection. Thus, completely new strains are less likely than the appearance of a virus following a change in animal reservoirs. This is particularly relevant to the modern world where the consequences of easy migration, deforestation, agricultural practices, dam building and urbanisation make and will continue to make a major impact on the ecology of animals.

To monitor the appearances of new viruses requires a number of detection strategies. The gold standard is viral culture. This is expensive and tedious since it is necessary to identify an appropriate host cell for growth and then suitable markers (cytopathic effect, viral product, immunochemical characterisation) must be defined. Electron microscopy is an additional approach which has the advantage that it is rapid and can identify unknown species. The hepatitis A virus and the rotaviruses were found in this manner. Immunological detection systems rely on antibodies or antigens either host-derived or prepared in vitro. Although very useful in identifying viruses they are less helpful in our understanding of the agent's life cycle and its pathogenic potential.

Nucleic acid based strategies have similar uses to the immunological techniques with the added potential that DNA amplification is rapid, it can be more specific than monoclonal antibody based assays and can detect very few of the infectious particles. Moreover, sequence data can be generated from the polymerase chain reaction. This information on the virus's genome is useful in determining the relatedness of viruses, factors which may be relevant to pathogenesis and potential targets for therapy. In the long term, our deficient knowledge of viral evolution will be greatly enhanced through comparative studies of their DNA/RNA sequences.

Recently discovered viruses

The human immunodeficiency viruses (HIV-1, HIV-2) as recently emerged human pathogens have been discussed previously. A better understanding of their epidemiology and pathogenesis will follow from characterisation at the DNA level of the HIV viruses in humans and other infected animals. The ability of some animals to adapt to this virus and not develop AIDS may have important implications in developing a strategy to control AIDS in humans. For example, the African Green monkey carries the simian immunodeficiency virus (SIV) but does not have AIDS. However, if this virus is given to unrelated monkeys, e.g. the macaque, an AIDS-like illness emerges. Similarly, a feline immunodeficiency virus (FIV) is present in the cat family including the domestic cat as well as many of the free range animals. It is intriguing that one of the cats (the puma) carries the virus but does not develop AIDS.

Another of the new arrivals is human herpesvirus-6 (HHV-6). This was first reported during 1986 in blood mononuclear cells from patients with lymphoproliferative diseases (lymphomas and leukaemias of lymphocyte origin). The virus is predominantly T cell lymphotropic although it infects other cells including B lymphocytes and brain glial cells. It was initially thought to belong to the herpes group of viruses. Comparisons of DNA sequences would now suggest that human herpesvirus-6 is more closely related to cytomegalovirus. Seropositivity as high as 90% in the healthy adult population has been reported. Seroconversion appears to occur in early childhood.

Human herpes virus-6 is associated with disorders of the immune system in adults (e.g. lymphoproliferative disorders, AIDS) and is an important cause of acute febrile illnesses in children presenting to a hospital's emergency department. In one study, 14% of children in an emergency department had human herpes virus-6 cultured from their peripheral blood and, at the same time, viral-specific DNA identified by the polymerase chain reaction. The steps required to culture the virus are complex and would not be available in most laboratories since they involved the isolation of mononuclear cells, co-cultivation of these with cord blood cells, the addition of a number of growth factors and finally confirmation by cytopathic effects and immunofluorescent staining. In contrast, detection of viral-specific DNA by the polymerase chain reaction was faster and technically less demanding. The latter was as sensitive as culture in the acute phase and more sensitive during the convalescence phase. Serology was only helpful during convalescence with the majority of infected children demonstrating an elevation in their IgG antibody titres.

DNA characterisation of the human herpes virus-6 genome continues to provide a better understanding of its evolutionary origin and its apparent ability to remain dormant. A DNA-derived test will be a more rapid and effective diagnostic approach for what is proving to be a common infection in children.

Taxonomic classification

Continuous in vitro culture of human-derived *P. carinii* has not been possible. Therefore, its life cycle and metabolic processes are poorly understood. Indeed present taxonomic classification as a protozoan may be incorrect since rRNA sequence comparisons would place this organism closer to the fungi or even in a distinct lineage. Analysis of the *P. carinii* genome is presently underway and from this it will be possible to appreciate better its evolution and biology.

CONCLUSIONS

Despite the exciting prospects offered by recombinant DNA techniques in the infectious diseases, traditional approaches remain firmly entrenched in many of the diagnostic and research laboratories. The reasons are varied. It is still cheaper and faster to culture many of the pathogens. Culture can also provide additional information such as antibiotic sensitivity, virulence factors, antigen status and strain variation. However, alternative diagnostic methods remain necessary in situations for which conventional tests are unavailable, expensive, time-consuming or potentially dangerous.

Nucleic acid hybridisation techniques show promise, particularly the use of the repetitive rRNA-derived probes. Disadvantages associated with DNA hybridisation include the lack of automation and the requirement for skilled personnel. However, these will become problems of the past with the increasing availability of 'user-friendly' DNA hybridisation kits. These kits need not identify pathogens directly in clinical specimens but can be used to confirm cultures more rapidly than would otherwise be possible by conventional methods. Organisms detectable in this way include the mycobacteria, a number of fungi, many of the pathogens which cause gastrointestinal infections, *Staph. aureus*, *Legionella* spp. and others. Diagnostic kits are also available to detect some pathogens directly from clinical specimens, e.g. *C. trachomatis*, *N. gonorrhoeae*.

Although introduced only in 1985, DNA amplification by the polymerase chain reaction has emerged rapidly as a front runner for the identification and study of pathogens (Box 5.7). Automation is available and the technology is less demanding if probe labelling and hybridisation steps are not required. Multiplex DNA amplifications will increase cost-effectiveness by enabling simultaneous detection of a number of pathogens, e.g. in the sexually transmitted diseases, it would be possible to test for *C. trachomatis*, *N. gonorrhoeae* and other relevant organisms in the one reaction. Non-^{32}P labelled probes are very effective when hybridised to amplified DNA products.

Presently DNA amplification is not routinely used in many of the diagnostic laboratories because of two major drawbacks which result from its exquisite sensitivity. The first problem is **contamination**. Sources for contamination include other specimens, reagents, equipment and most importantly the amplified products themselves. Many strategies have been developed to avoid contamination. To some degree these have been successful but they are time-consuming and costly. Better procedures for control of contamination will make the polymerase chain reaction more acceptable, particularly in the routine diagnostic laboratory which is often crowded with a mixture of personnel who have a range of skills.

The second matter of concern is the high **sensitivity** which brings into question the biological significance of some positive results obtained through DNA amplification. For ex-

Box 5.7 PCR technology and the detection of microorganisms

DNA amplification by the polymerase chain reaction is an important new laboratory diagnostic tool.

Advantages

- **Speed, automation**
- **High degree sensitivity, specificity**
- **Crude extract satisfactory**
- **Inoculum need not be viable**
- **Target need not be infectious (e.g. provirus)**
- **Target can be DNA or RNA**

Disadvantages

- **DNA/RNA sequence must be known and unique**
- **Ease of contamination**
- **Difficult to quantify**
- **Colonisation versus invasion can be difficult to distinguish**
- **Cultures are not available for further characterisation, e.g. typing**

ample, the significance of detecting one or two copies of human herpesvirus-6 or cytomegalovirus in an adult remains unclear since these pathogens are frequently present in a latent form in normal individuals. As has already been discussed previously, is one human immunodeficiency virus-1 particle infectious? This problem may be resolved as the DNA amplification technology improves and an accurate quantification of amplified product is easier to achieve. In this circumstance, it would be possible to show a rising 'titre' of amplified product which might have more significance than a single finding.

It is important to note that DNA testing does not necessarily tell you anything more than that a certain nucleic acid sequence is present. Thus, culture may still be necessary if antibiotic sensitivity needs to be determined. Nevertheless, as recombinant DNA technology develops it will be possible to obtain more comprehensive DNA/RNA profiles. For example, multiplex polymerase chain reactions could be designed to identify an organism as well as its virulence-specific sequences or antibiotic resistance plasmids simultaneously. The use of prophylactic antibiotics in at-risk situations, e.g. trimethoprim/sulphamethoxazole for *P. carinii* infection in patients with AIDS will become less of a necessity if a rapid diagnostic test is available to identify life-threatening situations.

The potential role of recombinant DNA in the areas of infection control and epidemiological studies has had a lower profile compared to its utility in diagnostic testing. Nevertheless, essential information to identify the source of an infection or the way it has spread is frequently unavailable after conventional approaches are exhausted. Knowledge of an organism's life-cycle, particularly newly-evolved pathogens or those which assume significance in association with the immunocompromised host, may be obtained more easily through DNA approaches. These will give more representative in vivo changes compared to the in vitro culture conditions often required for research strategies.

To date many comparative studies evaluating the role of DNA tests in the infectious diseases have utilised cultured organisms as sources of DNA. These studies will need to be repeated in the in vivo situation since inhibitors may be present in clinical specimens and the effects of other organisms/DNA species may be significant. Thus, in the infectious diseases, the clinical sample and the means used to isolate DNA are important considerations in recombinant DNA testing.

A balance between what is possible by traditional approaches and what DNA tests offer will enable the establishment of priorities to identify which pathogens are best studied by recombinant DNA means. Commercial kits are becoming more readily available for the detection of a wide range of organisms. Kits will assist with the difficulties mentioned above and avoid the problem of finding laboratory personnel who are skilled in recombinant DNA technology. The

Table 5.7 Some microorganisms which can be detected by recombinant DNA tests

Bacteria or bacteria-like	*Mycobacterium tuberculosis* and atyptical mycobacteria, *M. leprae, Yersinia enterocolitica, Campylobacter* spp., *Aeromonas hydrophila, Shigella* spp., *Neisseria gonorrhoeae, Pseudomonas cepacia, Clostridium difficile, Borrelia burgdorferi, Haemophilus aphrophilus, Salmonella* spp., *Escherichia coli, Legionella pneumophila, Streptococcus pneumonia, Helicobacter pylori, Treponenema pallidum, Bordetella pertussis, Haemophilus ducreyi, Mycoplasma pneumoniae, Haemophilus influenzae, Neisseria meningitidis, Gardnerella vaginalis, Leptospira* spp., *Chlamydia* spp., *Coxiella burnettii, Rickettsia rickettsii, Nocardia asteroides*
Viruses	Human immunodeficiency viruses 1 and 2, human T cell lymphotropic virus-1, papillomavirus, herpes simplex virus, hepatitis B, C and delta viruses, Epstein–Barr virus, cytomegalovirus, enteroviruses, rhinovirus, parvovirus B19, rotaviruses A,B,C, picornavirus, dengue, herpes zoster virus, influenzae, respiratory syncytial virus, adenovirus, Lassa virus, rubella, measles, JC virus, molluscum contagiosum, human herpes virus 6
Fungi	*Aspergillus fumigatus, Blastomyces dermatitidis, Coccidioides immitis, Cryptococcus neoformans, Candida albicans, Histoplasma capsulatum*
Protozoans	*Pneumocystis carinii, Toxoplasma gondii, Naegleria fowleri, Entamoeba histolytica, Trichomonas vaginalis, Plasmodium falciparum, Leishmania* spp. *Trypanosoma* spp.
Metazoans	Filariae (*Brugia, Loa, Wuchereria, Onchocerca* spp.); *Echinococcus* spp., *Taenia* spp., *Fasciola* spp.

Table 5.8 Clinical infections for which rDNA diagnosis is presently of value

Infections	Type of rDNA test available	Advantages of rDNA testing
Meningitis/ encephalitis (diagnosis)	PCR: *M. tuberculosis*, herpes simplex virus rRNA: commercial kits for mycobacteria	Avoids long-term culture, likelihood negative microscopy (TB). Avoids brain biopsy, provides early detection system (HSV) Commercial rRNA kits: very promising results
Leprosy (diagnosis)	PCR: *M. leprae*	*M. leprae* cannot be cultured; few cells present in tuberculoid leprosy
Cytomegalovirus in immuno-compromised (diagnosis)	PCR: CMV	Reactivation of latent infection/transferred CMV with organ transplants requires assessment of CMV status in transplantation situation. Viral culture expensive and time consuming. PCR results may also predict response to treatment
Sexually transmitted diseases (diagnosis), e.g. chlamydia and gonorrhoea	PCR; rRNA commercial kits	PCR still being evaluated to compare with immunofluorescence/ enzyme immunoassay approaches. rRNA kits compare favourably with above methods (*C. trachomatis*)
Lyme borreliosis (diagnosis and epidemiology)	PCR: *B. burgdorferi*	Few organisms, difficult to culture, atypical clinical presentations, serological responses slow to develop
Legionella (diagnosis and epidemiology)	rRNA commercial kits; PCR, PFGE: *L. pneumophila*, *L. longbeachae*	Fastidious grower; typing for epidemiological purposes
Intestinal infections (diagnosis and virulence)	rRNA; PCR: *Campylobacter* spp. PCR: rotavirus, A, B, C	Fastidious growers (Campylobacter) Avoidance immunoassay, electron microscopy or RNA characterisation (rotavirus)

latter becomes an important consideration when a laboratory's future needs and directions are being planned.

Recombinant DNA approaches have now been described for detection of a wide range of infectious agents (Table 5.7). Some remain research activities, others are more appropriate to the routine diagnostic laboratory (Table 5.8). The emphasis in this chapter has been on DNA amplification by the polymerase chain reaction since this technology has a real potential to change significantly the future approach to the practice of medical microbiology particularly in relation to the routine diagnostic laboratory. Already research laboratories have incorporated DNA technology into a number of their protocols. The prospects for molecular medicine in microbiology are vast and will have profound long-term effects in laboratory and clinical practice.

FURTHER READING

Barker D C 1989 Molecular approaches to DNA diagnosis. Parasitology 99: S125–S146
Brisson-Noel A, Aznar C, Chureau C et al 1991 Diagnosis of tuberculosis by DNA amplification in clinical practice evaluation. Lancet ii: 364–366
Char S, Farthing M J G 1991 DNA probes for diagnosis of intestinal infection. Gut 32: 1–3
Evans D M A, Dunn G, Minor P D et al 1985 Increased neurovirulence associated with a single nucleotide change in a noncoding region of the Sabin type poliovaccine genome. Nature 314: 548–550
Hopkin J M, Wakefield A E 1990 DNA hybridization for the diagnosis of microbial disease. Quarterly Journal of Medicine 75: 415–421
Jackson J B 1991 Polymerase chain reaction assay for detection of hepatitis B virus. American Journal of Clinical Pathology 95: 442–444
Kakaiya R, Miller W V, Gudino M D 1991 Tissue transplant-transmitted infections. Transfusion 31: 277–284
Kandolf R, Hofschneider P H 1989 Enterovirus-induced cardiomyopathy. In: Notkins A L, Oldstone M B (eds) Concepts in viral pathogenesis III. Springer-Verlag, New York, p 282–290
Krause R M 1992 The origin of plagues: old and new. Science 257: 1073–1078.
Lanser J A, Adams M, Doyle R, Sangster N, Steele T W 1990 Genetic relatedness of *Legionella longbeachae* isolates from human and environmental sources in Australia. Applied and Environmental Microbiology 56: 2784–2790
Liang T J, Hasegawa K, Rimon N, Wands J R, Ben-Porath E 1991 A hepatitis B virus mutant associated with an epidemic of fulminant hepatitis. New England Journal of Medicine 324: 1705–1709
McLaughlin G L, Vodkin M H, Huizinga H W 1991 Amplification of repetitive DNA for the specific detection of *Naegleria fowleri*. Journal of Clinical Microbiology 29: 277–230
Morse S S, Schluederberg A 1990 Emerging viruses: the evolution of viruses and viral diseases. Journal of Infectious Diseases 162: 1–7

Ott M, Bender L, Marre R, Hacker J 1991 Pulsed field electrophoresis of genomic restriction fragments for the detection of nosocomial *Legionella pneumophila* in hospital water supplies. Journal of Clinical Microbiology 29: 813–815

Pays E 1991 Genetics of antigenic variation in African trypanosomes. Research in Microbiology 142: 731–735

Pfaller M A, Wakefield D S, Hollis R, Fredrickson M, Evans E, Massanari R M 1991 The clinical microbiology laboratory as an aid in infection control. The application of molecular techniques in epidemiologic studies of methicillin-resistant *Staphylococcus aureus*. Diagnostic Microbiology and Infectious Disease 14: 209–217

Phelps C J, Lyerly D L, Johnson J L, Wilkins T D 1991 Construction and expression of the complete *Clostridium difficile* toxin A gene in *Escherichia coli*. Infection and Immunity 59: 150–153

Pruksananonda P, Breese Hall C, Insel R A et al 1992 Primary human herpesvirus 6 infection in young children. New England Journal of Medicine 326: 1445–1450

Rosa P A, Hogan D, Schwan T G 1991 Polymerase chain reaction analyses identify two distinct classes of *Borrelia burgdorferi*. Journal of Clinical Microbiology 29: 524–532

Rotbart H A 1991 Nucleic acid detection systems for enteroviruses. Clinical Microbiology Reviews 4: 156–168

Shankar P, Manjunath N, Mohan K, Prasad K, Behari Shriniwas M, Ahuja G 1991 Rapid diagnosis of tuberculous meningitis by polymerase chain reaction. Lancet i: 5–7

Sheppard H W, Ascher M S, Busch M P et al 1991 A multicenter proficiency trial of gene amplification (PCR) for the detection of HIV-1. Journal of Acquired Immune Deficiency Syndromes 4: 277–283

Stull T L, LiPuma J J, Edlind T D 1988 A broad-spectrum probe for molecular epidemiology of bacteria: ribosomal RNA. Journal of Infectious Diseases 157: 280–286

Syrjanen S M 1990 Basic concepts and practical applications of recombinant DNA techniques in detection of human papillomavirus (HPV) infections. Acta Pathologica, Microbiologica et Immunologica Scandinavica 98: 95–110

Tannich E, Burchard G D 1991 Differentiation of pathogenic from nonpathogenic *Entamoeba histolytica* by restriction fragment analysis of a single gene amplified in vitro. Journal of Clinical Microbiology 29: 250–255

Telenti A, Imboden P, Marchesi F et al 1993 Detection of rifampicin-resistance mutations in *Mycobacterium tuberculosis*. Lancet i: 647–650

Wellems T E 1991 Molecular genetics of drug resistance in *Plasmodium falciparum* malaria. Parasitology Today 7: 110–112

Wilson S M 1991 Nucleic acid techniques and the detection of parasitic diseases. Parasitology Today 7: 255–259

6

MEDICAL ONCOLOGY

INTRODUCTION

Nomenclature used in molecular oncology can be cumbersome and so a number of terms need definition. *Oncogenes* (onkos is the Greek word for mass or tumour) are genes which are associated with neoplastic proliferation following a mutation or perturbation in their expression. Oncogenes have normal counterparts in the genome. These antecedent genes or *proto-oncogenes* play an essential physiological role in normal cellular proliferation and differentiation. v-*onc* or c-*onc* are abbreviations for oncogenes found in retroviruses (v) or their cellular equivalents (c). *Tumour suppressor genes* (also called recessive oncogenes, anti-oncogenes, growth suppressor genes) are genes which also play a role in normal cellular proliferation and differentiation. Loss or inactivation of tumour suppressor genes can lead to neoplastic changes. *Transformation* refers to the acquisition by normal cells of the neoplastic phenotype. *Transfection* refers to the procedure of gene transfer in which segments of DNA are incorporated into eukaryotic cells. *Transduction* describes the transmission of a gene from one cell to another by viral infection. In the context of the oncogenes, transduction has been used to describe the origin of retroviral oncogenes from cellular proto-oncogenes.

'Cancer is essentially a genetic disease at the cellular level' (W F Bodmer). Evidence for a genetic component in cancer includes:

- An increased risk for tumour development in those with a genetic defect in DNA repair
- Chemicals or physical agents which mutate DNA also elicit tumours in animals
- Some structural chromosomal rearrangements in the germline predispose to tumour development
- Tumour cells can have somatically acquired DNA mutations which resemble the mutations seen in those with familial cancers
- The malignant phenotype can be conferred on normal cells by gene transfer studies with oncogenes.

For over 30 years, tumourigenesis has been hypothesised to represent a *multistep process*. It is only with the application of recombinant DNA techniques that evidence for this has become available and it has become possible to identify molecular (DNA) changes which may produce the initiation, promotion and progression of cancers. Understanding the pathogenesis of cancer has become a realistic goal within the framework of molecular medicine. The ability to define mutations at the DNA level has enhanced the accuracy of diagnosis in the oncological disorders. In the near future, therapeutic directions will be influenced by our knowledge of the DNA changes in cancer.

An individual's response to cancer is complex. It involves a number of factors such as the person's immunological state, his/her performance status, the extent of disease, its response to treatment and the development of resistance. In addition there are the effects of oncogenes or tumour suppressor genes and chromosomal changes. The last three will be discussed in this chapter. Treatment resistance will be reviewed in Chapter 7.

ONCOGENES

Retroviruses

Biology

The RNA tumour viruses (retroviruses) provided the first proof that genetic factors can play a role in carcinogenesis. Retroviruses have three genes (*env*, *gag* – coding for structural proteins – and the third, *pol*, which codes for reverse transcriptase) (Fig. 6.1). As discussed in Chapter 2, reverse trans-

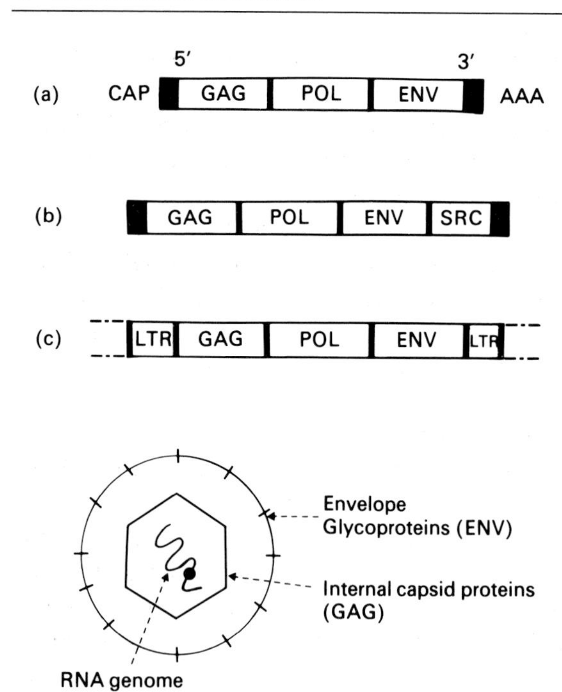

Fig. 6.1 The structure of retroviruses.
(**a**) RNA tumour viruses (retroviruses) have an RNA genome. This RNA has two features of eukaryotic mRNA, i.e. a capped 5′ end and a poly-A tail at the 3′ end. Retroviral RNA codes for three viral proteins: (1) a structural capsid protein (GAG) which associates with the RNA in the core; (2) an envelope glycoprotein (ENV) which is associated with the lipoprotein envelope of the virus and (3) the enzyme reverse transcriptase (POL). This is a DNA polymerase which makes a DNA copy of the RNA molecule and then a second DNA strand. Thus, it produces a double-stranded copy (cDNA) of the RNA genome. (**b**) Transforming retroviruses can have a fourth gene (oncogene). In the example here the oncogene is that of the Rous sarcoma virus (*src*). (**c**) Retroviruses are so named because they have a RNA genome and are able to replicate through formation of an intermediate (provirus) which involves integration of the retroviral genome into that of host DNA. The provirus has LTRs (long terminal repeats) on either side of the three RNA genes. The LTRs are several hundred base pairs in size and insert adjacent to smaller repeats derived from host DNA.

criptase is an enzyme that enables RNA to be converted into complementary DNA. In this way the retrovirus is able to make a DNA copy of its RNA which can then become incorporated into the host's genome (Fig. 6.2).

An additional (fourth) gene gives retroviruses the ability to induce tumour growth in vivo or to transform cells in vitro. In the latter situation, cells lose their normal growth characteristics and acquire a neoplastic phenotype. An example of this would be loss of contact inhibition so that instead of growth in a single cell layer in vitro, there is unregulated proliferation into clumps. DNA sequences in retroviruses which give them their transforming properties were called viral oncogenes (v-onc). Names for these are derived from the tumours in which they were first described. For example, v-*sis*, **Si**mian **s**arcoma virus; v-*abl*, murine **Ab**elson **l**eukaemia virus; v-*mos*, **Mo**loney **s**arcoma virus.

Viral oncogenes were subsequently shown to have cellular homologues called cellular oncogenes (c-onc). The term proto-oncogene was coined to describe cellular oncogenes which do

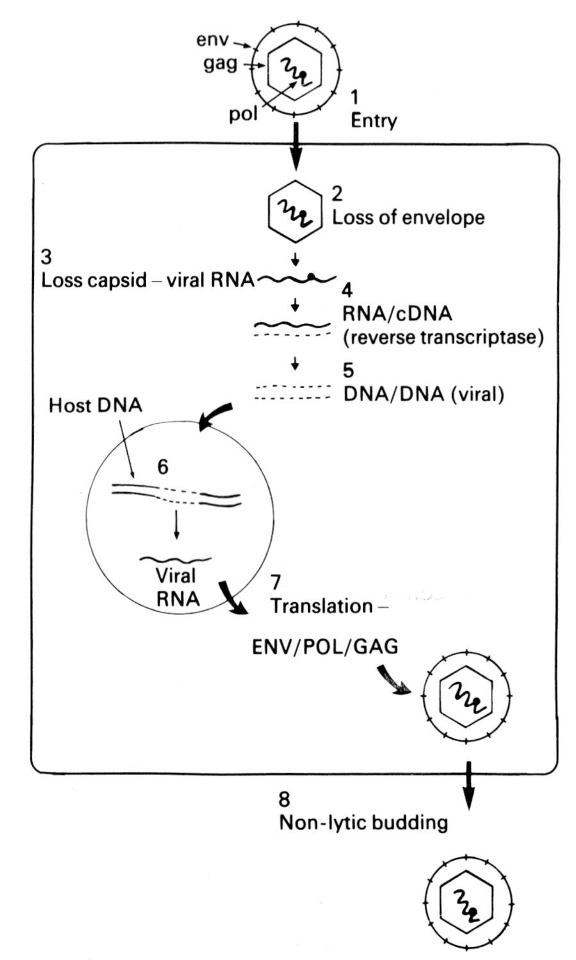

Fig. 6.2 Life cycle of a retrovirus.
(**1**) The envelope protein enables the retrovirus to bind to the surface of host cells on infection. (**2–5**) Double-stranded DNA derived from viral RNA and the action of reverse transcriptase is required before the retroviral genome can be integrated into that of the host. (**6–8**) The provirus so formed replicates to produce mature viral particles which are extruded from the cell by non-lytic budding.

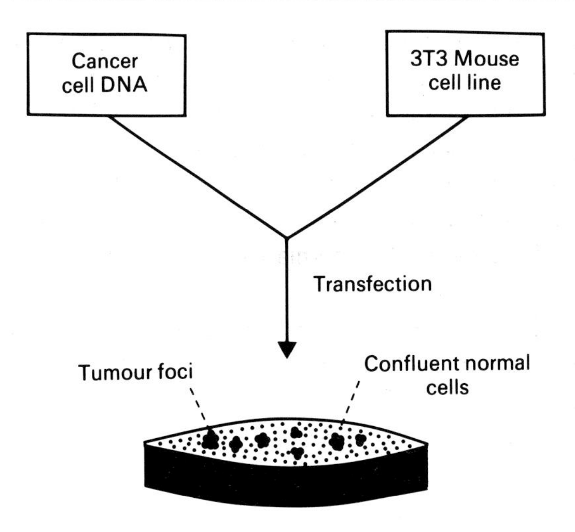

Fig. 6.3 Gene transfer to identify oncogenes.
DNA derived from cancer cells is transfected into a susceptible cell line such as the NIH 3T3 fibroblast mouse line. Transfection can be induced by co-precipitation of DNA with calcium phosphate. An alternative and more efficient method involves electroporation, i.e. DNA uptake occurs through permeabilising the cell to high molecular weight molecules by exposing it to an intense electrical field. Gene transfer studies enable new oncogenes to be identified and the role of oncogenes to be studied. Fractionating the DNA enables the oncogene to be cloned.

not have transforming potential to form neoplasms in their native state. Proto-oncogenes have now been shown to play important roles in control of cell growth and will be discussed in more detail below.

The initial observation implicating viruses and cancer came in 1910 when Roux demonstrated that a filterable agent (virus) was capable of inducing cancers in chickens. 56 years later, he was awarded the Nobel Prize for this work. Retroviruses have since been identified as the cause of cancers in many other species including mammals, although rarely in humans. Nevertheless, the direct relevance of the retroviruses to human cancers remained unclear until alternative lines of investigation became available. These included the characterisation at the molecular level of chromosomal translocations and amplified DNA and gene transfer studies.

Gene transfer

Gene transfer is illustrated in Figure 6.3. In brief, DNA derived from a malignant tumour is transfected or transferred into a cell line. In normal circumstances, cell lines grow as confluent monolayers with each cell remaining discrete from its neighbour. In the presence of a transforming oncogene, cells can form clumps. Additional evidence that the cells in clumps have acquired a neoplastic phenotype is obtainable by inoculation of immunologically-deficient 'nude' mice. Normal cells would not grow in these circumstances. Transformed cells will form tumours.

Gene transfer studies have demonstrated the relevance of the oncogenes to human cancers such as bladder, colon, lung and many others. New oncogenes have been defined in this way. From the type of approach described above, it is possible to fractionate DNA so that eventually the oncogene can be cloned. This approach was first

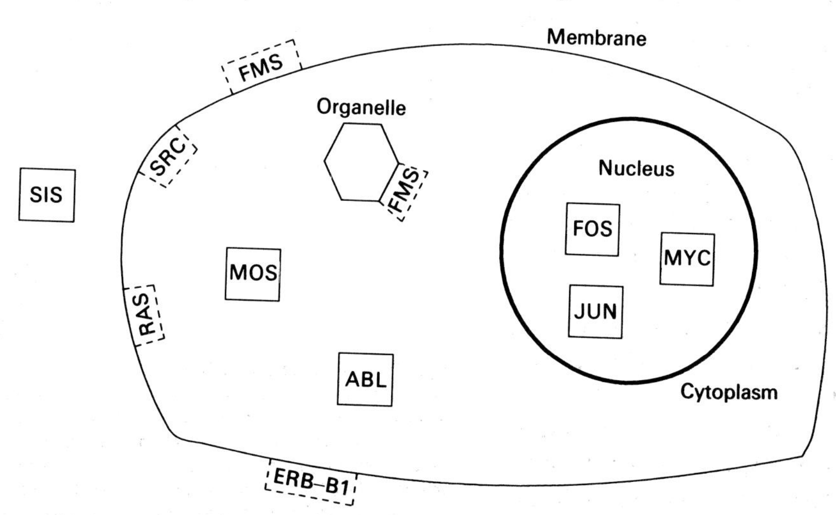

Fig. 6.4 The cellular localisations of proto-oncogene products.

Table 6.1 Some proto-oncogenes and their modes of action. Proto-oncogenes are normal genes involved in cellular proliferation and differentiation

Proto-oncogene	Example	Function
Growth factors	*sis*	B chain of PDGF
	int-2, *hst*	Fibroblast growth factor-related
Growth factor receptors	*erb*B	Epidermal growth factor receptor
	fms	CSF-1 receptor
Protein kinases	*abl*, *src*, *fes*,	Protein tyrosine kinases
	mos, *raf*	Protein serine kinases
G proteins	H, K and N-*ras*	GTP-binding/GTPase
Nuclear transcription factors	*myc*, N-*myc*	DNA-binding protein
	myb	DNA-binding protein
	jun, *fos*	DNA-binding protein
	*erb*A	DNA-binding protein (thyroid hormone receptor)

successful with a human bladder cell line called T24. The oncogene isolated was H-*ras*.

Proto-oncogenes

There are now over 60 proto-oncogenes described. Their roles in terms of cellular growth and differentiation are complex and may involve interactions between a number of the proto-oncogenes. The sites and modes of action of many proto-oncogenes have been defined. They include growth factors, growth factor receptors, membrane and cytoplasmic transducers as well as DNA binding proteins (Fig. 6.4, Table 6.1).

Growth factors

Growth factors are molecules which act via cell-surface receptors to induce cellular division. Some growth factors will lead to transformation of normal cells in vitro. Alternatively, the process of transformation itself can produce growth factors. The scenario of self-perpetuating, positive-feedback loops is one mechanism by which the neoplastic phenotype can arise.

Growth factor receptors

Several proto-oncogene-derived proteins form components of growth factor receptors at the cell surface. An example of this is the *fms* proto-oncogene which forms the CSF-1 (colony stimulating factor) receptor in differentiating macrophages. Binding of growth factors and their respective membrane receptors is the first step in the delivery of mitogenic signals to the cell's interior to initiate cell division.

Intracellular signal transduction pathways

Once an extracellular growth factor has bound to its cell surface receptor it is able to transfer (transduce) its signal to the nucleus by a number of processes some of which may involve 'second messengers'. GTP (guanosine triphosphate) and proteins that bind GTP (known as G proteins) play an important role in intracellular signal transduction. GTP is converted to GDP (guanosine diphosphate) by the GTPase activity of the G proteins.

The *ras* proto-oncogenes (viz. H-*ras*-1, K-*ras*-2, N-*ras*) comprise a multigene family which is present in eukaryotic organisms as divergent as yeast and humans. This type of conservation during evolution would suggest that the *ras* genes have functional significance. *Ras* proto-oncogenes encode related proteins of 21 kDa size called p21. Proteins derived from the *ras* proto-oncogenes have GTPase, GTP and GDP binding activities, i.e. they are related to G proteins. One physiological function of the *ras* proto-oncogenes is in modulation of cellular proliferation via the transduction of signals from the cell surface to the nucleus. Wild-type *ras* p21 binds GTP and thence becomes inactive following conversion of GTP to GDP (see oncogenes below for a continuation of the G protein story involving mutant *ras*).

Nuclear proteins

An increasing number of proto-oncogenes have been shown to encode nuclear binding proteins. Some of these are previously described transcription factors.

Oncogenes

From molecular studies, a number of mechanisms have been defined in which the normal products of proto-oncogenes can be disrupted to produce

Table 6.2 Changes in the DNA of proto-oncogenes and cancer development.
Some molecular rearrangements found in the proto-oncogenes and their neoplastic associations (from Bishop 1991)

Proto-oncogene	DNA abnormality	Human neoplasm(s)
abl	Translocation	Chronic granulocytic leukaemia
myc	Translocation	Burkitt's lymphoma
*erb*B	Amplification	Squamous cell carcinoma, astrocytoma, carcinoma oesophagus
neu	Amplification	Adenocarcinomas of breast, ovary,stomach
myc	Amplification	Carcinomas of breast, lung, cervix, oesophagus
N-*myc*	Amplification	Neuroblastoma, small cell carcinoma of lung
int-2	Amplification	Carcinoma oesophagus
K-*ras*	Point mutations	Carcinomas of colon, lung, pancreas, melanoma
N-*ras*	Point mutations	Acute myeloid and lymphoblastic leukaemias, carcinoma of thyroid, melanoma
H-*ras*	Point mutations	Carcinomas of genitourinary tract, thyroid
ros	Unknown	Astrocytoma
sis	Unknown	Astrocytoma
src	Unknown	Carcinoma of colon

neoplastic transformation (summarised in Table 6.2).

Gene translocation

Relocation of oncogenes within the human genome can alter their function. Chromosomal breakpoints, which produce the Philadelphia chromosome found in chronic granulocytic leukaemia, involve a translocation of the c-*abl* oncogene on chromosome 9 to a gene on chromosome 22 which is called the breakpoint cluster region (bcr). At the same time, the *sis* proto-oncogene on chromosome 22 is translocated to chromosome 9 (Fig. 6.5). The hybrid bcr-*abl* transcript produces a novel protein (p210) with tyrosine kinase activity. Transfection of murine stem cells with the bcr-*abl* fusion gene produces in some irradiated mouse recipients a phenotype which is similar to the human leukaemia.

Gene amplification

Quantitation of gene numbers is possible by DNA mapping. Amplified c-*myc* sequences have been described in a number of human cancers. In the childhood tumour neuroblastoma, the N-*myc* gene may become amplified up to 300 times. The mechanisms whereby gene amplification can lead to transformation remain to be defined.

Point mutations

Single base changes in the DNA sequence of proto-oncogenes have been consistently observed in the different *ras* genes. A wide range of malignancies has been shown to have *ras* mutations. How these mutations produce their effects remains unknown although it is noteworthy that the sites for mutations in *ras* are limited to codons 12, 13 and 61 which are located within the regions coding for binding of GTP/GDP. Therefore, failure by *ras* to convert the active complex (GTP-*ras*) to inactive GDP-*ras* would produce an excess of stimulatory activity leading to unregulated cellular proliferation.

Viral insertion

Another mode by which proto-oncogene function can be perturbed is via insertion of viral elements. An example of this is the hepatitis B virus which is discussed in more detail below (p. 136).

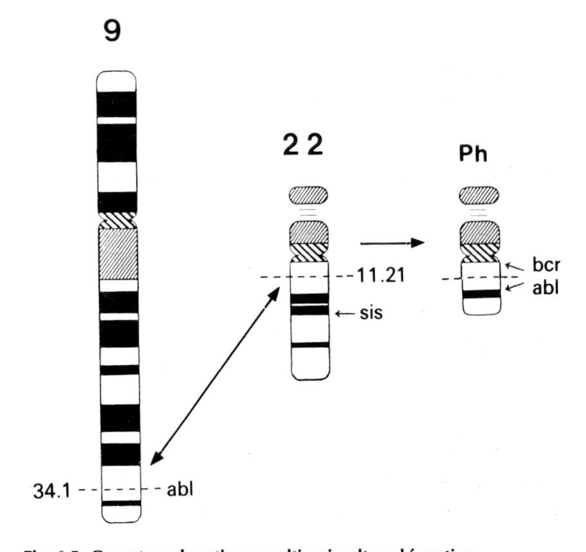

Fig. 6.5 Gene translocation resulting in altered function.
A reciprocal translocation between chromosomes 9 and 22 which produces the Philadelphia chromosome found in chronic granulocytic leukaemia. The *sis* proto-oncogene is not considered to have a functional effect from this translocation because it is located at some distance (22q12.3-q13.1) from the actual chromosome 22 breakpoint (22q11.21). ----- = breakpoints.

TUMOUR SUPPRESSOR GENES

Perturbations in genes can produce stimulatory or inhibitory stimuli to cell growth. These differences are also described as dominant or recessive. Examples of *dominant* or stimulatory effects are the proto-oncogenes described above. In these situations, mutations produce stimulatory signals leading to uncontrollable proliferation. On the other hand, oncogenesis can result through loss of inhibitory function which is associated with *recessive* lesions involving another group of DNA sequences known as the *tumour suppressor genes*. Evidence for the latter in the pathogenesis of cancer has come from three observations: cell hybrids, inherited cancer syndromes and loss of heterozygosity in tumours.

Experimental evidence

Somatic cell hybrids

In the late 1960s, murine cell hybrids formed by fusions between normal and tumour cells were found to revert to the normal phenotype. Subsequently, as the hybrid clones were propagated in culture the tumour phenotype became re-established. This effect was seen in a wide range of tumour lines and was considered to indicate the influence of a tumour suppressor gene derived from the normal cells. Subsequent loss of chromosomes, which occurred on serial passage of cell lines, enabled reversion to the neoplastic phenotype when the tumour suppressor gene(s) were lost. Today, micro-cell transfers allow one or a few chromosomes within a reconstituted membrane from normal cells to be delivered to recipient tumour cells. In this way, it has been possible to identify tumour suppressor genes in a variety of chromosomes.

Inherited cancer syndromes

Familial recurrence is seen in most tumours although the proportion is often low. One exception is retinoblastoma in which the inherited form comprises up to 40% of cases. Retinoblastoma is the most common intraocular malignancy in children with an incidence of about 1 in 14 000 live births. Retinoblastoma can occur as a sporadic or heritable disorder. In the early 1970s, epidemiological studies of both retinoblastoma

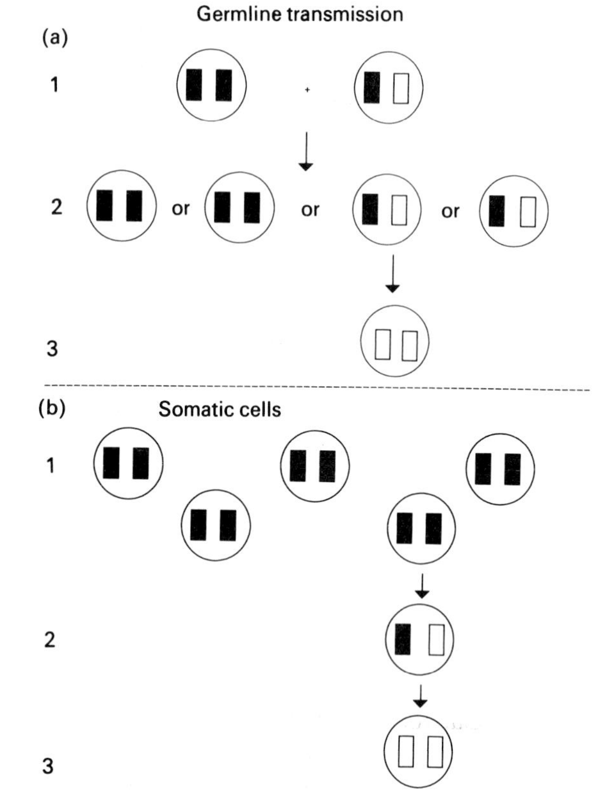

Fig. 6.6 Two-hit model for tumourigenesis at the gene level. The first or predisposing event can be inherited through the germline or arise in somatic cells. A second somatic event is required to inactivate the remaining normal allele. (**a**) First event occurring in the germline. **1**. Tumour suppressor genes for one particular locus are shown as ■. The first hit affects one of the germ cells to inactivate one tumour suppressor gene □. **2**. Offspring have a 50% chance of inheriting the mutant tumour suppressor gene (dominant inheritance). **3**. A second hit is required to inactivate the remaining tumour suppressor gene. Because the germ cells were initially involved all cells will be predisposed. Therefore, the second hit is more likely to involve the appropriate cell, e.g. retinal cells in retinoblastoma. This produces multifocal tumours which present at an earlier age. Other members of the family can become affected. (**b**) Sporadic tumour formation from two separate somatic events. **1,2**. If the first hit affects a somatic cell it inactivates the tumour suppressor gene in that cell and its progeny. To get both tumour suppressor genes inactivated the second hit (**3**) must involve the predisposed line. This is less likely to occur than the situation in (a). Tumour formation is delayed. If it occurs it is sporadic and unifocal.

and Wilms' tumour led Knudson to propose the two-hit model of tumourigenesis (Fig. 6.6).

In the above model, the *first* or predisposing event can be inherited through the germline or arise in somatic cells. In the case of a familial tumour, the affected individual will inherit via the germline a mutant non-functional allele from one parent (who may be asymptomatic) and then a *second* somatic event inactivates the normal allele inherited from the other parent. The frequency of somatic mutations is sufficiently high that those who have inherited the first mutation are likely to develop one or more tumours. This mode of inheritance shows up as a dominant effect with incomplete penetrance or appearance in pedigree analysis. However, at the cellular level it is recessive since loss or inactivation of both alleles is required for expression in an affected cell. A cell containing only one of its two normal alleles will usually produce enough tumour suppressor gene product to remain non-transformed. Loss of the cell's remaining normal allele exposes it to uncontrolled proliferation. Incomplete penetrance reflects the requirement for the second (somatic) mutational event and explains the 'skipping' of some generations. On the other hand, sporadic forms of the tumour involve two separate *somatic* events. The second hit must also occur in the same cell lineage that has experienced the first or predisposing hit. The probability of this is relatively rare and so sporadic forms of the tumour occur later in life and have the additional features of being unifocal and unilateral. Retinoblastoma is a useful model for the tumour suppressor genes and will be discussed in more detail below (p. 131).

Loss of heterozygosity

In the inherited tumours, the somatic event that affects the second (normal) allele and so exposes the recessive mutation can be chromosomal loss or molecular abnormalities such as a deletion. Chromosomal rearrangements, losses or aneuploidy (abnormal chromosome numbers) can be detected by studying constitutional (e.g. lymphocytes or fibroblasts) and tumour cells. However, microdeletions will be difficult to find. In most cases, cytogenetic analysis does not allow the parental origin for individual chromosomes to be determined.

The availability of DNA polymorphic markers has meant that individual parental contributions in normal and tumour cells can be distinguished (Fig. 6.7). DNA markers flanking a locus which contains a tumour suppressor gene will thus demonstrate loss of heterozygosity if that segment is lost. The same approach is useful in

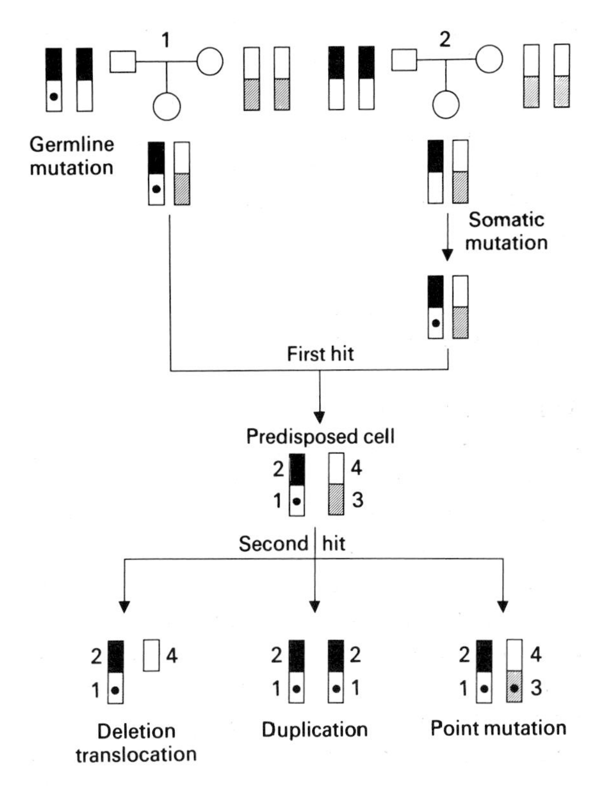

Fig. 6.7 Molecular defects which produce a second (somatic) loss at a tumour suppressor gene locus.
The mutation in the tumour suppressor gene is depicted by ●. (**1**.) There is inheritance of a germline mutation. (**2**.) The germline is normal but a somatic mutation occurs. In both situations the end result is a predisposed cell. Loss of the remaining (normal) tumour suppressor gene via a second hit can occur by a number of different mechanisms, e.g. deletion or an unbalanced translocation; a duplication or a more discrete abnormality such as a point mutation. Loss of heterozygosity at the DNA level can be illustrated if it is presumed that there are four informative DNA markers which distinguish the two parental contributions and are situated at the tumour suppressor location. For simplicity, the polymorphic markers are depicted as 1 (mutant); 2,3,4 (normal) which would give two haplotypes of 1,2/3,4. Loss of heterozygosity is seen as 1,2/–4 (deletion) and 1,2/1,2 (duplication). Loss of heterozygosity will not be found in the point mutation situation since the normal locus (identified by the '3' marker) is present but non functional since it has acquired a discrete mutation.

Table 6.3 Loss of heterozygosity in some human tumours.
A second somatic event will result in loss of the remaining normal gene activity

Tumour	Implicated chromosome(s)
Breast carcinoma	1p, 1q, 3p, 11p, 13q, 17p, 17q, 18q
Lung carcinoma	3p, 13q, 17p
Colorectal carcinoma	5q, 17p, 18q
Cervical carcinoma	3p
Multiple endocrine neoplasia type 1, 2	11q, 1p respectively
Meningioma, acoustic neuroma	22
Astrocytoma	17p
Bladder carcinoma	11p,13q
Renal carcinoma	3p
Pheochromocytoma	1p, 22
Neurofibromatosis 1	17q
Retinoblastoma	13q
Osteosarcoma	13q, 17p
Neuroblastoma	1p
Wilms' tumour	11p13, 11p15.5
Melanoma	1p, 9p, 17p
Ovarian carcinoma	17p, 17q, 18q
Hepatocellular carcinoma	11p,13q

studying sporadic tumours with the base line or constitutional genotype determined from the affected individual's normal tissues and the two parental contributions from analysis of the pedigree. These can then be compared to DNA patterns in the tumour.

DNA studies to detect loss of heterozygosity in tumours have been used to identify putative tumour suppressor loci and from these attempts to clone and characterise candidate genes have been made. Two tumour suppressor genes, p53 on chromosome 17p and DCC (deleted in colon carcinoma) on chromosome 18q have been detected in this way (see p. 131, 134). Loss of heterozygosity has been described in many forms of cancer (Table 6.3). As is evident from Table 6.3, the same tumour suppressor regions may be involved in more than one malignancy.

Functions

The exact role(s) of the tumour suppressor genes in tumourigenesis remain(s) to be defined. Potential sites where these genes might inhibit the development of cancer include: cell proliferation, differentiation and senescence, cell to cell communication and chromosomal stability.

Cell proliferation

Two tumour suppressor genes (retinoblastoma and p53) have been shown to play a key role in regulation of the cell cycle. Products of both genes restrict cells at the G1-S phase of the cell cycle, preventing their entry into the proliferative phases of the cell cycle. Physiologically, this may be an essential step in allowing cells to repair damaged DNA before mitosis (Box 6.1).

Differentiation and senescence

Some oncogenes can block differentiation and stimulate cellular proliferation. Co-existent loss of a tumour suppressor gene which functions to inhibit proliferation would potentiate the oncogene effect leading to deregulated proliferation.

Cell to cell communication

Gap junctions are close linear appositions of adjacent plasma membranes which allow the transfer of small molecules between neighbouring cells. Experimental data indicate that cell to cell transfer of as yet undefined factors from normal

Box 6.1 Phases in the cell cycle

Growth and development of cells has been described in the form of a cell cycle which can be divided into five phases. G_1 (G for gap) is a pause after stimulation when the cell is involved in a lot of biochemical activity. S (synthesis) is the phase when the cell's DNA content is doubled in amount. G_2 is another pause and M is the phase involving mitosis when the cell divides into two. Cells can also stop cycling and enter a resting phase (G_o)

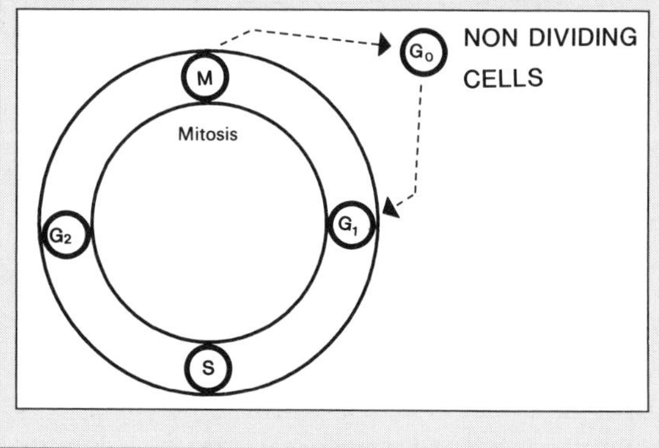

to neoplastic cells can suppress the transformed phenotype.

Chromosomal stability

Cancer-prone genetic disorders such as xeroderma pigmentosa, Fanconi anaemia, ataxia telangiectasia and Bloom syndrome have the common feature of abnormal repair of damaged DNA. A consistent finding in cancer cells is multiple abnormalities affecting the karyotype. A potential role for the tumour suppressor genes in DNA repair is illustrated by reference to p53.

p53

The p53 tumour suppressor gene is located on chromosome 17p12-13.3 and has been implicated in many inherited and sporadic forms of cancer in humans (Table 6.3). To date, mutations in p53 are the commonest genetic change in these cancers. p53 functions as a tumour suppressor gene since it inhibits the transformation of cells in culture by oncogenes as well as the formation of tumours in animals. A key feature of tumour suppressor genes, i.e. loss through chromosomal or DNA rearrangement, is seen with the p53 locus.

The 53 kDa protein encoded by p53 is a nuclear phosphoprotein which is involved in the regulation of cellular proliferation. The function of p53 in restricting or holding cells at the G_1-S phase of the cell cycle, allowing DNA repair to occur, has led to it being termed the 'genome policeman'. The mutant forms demonstrate altered growth regulatory properties and can also inactivate normal (wild-type) p53 protein. The latter phenomenon is called a '**dominant negative**' effect since inactivation of one of the two tumour suppressor loci can produce what appears to be a dominant effect if the mutant protein inhibits the product from the remaining normal allele.

Cancers in which there have been mutations affecting the p53 gene include colon, lung, brain, breast, melanoma, ovary and chronic myeloid leukaemia in blast crisis. Defects observed lead to loss of both p53 alleles in 75–80% of cases with one defect often a deletion and the second a missense point mutation. The latter leads to production of an abnormal protein. The loci for mutations in p53 involve predominantly four regions of the protein which presumably have functional roles since they are highly conserved amongst different species.

The *Li Fraumeni syndrome* is a rare inherited predisposition to cancer including various types of sarcomas in children and young adults and premenopausal breast cancer in relatives. Recently, it has been shown that some affected families have one mutant p53 gene in their *germline*. Malignant cells, on the other hand, have abnormalities affecting both alleles. These data reinforce the two-hit hypothesis for tumourigenesis and implicate the p53 tumour suppressor gene as one component in the pathogenesis of the Li Fraumeni syndrome. What other genetic defects are involved in this disorder remain to be determined.

GENETIC FACTORS IN CANCER

Retinoblastoma

Clinical features

Retinoblastoma is an important intraocular malignancy in children. The disease can be detected at birth although the usual time is within the first 3 years of life. Since treatment with surgery or radiotherapy is potentially curative, it is essential to make an early diagnosis. *Sporadic* retinoblastoma is usually a unilateral tumour which has a single focus. It occurs later in life than the *heritable* form which is frequently bilateral and multifocal. With increasing survival of treated patients it has become apparent that those with the heritable form are at increased risk for other tumours, e.g. osteosarcoma.

An important breakthrough in our understanding of retinoblastoma came with the finding of a deletion involving chromosome 13q14 in the blood of patients with the heritable type and in the tumour cells of both the sporadic and heritable forms. Subsequently, DNA polymorphic markers showed that the *normal* allele on the chromosome 13 inherited from the unaffected parent was the one lost in the tumour cells, indicating that loss of heterozygosity brought about its effect by uncovering a mutation in the germline. This was consistent with the two-hit model for cancer described earlier in this chapter.

Positional cloning in retinoblastoma

Using positional cloning, a candidate gene at the retinoblastoma locus was identified. A 4.7 kb mRNA transcript which encoded for a 928 amino acid protein (called p105-RB) was isolated. Surprisingly, the protein was expressed in most tissues not only in the retina. Nevertheless, confirmation of the above as a likely candidate gene came with the finding that the normal 4.7 kb mRNA transcript was not present in retinoblastoma cell lines. Mutations involving the gene, i.e. deletions, point mutations, were also detected in DNA from tumours. An apparent increased frequency of osteosarcoma in patients with retinoblastoma was strengthened when similar changes in the retinoblastoma gene were found in osteosarcomas. Surprisingly, sporadic cases of osteosarcoma which had occurred unrelated to retinoblastoma also had mutations involving the retinoblastoma gene.

DNA changes in retinoblastoma

The tumour suppressor gene involved at the retinoblastoma locus and its mode of action remain to be defined. However, DNA probes are now available to identify those who are predisposed. It is estimated that in approximately 10% of cases the disease can arise by inheriting a germline mutation from a carrier parent or through the occurrence of a de novo germline mutation (~30% of cases). This leads to bilateral heritable disease. Progeny of those with bilateral heritable disease have a 50% incidence of retinoblastoma.

Infants at-risk require repeated ophthalmological examinations under anaesthesia to detect tumours at an early stage. DNA linkage analysis using retinoblastoma-specific DNA locus probes enables the wild-type to be distinguished from the mutant 13q14 loci in a pedigree analysis. Those at-risk can be assessed and their risk modified depending on their DNA polymorphism patterns. An absolute exclusion or confirmation as susceptible is not possible unless a mutation can be detected since DNA polymorphisms located outside the retinoblastoma locus can lead to error through recombination.

Leukaemias and lymphomas

Haematological malignancies either become more malignant during the course of their natural history or present in the first instance as an aggressive disorder. Access to abnormal cells in the peripheral blood or bone marrow is possible. Therefore, they provide convenient models with which to study changes in DNA during various stages of a malignancy.

Chronic granulocytic leukaemia

A preliminary description of the characteristic chromosomal change (the Philadelphia chromosome) in chronic granulocytic leukaemia has already been given (p. 127). This leukaemia affects young adults and is a malignant clonal disorder involving a pluripotential haematopoietic stem cell. It usually presents in chronic phase and within 3 to 4 years develops into an accelerated or acute phase called *blastic transformation*. Over 95% of cases have the Philadelphia chromosome. The fusion gene product formed following the translocation between chromosomes has tyrosine kinase activity and is implicated in the pathogenesis of the chronic phase (Fig. 6.8). During development of blastic transformation, additional DNA changes are seen. One of these produces deletions or rearrangements involving the tumour suppressor gene p53. Occasionally, a Philadelphia chromosome is detected in some forms of acute lymphoblastic leukaemia. In this situation, the acute leukaemia has a poorer prognosis.

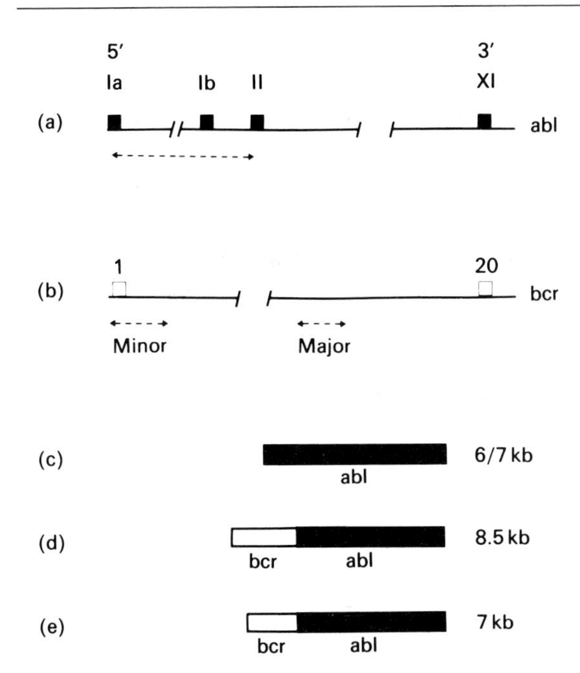

Fig. 6.8 Proto-oncogenes involved in the translocation which produces the Philadelphia chromosome.
(**a**) The *abl* gene (■) on chromosome 9 showing two alternatives for exon I (Ia and Ib) which are situated 5′ to exon II. The translocation break in chromosome 9 usually occurs 5′ to exon II. Thus, *abl* exons II–XI (occasionally exon I is included) are translocated onto chromosome 22. (**b**) There are two breakpoint regions in the *BCR* gene (□) which has 20 exons. That closest to exon 1 is termed the minor region and is involved in the Philadelphia chromosome which is occasionally found in acute lymphoblastic leukaemia in adults. The major breakpoint cluster region (bcr) is situated further 3′ and is the site for the translocation break in chronic granulocytic leukaemia and some cases of adult acute lymphoblastic leukaemia. (**c**) The normal *abl* mRNA transcript is 6–7 kb long depending on which exon I is utilised. (**d**) The bcr-*abl* fusion gene involving the major bcr produces a transcript of 8.5 kb. (**e**) The bcr-*abl* fusion gene involving the minor bcr gives a smaller 7 kb transcript.

Table 6.4 Proto-oncogenes and tumour suppressor genes implicated in the pathogenesis of leukaemias and lymphomas (from Cline & Ahuja 1991)

Neoplasm	Proto-oncogenes	Tumour suppressor genes
Chronic granulocytic leukaemia	*abl*	p53
Acute myeloid leukaemia	N-*ras* (also H-*ras*, K-*ras*), *myc*	p53, 11p
Pre-leukaemias	N-*ras*, H-*ras*, K-*ras*	–
Acute lymphoblastic leukaemia	*abl*, *myc*, N-*ras* (also H-*ras*, K-*ras*)	–
Follicular lymphoma, chronic lymphocytic leukaemia	bcl 1, bcl 2	–
T cell and B cell lymphomas	tcl 1,2,3*, *myc*, N-*ras*	–

* putative oncogenes

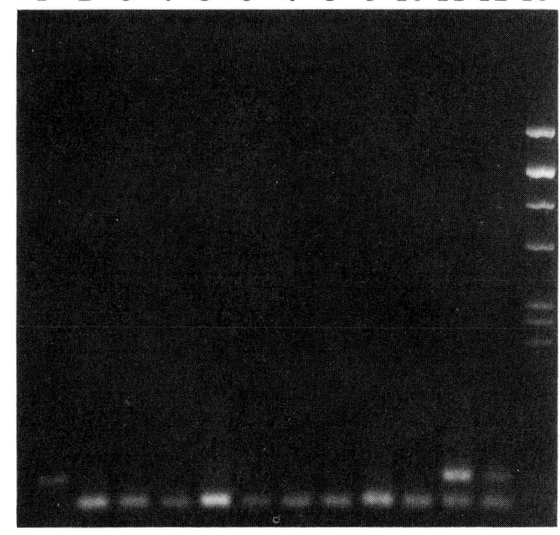

Fig. 6.9 Mutations in the N-*ras* gene present in acute granulocytic leukaemia.
Lane 1. Control DNA fragment containing a codon-12 mutation in the N-*ras* proto-oncogene. Lane 13. Molecular weight marker. Lanes 2–12. Peripheral blood from different patients with acute granulocytic leukaemia. The codon-12 mutation detected by DNA amplification is present in the leukaemic cells tested in lanes 11 and 12 (example courtesy of Dr A Todd and Dr H Iland, Kanematsu Laboratories, Royal Prince Alfred Hospital).

Acute granulocytic leukaemia

This is a leukaemia of primitive haematopoietic cells and in contrast to chronic granulocytic leukaemia, it presents as an acute disorder which progresses rapidly without therapeutic intervention. *Ras* proto-oncogene abnormalities, which are predominantly N-*ras*, occur in 20–60% of cases (Fig. 6.9). However, mutations in the N-*ras* proto-oncogene are not considered sufficient to explain the acute leukaemic phenotype. Additional DNA changes detected in this and other leukaemias occur in the p53 tumour suppressor gene (~50%) and other proto-oncogenes, e.g. *myc* (Table 6.4).

Cancer of the colon

Colonic cancer has been used to study evolution of a *solid* tumour along similar lines described for the haematological malignancies. This is possible because there is a model for colon cancer that allows progression of cancer to be followed from the pre-malignant stage to the advanced tumour and then the invasive (metastatic) tumour.

Clinical features

Colon cancer is one of the most common cancers in western countries. Features which distinguish it from other frequently occurring malignancies include a defined precancerous state (the adenomatous polyp) and a rare but well described genetic variant called familial adenomatous polyposis. The latter is inherited as an autosomal dominant disorder with high penetrance. Familial adenomatous polyposis is characterised by many polyps in the colon with a probability close to 100% that one or more will become malignant. Since there is high risk of cancer in familial adenomatous polyposis, treatment involves prophylactic (preventative) removal of the colon in those who are predisposed.

Positional cloning in familial adenomatous polyposis coli

The various landmarks achieved in looking for the gene(s) causing this genetic form of colon cancer are summarised (Box 6.2). As has been seen in other examples, a key early finding was the observation of a deletion involving the long arm of chromosome 5. From linkage studies of affected pedigrees, the familial adenomatous polyposis locus was next localised to the 5q21-22 region and DNA probes were soon isolated. During the next 3 years it became apparent that in both *familial* and *sporadic colon cancers* there were other frequently detected abnormalities. These included the p53 gene on chromosome 17p; a tumour suppressor locus on chromosome 18q and the *ras* proto-oncogenes. In 1990, a tumour suppressor gene on chromosome 18q was isolated and called DCC (deleted in colon cancer). In the following year, the MCC gene (mutated in colon cancer) was found at the chromosome 5q locus linked to familial adenomatous polyposis. MCC thus became a good candidate for the familial genetic defect but a unique role was subsequently excluded when a number of affected families showed no defects in this gene. Further DNA searches then identified, not far from MCC, another gene called APC (adenomatous polyposis coli). APC is a better candidate since at the time of writing all families studied have demonstrated point mutations (stop codons, frameshifts) in this gene. Domain structure in APC suggests it may have G protein like activity.

Box 6.2 Historical developments in the colon cancer story

1986	Familial adenomatous polyposis: cytogenetic deletion on chromosome 5q observed
1987	Genetic linkage located defect to 5q21-22
1987–9	Mutations / deletions involving chromosomes 17p (p53 gene); 18q and the K-*ras* proto-oncogene
1990	DCC* gene (chromosome 18q21.3) isolated
1991	MCC* gene (chromosome 5q21)
1991	APC* gene (chromosome 5q21)

* DCC = deleted in colon cancer; MCC = mutated in colon cancer, APC = adenomatous polyposis coli

DNA changes in colon cancer

DNA changes which are frequently found in colon cancer are summarised in Table 6.5. From these, a number of observations were made by Fearon and Vogelstein:

- Both proto-oncogene and tumour suppressor genes are involved, with the latter having a major role
- Multiple mutations (e.g. more than 4) are required: the numbers present correlate approximately with the stage of evolution, i.e. there are fewer changes in adenomatous polyps compared to carcinoma
- There is no single genetic mutation (or combination of mutations) which predictably leads to a specific phenotype
- The dominant-negative effect of tumour suppressor genes, where the mutant protein product inhibits the activity of the normal gene product, is important.

It should also be noted that similar DNA muta-

Table 6.5 Genetic changes present in colon cancers (from Fearon & Vogelstein, 1990).

Locus/gene(s)	Defects and features
Chromosome 5q, tumour suppressor gene	Allele loss: 20–50% of carcinomas and ~30% of adenomas except in the familial cases where 5q changes in adenomas are rare
Chromosome 17p, p53 tumour suppressor gene	Allele loss: ~75% of carcinomas but rare in adenomas
Chromosome 18q, tumour suppressor gene: DCC	Allele loss: ~70% of carcinomas and ~50% of advanced adenomas
ras proto-oncogene	Point mutations: ~50% of carcinomas and advanced adenomas; <10% in small adenomas

tions to those found in the genetic form of colon cancer are also being detected in the more common but sporadic cases. Thus, knowledge about the latter is indirectly being obtained by studying the tumours which have an obvious genetic component.

Model for tumourigenesis in colon cancer

A number of hypotheses have been advanced to explain the clinical and genetic findings in colon cancer. One suggestion for the (precancerous) adenomatous polyp formation involves a dose effect from the APC gene such that a single mutation is sufficient to initiate the growth of a polyp. This, as well as a number of additional superimposed genetic changes, enable the slow but steady progression to carcinoma. Important components allowing tumour progression include genetic instability and clonal growth advantage. Invasion and metastasis then require additional genetic changes (Fig. 6.10). The net effect is multiple well-recognised mutations in key regions superimposed on which are other effects (genetic or environmental in origin). Clinical heterogeneity (e.g. invasiveness, the number of adenomatous polyps present) reflects the various genetic combinations or types of mutations present. Similar changes in oncogene and tumour suppressor gene loci are seen in other malignancies suggesting there are common pathways in tumourigenesis.

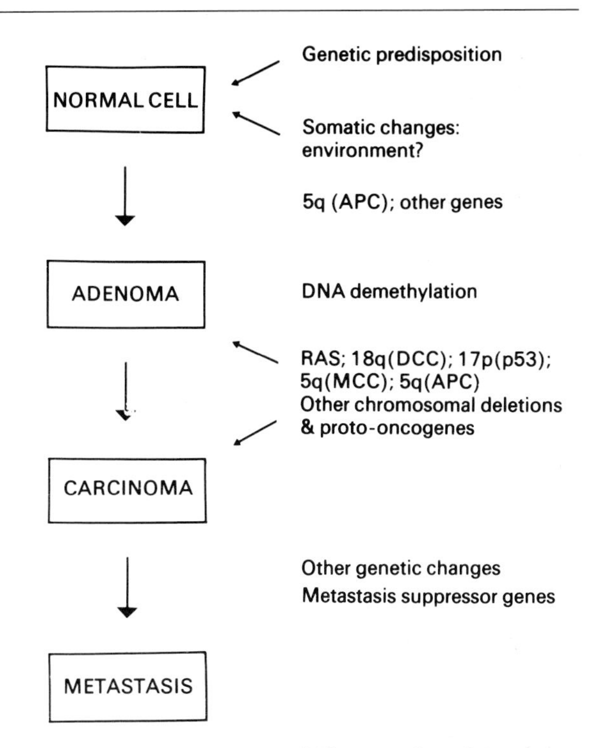

Fig. 6.10 A multistep genetic model for tumourigenesis in colon carcinoma.
Progression will depend on genetic and environmental effects imposed on an initial mutation. APC–adenomatous polyposis coli gene; DCC–deleted in colon cancer gene; MCC–mutated in colon cancer gene (from Fearon & Vogelstein 1990).

Cancer of the breast

Clinical features

In contrast to the inherited cancer syndromes such as retinoblastoma described above, *familial cancers* refer to neoplasms that cluster in families. However, because of inadequate markers to detect the predisposed phenotype, it is difficult to ascertain who are at-risk. Many types of familial cancers have been reported but the sites most commonly involved are breast, ovary, melanoma, colon, blood and brain. Clinical features which would suggest a familial cancer include: two or more close relatives affected; multiple or bilateral cancers in the same person; early age of onset and clustering (e.g. occurrence of both cancer of the breast and cancer of the ovary).

Breast cancer is the commonest cause of cancer death in women. Its pathogenesis is complex and involves physiological, environmental and genetic factors. A positive family history for breast cancer is a significant risk factor which can exceed by 50% the life time risk if the affected are first degree relatives with early and bilateral breast involvement.

DNA changes in breast cancer

The potential role of tumour suppressor genes in the pathogenesis of human breast cancer is complex with many loci demonstrating loss of heterozygosity (Table 6.3). Researchers are presently attempting to define genetic factors involved in the familial cases since it is more likely that the underlying defects will be identifiable in these situations. Once defined, the same genetic mutations can be sought in examples of sporadic disease.

To date, the gene for premenopausal familial breast cancer has been mapped to the long arm of chromosome 17. It has also been observed that tissues examined from sporadic breast cancers demonstrate loss of heterozygosity involving similar loci to those described in familial cases.

Viral-induced cancer

Clinical features

The hepatitis B virus genome was cloned and sequenced in 1979. 2 years later an association between chronic hepatitis B infection and cancer of the liver (called hepatocellular carcinoma) was described. Evidence for the above association came from epidemiological studies and animal models.

Hepatitis B virus carriers are estimated to have a 200-fold greater risk of developing hepatocellular carcinoma compared to uninfected individuals from the same region. Male carriers of the hepatitis B virus are particularly at risk of developing liver cancer with a latency period which ranges from 20–40 years.

DNA changes

The hepatitis B virus is a DNA virus. Therefore, unlike retroviruses (see Fig. 6.2) it does not have to integrate into the host genome to complete its replication cycle. However, hepatitis B viral integration does occur during persistent infection. For example, the gene for the hepatitis B virus X protein appears to be incorporated into the genome in such a fashion as to alter the expression of growth regulatory genes. The hepatitis B virus surface antigen (HBsAg) also acts to induce a chronic inflammatory response which is an important predisposing factor to abnormal mitosis and so cancer. Approximately 50% of hepatocellular cancers have loss of heterozygosity at chromosomal loci 11p and 13q. In both these regions there are tumour suppressor genes.

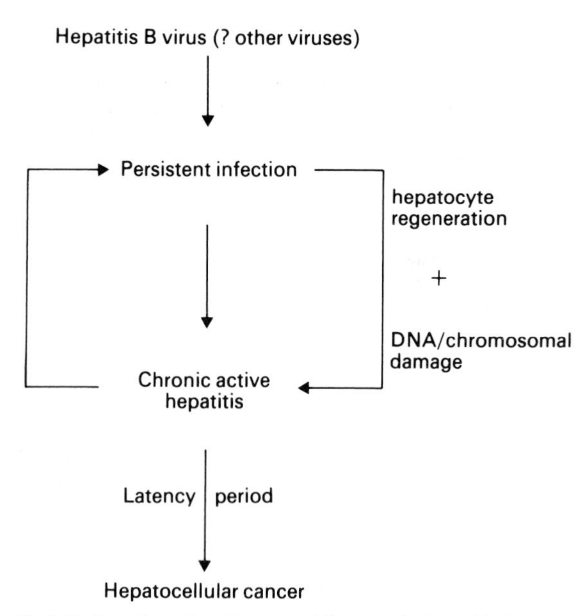

Fig. 6.11 Genetic and environmental factors which could play a role in hepatocellular carcinoma.

At present, two important components have been implicated in the hepatitis B virus-induced hepatocellular carcinoma: (1) the presence of active liver cirrhosis and (2) viral DNA effects on the host genome (Fig. 6.11). However, it is important to note that infection with the hepatitis B virus is not a necessary event for hepatocellular carcinoma (see cancer and the environment below).

Cancer and the environment

Experimental models

Chemical and physical agents which produce tumours also cause characteristic DNA-associ-

Table 6.6 DNA changes and carcinogens (from Jones et al 1991)

Carcinogens	Changes in DNA
Some alkylating agents	G → A transitions[1]
Benzo[α]pyrene	G → T transversions[2]
Melphelan	A → T transversions
Ionising radiation	Deletions

[1] transition = change of a purine to a purine or a pyrimidine to a pyrimidine
[2] transversion = change of a purine to a pyrimidine or vice versa

ated defects (Table 6.6). Animal models of carcinogen-induced tumourigenesis confirm that predominant mutations in tumour cells mirror the type of chemical agent used for induction. For example, breast cancers in rats which are induced by the carcinogen (cancer producing agent) nitrosomethylurea contain G to A mutations in the H-*ras* proto-oncogene. In comparison are skin tumours induced by dimethylbenzanthracene in mice which have A to T mutations in DNA.

Environmental carcinogens and human cancer

Mutations found in the south African or east Asian-associated hepatocellular carcinoma have been shown to be predominantly G to T mutations involving codon 249 of the p53 gene. This finding is consistent with an environmental agent (for example, aflatoxin B_1 a food contaminant present in the above regions) as an alternative or contributing risk-factor for hepatocellular cancer (see also hepatitis B virus above). In experimental mutation assays, aflatoxin B_1 binds preferentially to guanine residues in GC rich regions and induces the same G to T changes.

Another example of environmentally-related tumourigenesis is skin cancer. Ultraviolet light induced mutations have characteristic in vitro changes which include frequent C to T mutations at dipyrimidines. Also diagnostic of ultraviolet damage are CC to TT dinucleotide changes. The above mutations are frequently found in the p53 tumour suppressor gene from squamous cell carcinomas of the skin. Objective evidence is thus provided of the effect that sunlight has on the development of skin-related tumours.

Imprinting

Reference was made in Chapter 3 to parent of origin effects (imprinting) and the possible role these play in genetic disorders. It is now becoming apparent in some cancers that the parental origin (paternal versus maternal) of the mutated gene may affect subsequent expression of that gene. Three tumours (osteosarcoma, Wilms' tumour and glomus tumour) demonstrate features which are consistent with imprinting (Table 6.7).

The molecular basis for imprinting is unknown. It is intriguing, however, that methylation differences are found in association with imprinting in transgenic mice, i.e. the functional allele is hypomethylated and the inactive allele is methylated. Whether methylation patterns are primary or secondary effects remains to be determined. Nevertheless, differential methylation of alleles due to imprinting might lead to alleles having dissimilar susceptibilities to mutation. It is also interesting that imprinting per se can be one effect by which the 'second hit' occurs. If a locus is imprinted, only one of the two alleles would be

Table 6.7 Evidence from inherited cancer syndromes that DNA imprinting occurs.
The parental origin of the mutated gene affects its subsequent expression

Locus	Phenotype	Parent of origin effect
Retinoblastoma 13q14	Inherited form of retinoblastoma	Germline mutations occur more frequently during spermatogenesis than oogenesis*
	Somatic mutations in sporadic retinoblastoma	Both maternal and paternal 13q14 loci are affected
	Somatic mutations in sporadic osteosarcoma	Predominantly involve maternal allele loss
Wilms' tumour 11p13	Somatic mutations in sporadic Wilms' tumour	Predominantly involve maternal allele loss
11q23-qter	Hereditary glomus tumour (also called paraganglioma)	Autosomal dominant transmission when paternally inherited; offspring of female patients are not affected

* not an imprinting effect but a higher susceptibility to mutation in sperm compared to ova

functional. In this circumstance, it would only require one additional 'hit' to inactivate the remaining functional allele.

What then is the significance of imprinting in the neoplastic disorders? First, it is important to consider the possibility of imprinting so that counselling given to families with inherited or familial cancers is accurate. The potential for imprinting to confuse the inheritance patterns may explain why the genetic component of some tumours has remained obscure. In the long term, the role played by imprinting in cancers may be an additional epigenetic factor in their pathogenesis.

DIAGNOSTIC APPLICATIONS

Anatomical pathology

The standard approach in anatomical pathology is examination of a tissue section stained with a dye such as haematoxylin and eosin. On the basis of tissue morphology and staining characteristics, a diagnosis is made. Additional stains which help to identify specific tissues or components in cells are used to obtain further information. A greater level of resolution than light microscopy is possible with electron microscopy. Immunophenotyping to identify specific antigens with monoclonal or polyclonal antibodies has been an important additional development.

In situ hybridisation

To the above investigative approaches can now be added in situ hybridisation using DNA probes to identify specific sequences. An illustration of the value of in situ hybridisation is provided by the human papillomaviruses. A description of these viruses, their detection and association with genital tract lesions has been given in Chapter 5. Diagnosis and typing of such viruses has become simplified with tests such as Southern blotting and DNA amplification. In the research laboratory, the availability of in situ hybridisation has allowed tissue and cellular localisation of mRNA transcripts from these potentially oncogenic viruses.

Flow cytometry

Procedures described above provide details on the type and geographical localisation of a tumour. The distribution of oncogenes or defects in DNA can also be determined. However, the studies described to date are often inadequate when it comes to quantification, e.g. is an oncogene or its product amplified? Quantification is possible with flow cytometry. Components in this technique include: a laser which permits measurements of large numbers of individual cells in a short time period and monoclonal antibodies which are labelled with fluorochromes. The immunological (antigenic) properties of the cells being assayed are enhanced further by measurements of additional parameters such as size and granularity. More sophisticated flow cytometers are able to sort out cell populations for further study. In this way, isolated nuclei, even from archival specimens, can be examined individually.

Just as the p53 tumour suppressor gene is the most common DNA abnormality in tumour cells, so is *aneuploidy* (any chromosome number other than the normal 46) the usual cytogenetic finding. Determination of the DNA content of tumour cells is undertaken by flow cytometry. In this approach, nuclei prepared from tumour tissues are stained with DNA-binding fluorescent dyes such as propidium iodide. Flow cytometric analysis enables the various phases of the cell cycle (G_0, G_1, S, G_2 and M) to be defined (Box 6.1). Aneuploidy is assessed from the DNA content of the cell's G_0/G_1 phases relative to the DNA content of normal diploid cells. Aneuploid populations are seen as separate peaks from the normal diploid peak. Additional calculations are possible from the flow cytometry profile, for example the percentage of cells in S phase.

Cell cycle parameters described above are being used to attempt prediction of disease or treatment outcomes in a number of malignancies such as cancers of the breast, ovary, colorectum and many others. Results can have both prognostic and therapeutic implications, e.g. patients with aneuploid ovarian carcinoma have a far worse prognosis than those with diploid tumours, irrespective of the extent of disease or histological appearance.

Fine needle aspiration and DNA amplification

A useful source of tissue for histopathological examination is obtainable through fine needle aspiration. This is a rapid, less traumatic procedure than conventional biopsy and carries negligible risk of tumour dissemination. The problem of whether sufficient material can be obtained for histological assessment has now been overcome to some extent by the polymerase chain reaction. The utility of the above approach can be illustrated by reference to nasopharyngeal carcinoma. In this cancer, conventional biopsies can miss the primary lesion because the nasopharynx is difficult to visualise and the tumour infiltrates submucosally. In such circumstances, the Epstein–Barr viral genome can be sought in tumours by fine needle aspiration and DNA amplification. Finding the virus enables the diagnosis of nasopharyngeal carcinoma to be made as well as distinguishing metastatic lesions from other head and neck tumours. Another advantage of the polymerase chain reaction technique is that archival materials such as formalin fixed, paraffin wax embedded tissue blocks are also suitable sources of DNA for amplification.

Haematological malignancies

Conventional diagnostic approaches

A similar strategy to that described above in anatomical pathology is usually followed in the haematological malignancies, e.g. leukaemia or lymphoma (Table 6.8). Tissues which can be studied include: peripheral blood, bone marrow or lymph nodes. Tissue appearances under the light or electron microscope are frequently diagnostic. Examination of the cells' chromosomal con-

Table 6.8 Approaches used in the diagnosis and classification of the haematological malignancies

Test	Information provided
Tissue aspiration or biopsy Peripheral blood	Morphological appearances of cells can be seen. Staining by conventional means such as May–Grunwald–Giemsa can be supplemented by specific stains, e.g. esterase detection distinguishes leukaemias which are not lymphoid in origin
Cytogenetics	Analysis of the cell's chromosome constitution will detect translocations, e.g. the Philadelphia chromosome (chromosomes 9 and 22) and that seen in Burkitt's lymphoma (chromosomes 8 and 14 usually)
Immunophenotyping	Monoclonal antibodies to antigens formed during differentiation are particularly useful to distinguish leukaemic cells as myeloid or lymphoid in origin or typing as a primitive versus differentiated cell

Box 6.3 Translocations in lymphoproliferative disorders

Translocation is the movement of a segment of a chromosome to another chromosome. The best known chromosomal translocation in the lymphoproliferative disorders is the Philadelphia chromosome which produces a chromosome 9 and chromosome 22 reciprocal translocation. Two oncogenes, *abl* and *sis* are involved in this translocation. In follicular lymphomas, the usual translocation includes chromosomes 14 and 18. The two critical regions in this rearrangement are the immunoglobulin heavy chain gene (14q32) and bcl-2. In Burkitt's lymphoma, the usual translocation involves chromosomes 14 and 8. Less frequently the chromosome 14 locus, which contains the *myc* oncogene, translocates to chromosomes 2 or 22 which are the loci for the immunoglobulin light chain genes κ and λ respectively. Chronic lymphocytic leukaemia and some of the diffuse lymphomas have a translocation affecting the immunoglobulin heavy chain locus on chromosome 14 and a gene designated *bcl*-1 on chromosome 11. Translocations in the T cell lymphoproliferative disorders are less well characterised although in some cases they involve the immunoglobulin locus.

stitution will enable translocations to be detected and these can be specific for a number of disorders (Box 6.3).

Immunophenotyping by flow cytometry with the extensive range of monoclonal antibodies now commercially available has become an important investigative tool. Monoclonal antibodies to cell-surface antigens enable cell lineages to be identified. The level of differentiation for neoplastic clones can also be determined. The potentially confusing nomenclature arising from the vast array of monoclonal antibodies has been addressed by four International Workshops on Leukocyte Differentiation Antigens. Following on from these workshops, 78 major clusters of differentiation ('CD') have been defined. For example, primitive leukaemic cells (called blasts) can be classified on the basis of their CD antigens into the three major lineages as B lymphocytes (positive immunophenotyping to CD19, CD22), T lymphocytes (CD7, CD3 positive) and myeloid (CD13, CD33 positive but CD19, CD22, and CD3 negative).

Oncogenes and the leukaemias and lymphomas

The function of a proto-oncogene can be altered following its mutation or translocation onto another chromosome. Thus, detection of the Philadelphia chromosome or its product, the hybrid bcr-*abl* p210 protein or mRNA, provides important diagnostic information. Both DNA mapping (Southern blotting) and polymerase chain reaction-based techniques can be used to detect chromosomal translocations such as the Philadelphia chromosome and also determine what type of rearrangement has occurred. Compared to conventional cytogenetic tests, the sensitivity of the molecular approaches (particularly DNA amplification) is high with claims that even 1 cell in 1000 with the Philadelphia chromosome will be identified. This is discussed below under Minimal residual disease (p. 143). Different regions of the bcr (breakpoint cluster region) can be involved in the translocation (Fig. 6.8). The significance of this heterogeneity is unknown. To date there is no consistent evidence to suggest that these differences influence the prognosis.

Immunoglobulin and T cell receptor gene families

Lymphocytes are unique cells since they are able to undergo somatic rearrangements of their immunoglobulin or T cell receptor genes. This is essential to generate immune receptor molecules of sufficient diversity to enable recognition of the vast array of antigens in the external environment. Should one such lymphoid cell form a malignant clone, all sister cells will carry the hallmark of its unique gene rearrangement. This can be of considerable value when investigating patients with haematological malignancies.

Immunoglobulin molecules consist of two identical heavy chain genes (either μ, δ, γ. ε, or α) and two identical light chain genes (κ or λ). The variable or antigen recognition portion of the immunoglobulin molecule (called Fab, **f**ragment **a**ntibody **b**inding) is located at the amino terminal end. At the carboxyl end is the constant region (called Fc, **f**ragment **c**rystalline) which defines the isotype or genetic subgroup of the immunoglobulin molecule: IgM (μ), IgG (γ), IgE (ε), IgD (δ) or IgA (α) (Fig. 6.12). There are two types of T cell receptors: one is a heterodimer of α and β chains and the second is a $\gamma\delta$ heterodimer. Genes for the immunoglobulin and T cell receptor loci are

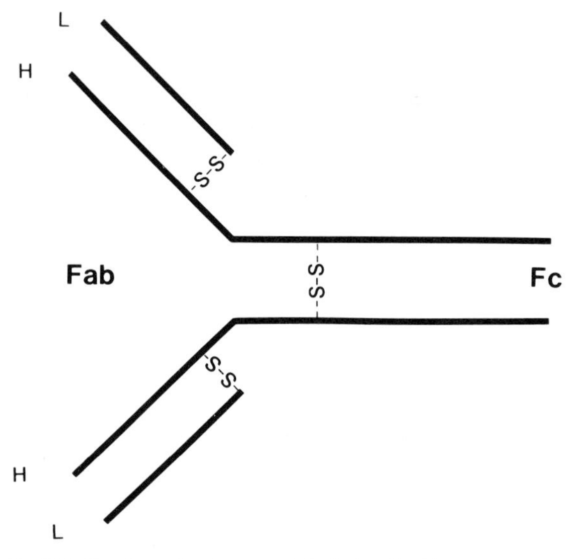

Fig. 6.12 Structure of an immunoglobulin (antibody). Disulphide bonds (–S–S–) hold together the two heavy chains (H) and the two light chains (L). Fab = fragment antibody binding; Fc = fragment crystalline.

Table 6.9 Genes which comprise the immunoglobulin and T cell receptor loci

Type	Chromosome locus	Component genes*
Immunoglobulin heavy chains	14q32	$V_H–D_H–J_H–C_H$ (μ, δ, γ_3, γ_1, α_1, γ_2, γ_4, ε, α_2)
Immunoglobulin κ light chains	2p11	$V_\kappa–J_\kappa–C_\kappa$
Immunoglobulin λ light chains	22q11	$V_\lambda–(J_\lambda C_\lambda)_7$
T cell α, δ chains	14q11	$V_\alpha–V_\delta–V_\alpha–V_\delta–D_\delta–J_\delta–C_\delta–V_\delta–J_\alpha–C_\alpha$
T cell β chains	7q34	$V_\beta–D_\beta–J_\beta–C_\beta–D_\beta–J_\beta–C_\beta$
T cell γ chains	7p15	$V_\gamma–J_\gamma–C_\gamma–J_\gamma–C_\gamma$

*V = variable, D = diversity, J = joining, C = constant

located on a number of different chromosomes and comprise various combinations of the following: variable (V); diversity (D); joining (J) and constant (C) genes (Table 6.9). Gene families encoding the immunoglobulin and T cell receptor molecules are arranged in two distinct configurations: (1) a functionally inactive or germline state and (2) a functionally active or rearranged configuration (Fig. 6.13). During development of a stem cell into a B or T lymphocyte, there is rearrangement of its germline immunoglobulin or T cell receptor genes. Where present, there is first a combination of a D (diversity) and a J (joining) segment (DJ rearrangement). The next step involves combination between the DJ and a V (variable) gene. This comprises the *variable* unit of the molecule. The fully rearranged immunoglobulin or T cell receptor gene requires joining to a C (constant gene) (Fig. 6.13). Each of the immunoglobulin or T cell receptor gene rearrangements is *unique*. The incredible diversity which can be generated explains in part the immunological repertoire which an organism can develop. Additional superimposed somatic mutations produce further diversity.

At the DNA level, the germline state will be present on non-lymphoid cells. This can be detected by DNA mapping using Southern blotting. Within a polyclonal population of lymphocytes, such as those normally circulating in the peripheral blood, rearranged genes cannot be identified because the number of possible rearrangements is so many that individual ones are

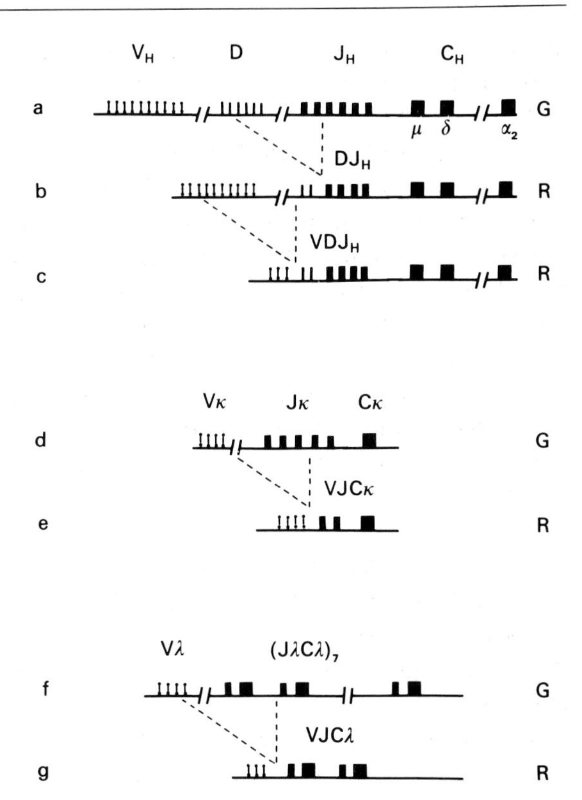

Fig. 6.13 Immunoglobulin genes in their germline and rearranged configurations.
During development of a stem cell into a B or T lymphocyte, there is rearrangement of the germline immunoglobulin or T cell receptor genes. This rearrangement generates the diversity in immune proteins necessary for effective antigen recognition. The different genes are : V = variable; D = diversity; J = joining; C = constant. A break in the base-line indicates that the full gene complement has not been given. G = germline; R = rearranged. (**a**) The immunoglobulin heavy chain locus. (**b**) The first recombination in the heavy chain locus involves a D to J step. (**c**) This is then followed by V to D-J recombination. (**d**) The germline κ light chain gene locus which becomes rearranged in (**e**). The germline λ light chain gene locus (**f**) and its rearrangement in (**g**).

too few to be detectable. On the other hand, a large number of cells which bear the identical rearranged pattern, i.e. a monoclonal population such as that found in leukaemia or lymphoma, can be detected by DNA mapping or DNA amplification techniques.

Rearranged immunoglobulin and T cell receptor genes

A rearranged immunoglobulin gene locus indicates that a monoclonal population of B lymphocytes is present. This means there is a haematological or immunological disorder involving these cells. The corresponding T lympho-

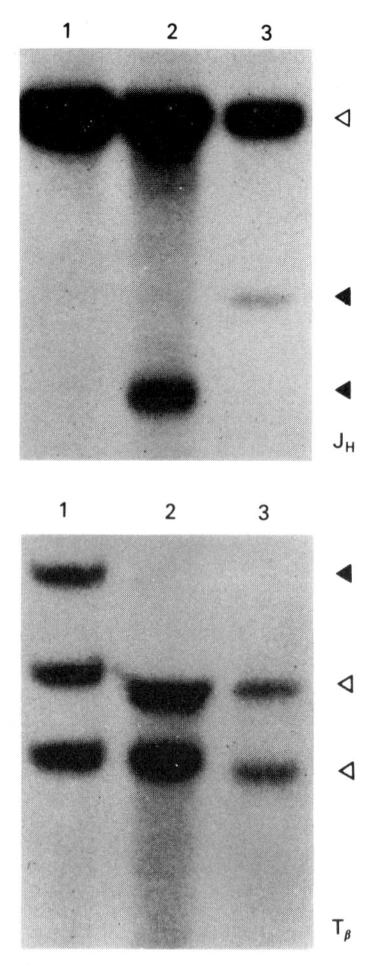

Fig. 6.14 Monoclonal population of lymphocytes indicative of a disease state.
The normal circulating lymphocytes will be polyclonal reflecting the somatic rearrangements of their genes. Detection of a monoclonal population will mean there is a haematological or immunological disorder involving these cells. DNA mapping patterns illustrating monoclonal populations in B and T lymphocytes. Immunoglobulin (J_H) and T cell receptor β chain patterns are shown. Germline bands are indicated by open triangles and rearranged bands by ◀. Sample 1, T cell rearrangement; samples 2 and 3, different B cell rearrangements.

cyte-derived leukaemias and lymphomas demonstrate monoclonal patterns associated with the T cell receptor β and to a lesser extent the γ and δ genes (Fig. 6.14). Very occasionally both immunoglobulin and T cell receptor genes are rearranged. This is called 'lineage infidelity' and is thought to reflect the activity of the enzyme recombinase which is involved in joining of the discontinuous segments from the various genes. Mixed rearrangements as described above are often incomplete, e.g. there is only one of the two immunoglobulin heavy chain alleles rearranged. Such aberrant rearrangements occur more often with B lymphocytes

The Southern blotting technique to detect immunoglobulin or T cell receptor gene rearrangements can distinguish, at a conservative estimate, approximately 10% of a cell population which is monoclonal. This relative insensitivity (although more objective and more sensitive than microscopy) as well as the technically long and demanding steps in DNA mapping has meant that DNA amplification by the polymerase chain reaction is now a more attractive option. The exquisite sensitivity of the polymerase chain reaction enables a monoclonal lymphocyte population of around 1–5% to be detected. This is better than Southern blotting but less than would normally be expected of the polymerase chain reaction (see minimal residual disease, p. 143) because the amplification products of *coexistent polyclonal* lymphocyte populations will make it difficult to distinguish the *amplified monoclonal* product.

The major disadvantage of the polymerase chain reaction technique to detect immunoglobulin or T cell receptor gene rearrangements is the problem of false *negative* results. These do not reflect a deficiency in the technology but the underlying gene rearrangements which are being sought. For DNA amplification it is necessary to design primers which are specific in terms of DNA sequence for a particular region. Furthermore, the amplification primers cannot be too far apart (an approximate distance would be 0.5–1.0 kb) or amplification will not occur or be inefficient. Designing DNA-specific primers for amplification in the case of immunoglobulin or T cell receptor genes is particularly difficult for the following reasons: (1) the complexity of the germline structure, (2) the occurrence of DNA rearrangements, (3) additional somatic mutations once rearrangement has occurred and (4) the possibility of incomplete or aberrant gene rearrangements. To get around the problem, primers to sequences which tend to be conserved (e.g. regions 3′ to the J_H genes or 3′ to C_μ in the heavy chain genes) can be mixed with a number of primers derived from

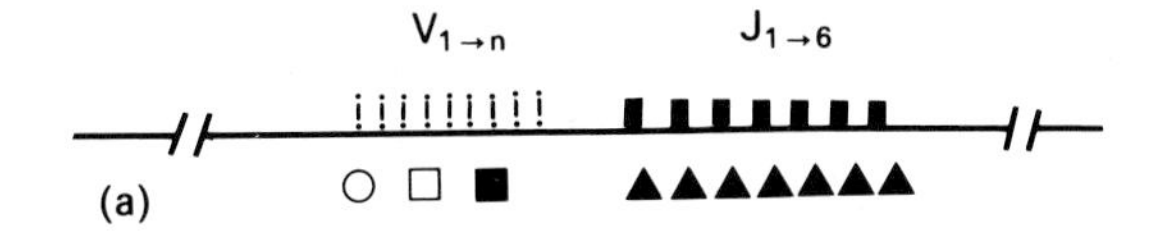

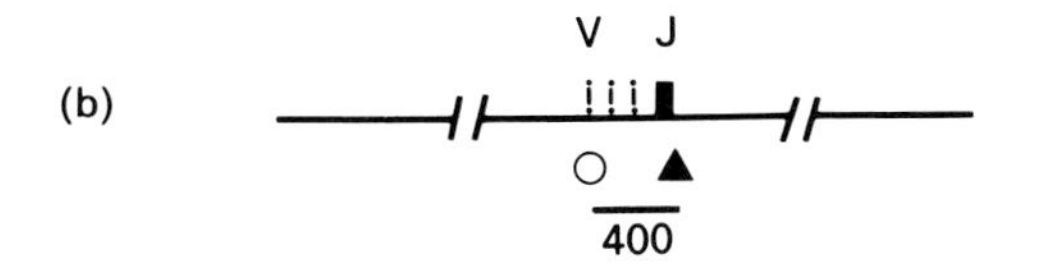

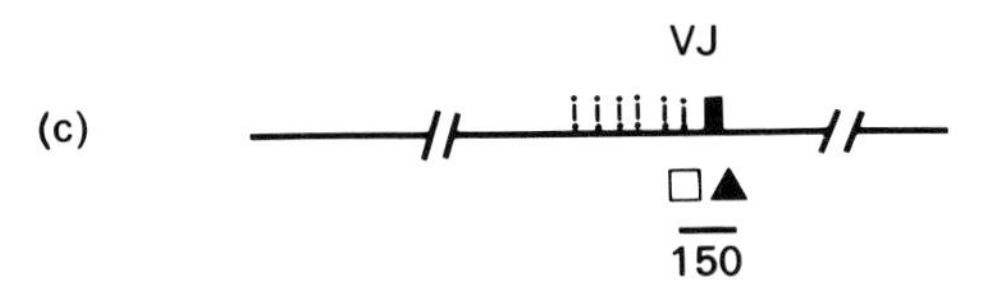

Fig. 6.15 DNA amplification-based strategy to identify immunoglobulin gene rearrangements such as those occurring in lymphoproliferative disorders.
Primers to conserved sequences are mixed with primers from variant sequences. Amplification will occur from the conserved primer to one of the variant primers. (**a**) The immunoglobulin heavy chain locus showing an unlimited number of V genes and the 6 J genes. DNA amplification primers designed from the 3′ ends of the J genes are constant (▲) whilst those from the V loci are different (○, □, ■) and reflect a particular V gene. (**b**,**c**) Two rearrangements have occurred involving V and J genes. The 3′ constant primer will form an amplified product provided there is a nearby V-specific primer. The size of the amplified product will depend on the distance between the primers, e.g. 400 and 150 base pairs as illustrated here.

5′ sequences that are likely to be altered (e.g. V_H). In this way, it is anticipated that amplification will occur from the 3′ conserved region to one of the many 5′ primers found in the mixture. Amplification will not be obtained from germline sequences since these are too far apart (Fig. 6.15). Using the above approaches 50% or more of the gene rearrangements found in the lymphoproliferative disorders (leukaemias or lymphomas which involve the lymphocyte) are said to be detectable.

The DNA amplification strategy is more suitable for the identification of translocation breakpoints which involve the immunoglobulin genes (e.g. bcl-2 rearrangements) since these have more specific, invariant sequences associated with them and so the appropriate amplification primers are easier to design. It is important to note that the ability to detect a *monoclonal* population does not necessarily mean it is a *malignant* population. The final decision as to whether the cell population present represents a benign or malignant monoclonal expansion will require additional information based on clinical, laboratory data and perhaps long-term follow-up of an affected individual.

Minimal residual disease

High-dose chemotherapy and/or radiotherapy can be used to treat cancer and haematological malignancy provided there is rescue of bone marrow suppression by autologous (from the

Box 6.4 Assessment of minimal residual disease using the polymerase chain reaction

In contrast to the immunoglobulin and T cell receptor rearrangements which arise from multiple sites within the gene loci, chromosomal translocations are clustered within fewer and better defined sites. Thus, DNA amplification by the polymerase chain reaction can be used for detection since it is easier to construct amplification primers. Normal cells will not give rise to an amplified product because each primer comes from a different chromosome. In one study of B cell non-Hodgkin's lymphoma and its associated translocation involving chromosomes 14 and 18, DNA amplification was used to test for minimal residual disease in purged autologous marrow. Patients were treated with intensive chemotherapy and then their marrows were purged with monoclonal antibodies to B lymphocytes. Before purging, lymphoma cells (cells positive for the translocation) were detectable in the post-treatment bone marrow. After purging a proportion of patients no longer had lymphoma cells present in their marrows. These patients demonstrated an improved survival over individuals who still had residual lymphoma cells detected by DNA amplification after purging. This study illustrated both the value of ex vivo purging and the potential contribution of occult neoplastic cells to relapse in B cell non-Hodgkin's lymphoma (Gribben et al 1991).

same person) bone marrow transplantation and infusion of the haematopoietic growth factor G-CSF (see Ch. 7). The infusion of autologous bone marrow provides sufficient haematopoietic stem cells for the marrow to recover. However, there is concern that even after intensive treatment, residual neoplastic cells may remain or the autologous marrow itself may contain some malignant cells which then lead to recurrence of disease. For example, leukaemic cells which comprise $<1\%$ of the total marrow pool will not be detectable by conventional diagnostic approaches. This has led to the concept of 'minimal residual disease', i.e. disease which remains occult within the patient but eventually leads to relapse.

To attempt removal of minimal residual disease in autologous marrow, 'purging' with cytotoxic drugs or by immunological approaches (e.g. monoclonal antibodies to B cells in the leukaemias or lymphomas involving B lymphocytes) has been attempted. Whether it is essential to remove *all* neoplastic cells and how effective purging is in doing so cannot be determined unless a sensitive assay (such as the polymerase chain reaction) is available. This has increasingly been applied to assess the state of occult neoplastic cells in autologous marrow transplants (Box 6.4). Since the DNA amplification test has the potential to detect 1 in 10^5 to 1 in 10^6 tumour cells, it is important to study a wide range of tumours both haematological and solid to answer the above questions.

A variation of the minimal residual disease theme involves the monitoring of progress following bone marrow transplantation from another donor (allogeneic transfer). This type of treatment is used in a number of leukaemias and bone marrow aplasias. Early identification of engraftment (presence of donor cells) or relapse (presence of host cells) is important to optimise post-transplantation treatment. The DNA amplification technique is ideally suited for this if primers can be designed to detect DNA sequences which will differentiate host and donor cells. The extreme sensitivity of DNA amplification means that only a few cells are necessary for assay and so the dilemma of engraftment versus relapse can be resolved early and rapidly.

FUTURE DIRECTIONS

Cancer pathogenesis

As described above, the pathogenesis of cancer at the DNA level involves a number of sequential steps from a normal cell to a premalignant cell to a localised tumour which then invades and finally metastasises. Along this pathway a number of mutations affecting tumour suppressor genes and/or proto-oncogenes accumulate. The net effect is genetic instability and proliferation of a clone which has a growth advantage over normal cells.

A key issue emerging from molecular studies of cancer cells is whether the changes detected represent primary or secondary events. For example, changes in chromosomal number (aneuploidy) or gene amplification are frequently found in association with tumours. Both these defects have the same end-result, i.e. increased expression of mutated proto-oncogenes or tumour suppressor genes. Whether these changes are essential for cancer to develop and/or progress or whether they simply provide the neoplastic cell with a growth advantage over normal cells remains to be determined.

A consistent feature which emerges from the various examples given above is the considerable *variation* that is present at the molecular level. Thus, no single tumour suppressor gene or proto-oncogene per se is ultimately responsible for the cancer phenotype. Greater complexity arises when additional factors such as the environment or epigenetic events, as illustrated by imprinting, are involved in the gene's expression. Other proto-oncogenes and tumour suppressor genes will be discovered. In the long term, comparisons of genotypic changes with the biology and behaviour of the underlying cancer will make it possible to understand more fully how tumours arise and why they can behave differently.

Screening for cancer

Familial adenomatous polyposis is a good example to illustrate the types of preventative programmes which are presently available for those at-risk of cancer and the potential impact that DNA screening will have in this aspect of cancer management. Mortality directly related to familial adenomatous polyposis is associated with metastatic colorectal carcinoma (approximately 60%) or a number of extracolonic cancers which occur in this disorder. Even after prophylactic colectomy, many with familial adenomatous polyposis will still die from an extracolonic tumour. Thus, those who carry the mutant gene require life-long follow-up starting in childhood. One recommendation is annual flexible sigmoidoscopy from ages 10 to 35. Once the polyposis phenotype is detected, prophylactic colectomy is advised. Surveillance for the extracolonic malignancies must continue indefinitely. The gold standard remains sigmoidoscopy but the frequency of this procedure can be reduced if an at-risk individual in the family can be shown to have inherited the *normal* rather than the mutant allele from the affected parent (Fig. 6.16). One suggested regimen in this situation is an initial yearly sigmoidoscopy for 2–3 years which is then extended to examinations every 3 years until age 35 at which time sigmoidoscopies are ceased. Caution in formulating screening programmes based on DNA testing is being observed at present since recombination may lead to error in predicting those who are at-risk. Experience has been insufficient to be sure that all with familial adenomatous polyposis will have a mutation on chromosome 5q. Ultimately, as the genetic defects which produce the malignant phenotype are fully understood, it will be necessary to follow only those who carry the mutant gene.

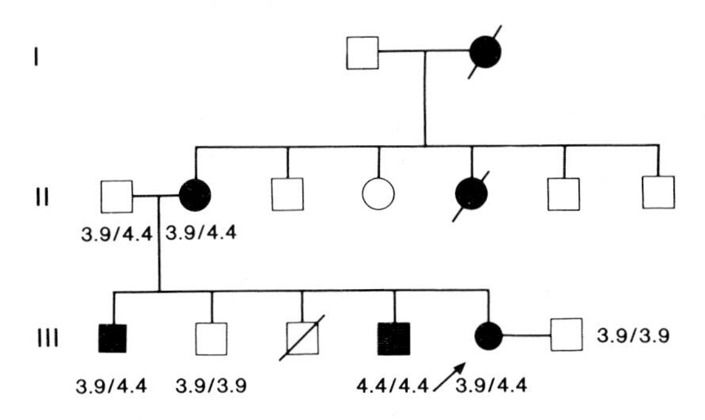

Fig. 6.16 Pedigree illustrating familial cancer.
Autosomal dominant inheritance is illustrated in this family with familial adenomatous polyposis (FAP). The consultand (→ who is also the proband) is shown to have an informative DNA polymorphism for a region on chromosome 5q which is associated with FAP. In this particular family, the disease-related polymorphism is 4.4. Her two siblings with haplotype patterns 4.4/4.4 and 3.9/3.9 help to establish linkage of the polymorphism and assign the 4.4 polymorphic marker to the mutant allele and 3.9 polymorphism to the normal one.

As further knowledge is gained about the types and frequencies of mutations in sporadic cancers it will be possible to develop screening programmes which are based on the identification of genetic defects. For example, with DNA amplification it might be possible to screen for a number of relevant mutations involving the APC, MCC and p53 genes in stool specimens to detect early signs of colon cancer. Similarly, cancer of the bladder (screening by urine cytology) and cervical cancer (screening by pap smears or cervicovaginal lavage) are additional areas in which testing for DNA mutations may prove more useful than waiting until histological features of cancer develop. Industry will also benefit since the mutagenic potential of chemicals, including food products, will be more objectively determined by looking for characteristic DNA changes.

Disease prognosis

Prognosis for cancer can be highly variable. In many situations, prognosis is determined on the histological appearance and/or staging based on surgical or investigative findings. DNA ploidy, the percentage of cells in S phase and hormone receptor status provide additional criteria in determining outcome. In some cases, clinical outcomes vary even within the categories defined by the above parameters. Additional prognostic indicators may thus be required to define disease subclasses. One parameter which will assume

Table 6.10 The Evans system for staging neuroblastoma. Prognosis and clinical staging correlate with a 75–90% chance of disease-free survival at 2 years in stages I and II and 10–30% in stages III and IV

Stage	Description
I	Tumour confined to the organ of origin
II	Tumour extends in continuity beyond the organ of origin but does not cross the midline. Regional ipsilateral lymph nodes may be involved
III	Tumour extends beyond the midline and may involve regional lymph nodes bilaterally
IV	Large primary tumour with distant metastases

increasing importance is an assessment of the underlying genetic defects. This is illustrated by reference to neuroblastoma, one of the commonest tumours in infancy and early childhood.

Neuroblastoma arises from immature, undifferentiated neuroblasts and is highly malignant. The tumour grows rapidly and metastasises early. There are a number of ways to stage the tumour, one of which is summarised in Table 6.10. Prognosis and clinical staging correlate with a 75–90% chance of disease-free survival at 2 years in stages I, II which falls to 10–30% in stages III and IV. An additional approach to determining prognosis in neuroblastoma is based on the amplification of the N-*myc* proto-oncogene. Amplification of this gene has been found in neuroblastomas, particularly those which are highly malignant. A number of studies have now confirmed that those with the greatest degree of amplification in terms of gene copy number have the worst prognosis. This is an easier and less invasive way to predict outcome of cancer.

Therapy

The decision whether to utilise intensive treatment for cancer depends to some extent on the prognosis of the underlying disorder. Similarly, some malignancies which are slow-growing will not respond to treatment but follow an indolent course over many years and so need not be treated aggressively. Better understanding of the biology of cancer, which is becoming increasingly more possible with knowledge of the molecular mechanisms involved, will be important in the future design of therapeutic regimens.

Potential targets for alternative forms of therapy in cancer will include the defective proto-oncogenes or tumour suppressor genes themselves. In these circumstances, a gene therapy-type strategy may be possible to correct or replace defective genes (see Ch. 7). At present this is highly speculative and a more realistic goal might be the potential to treat aberrant products of the above genes. This could involve inhibition of mutant proteins or the replacement of missing proteins. Whilst the above approaches will be difficult, given the multiple factors involved in carcinogenesis, it might be possible to interrupt one or two steps in the complex pathway to interfere with the tumour's overall progession.

FURTHER READING

Bishop J M 1991 Molecular themes in oncogenesis. Cell 64: 235–248

Black D M, Solomon E 1993 The search for the familial breast/ovarian cancer gene. Trends in Genetics 9:22–26

Campana D, Coustan-Smith E, Janossy G 1990 Immunophenotyping in haematological diagnosis. In: Cavill I (ed) Clinical haematology. Baillière Tindall, London, vol 3: 889–919

Carney D N, Sikora K 1990 Genes and cancer. Wiley, New York, p 1–7

Cline M J, Ahuja H 1991 Oncogenes and anti-oncogenes in the evolution of human leukemia/lymphoma. Leukemia and Lymphoma 4: 153–158

Deane M, Norton J D 1991 Detection of immunoglobulin gene rearrangement in B cell neoplasias by polymerase chain reaction gene amplification. Leukaemia and Lymphoma 5: 9–22

Fearon E R, Vogelstein B 1990 A genetic model for colorectal tumorigenesis. Cell 61: 759–767

Gribben J G, Freedman A S, Neuberg D et al 1991 Immunologic purging of marrow assessed by PCR before autologous bone marrow transplantation for B-cell lymphoma. New England Journal of Medicine 325: 1525–1533

Hansen M F, Cavenee W K 1988 Retinoblastoma and the progression of tumour genetics. Trends In Genetics 4: 125–128

Jones P A, Buckley, J D, Henderson B E, Ross R K, Pike M C 1991 From gene to carcinogen: A rapidly evolving field in molecular epidemiology. Cancer Research 51: 3617–3620

Kemshead J T, Patel K 1990 N-*myc* – its place as a diagnostic and prognostic indicator in neuroblastoma. In: Carney D, Sikora K (eds) Genes and cancer. Wiley, New York, p 43–53

Levine A J, Momand J, Finlay C A 1991 The p53 tumour suppressor gene. Nature 351: 453–456

Linder M E, Gilman A G 1992 G proteins. Scientific American 267: 56–65

Marshall C J 1991 Tumour suppressor genes. Cell 64: 313–326

Negrin R S, Blume K G 1991 The use of the polymerase chain reaction for the detection of minimal residual malignant disease. Blood 78: 255–258

Rogler C E, Hino O, Shafritz D A 1989 Mechanisms of hepatic

oncogenesis during persistent hepadna virus infection. In: Notkins A L, Oldstone M B A (eds) Concepts in viral pathogenesis III. Springer-Verlag, New York, p 255–267
Sager R 1989 Tumour suppressor genes: the puzzle and the promise. Science 246: 1406–1412
Vogelstein B, Kinzler K W 1992 Carcinogens leave fingerprints. Nature 355: 209–210
Worwood M, Wagstaff M 1990 Molecular biology and leukaemia diagnosis. In: Cavill I (ed) Clinical haematology. Baillière Tindall, London, vol 3: 949–976

7

THERAPEUTICS

RECOMBINANT DNA (rDNA)-DERIVED PRODUCTS

Blood coagulation factors

A review of the treatment options in haemophilia demonstrates the potential for molecular technology in the area of therapeutics. Haemophilia is an X-linked genetic disorder characterised by spontaneous bleeding into muscles, joints and mucous membranes. Deficiencies in two of the blood coagulation factors (factor VIII in haemophilia A; factor IX in haemophilia B or Christmas disease) are the underlying defects. Severity of haemophilia depends on the level of coagulation factor reduction with $<1\%$ of normal activity producing a potentially life-threatening disorder. The frequency of haemophilia A approximates 1 in 10 000 males and that of haemophilia B 1 in 50 000 males (see Ch. 3 for further discussion of haemophilia).

There are over 12 coagulation factors in the pathway involved with haemostasis, i.e. control of bleeding. The coagulation factors act as a cascade with one activating a second which then activates a third and so on. Factor VIII's role in the coagulation pathway is to accelerate the activation of factor X by activated factor IX in the presence of calcium and phospholipid. Factor VIII in plasma is also bound to another protein called von Willebrand factor. The latter is essential for the platelet's role in haemostasis and also stabilises the factor VIII molecule. Knowledge of the factor VIII's molecular structure was deficient until 1984 when the gene was cloned. From the gene's structure it was possible to predict potential protein domains and functions. The in vivo behaviour of factor VIII was further understood following expression of cDNA for factor VIII.

Structure and function of the factor VIII molecule

The factor VIII protein is encoded by a 186 kb gene containing 26 exons. mRNA specific for factor VIII is 9 kb in length. The protein is over 2000 amino acids in size and is synthesised as a single chain which is processed as it moves from the rough endoplasmic reticulum to the Golgi apparatus. As this occurs, the mid-portion of the molecule is excised. The heterodimers formed are held together by metal ions (Fig. 7.1). There are a number of domains associated with the 200 kDa amino terminal-derived heavy chain and the 80 kDa light chain which comes from the carboxy terminal. The A domains are very similar to a copper binding plasma protein called ceruloplasmin. The C domains appear to have phospholipid-binding function. The B domain has no known function or similarity to previously described proteins and does not appear to be essential for factor VIII coagulant activity.

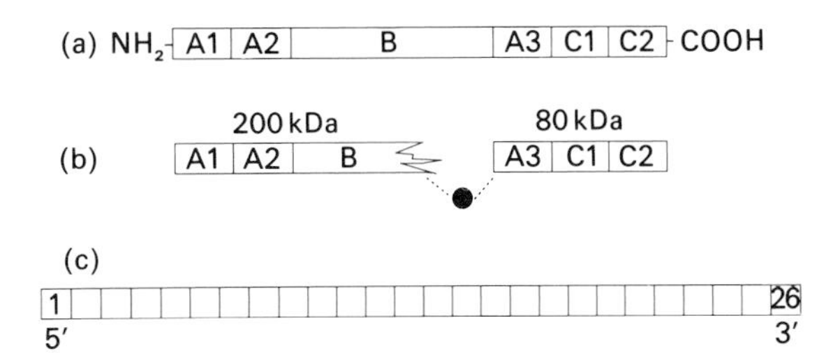

Fig. 7.1 Structure of the factor VIII protein and its gene.
(**a**) Protein domains. (**b**) Two heterodimers formed during synthesis. (**c**) the 26 exons which comprise the gene.

Expression studies of the recombinant DNA-derived product have now shown that optimal stability of factor VIII requires association with the von Willebrand factor. If clinical trials indicate that factor VIII stability is a problem, stability of the recombinant-derived product could be improved by co-expression of genes for factor VIII and von Willebrand factor. The recent observation that removal of the B-domain enables increased expression of recombinant factor VIII would also have potential commercial application if manipulation of the gene in this way enabled more efficient in vitro production.

Treatment of haemophilia with plasma-derived products

A summary of historical developments in the treatment of haemophilia is given in Table 7.1. Landmarks include the availability in 1965 of a specific factor VIII-enriched product known as

Table 7.1 Historical developments in the treatment of haemophilia A (from Recombinant Factor VIII 1991)

1840	Bleeding episode treated with normal fresh blood
1920s	Plasma rather than whole blood shown to be effective
1930s–50s	Fractionation of plasma producing various components with antihaemophiliac activity. 1937: factor VIII implicated in haemophilia A
1965	Cryoprecipitate isolated (cryoprecipitate is produced by allowing frozen plasma to thaw. When thawed a cold insoluble precipitate (cryoprecipitate) remains. This contains a high concentration of factor VIII.)
1970s	High potency freeze-dried factor VIII concentrates available (home therapy now feasible)
1984	Biotechnology companies Genentech (San Francisco) and Genetics Institute (Boston) clone and express the factor VIII gene
1980s	More effective viral inactivation steps incorporated in manufacture of factor VIII products
	Monoclonal antibody purified factor VIII becomes available
	Alternative haemostatic pathways used to bypass the effect of inhibitors
1988	Clinical trials using recombinant human factor VIII start
1990	First reports of multicentre international trials with recombinant factor VIII. Clinical and laboratory responses very satisfactory. Risk of inhibitor development still unclear. Potential effects of contaminants (DNA, protein) from the recombinant DNA process remain to be defined
1990s	Multicentred trial under way using recombinant factor VII to bypass factor VIII inhibitors
1980s–90s	Novel means to produce recombinant DNA-derived products are reported, e.g. transgenic expression of factor IX in sheep milk; autologous transplantation with factor IX gene in fibroblasts

Table 7.2 Some problems associated with conventional haemophilia treatment options

Infection	Hepatitis B virus Hepatitis C virus (60%–80% seropositive) HIV-I* (34% seropositive) parvovirus** unknown
Liver disease	Up to 20% of haemophiliacs have progressive and potentially fatal liver disease
Immunosuppression	Contaminating proteins in factor VIII concentrates (including relatively pure ones) are implicated as the cause for immunosuppression which is independent of infection with the human immunodeficiency virus; both B and T cell lymphocyte functions are impaired
Inhibitor development	Exposure to neoantigens leads to the risk of antibodies developing against the coagulation factor(s). This complication is not excluded in the monoclonal antibody purified (or even recombinant) products
Cost	Purification; viral inactivation
Availability	Conventionally produced plasma factors will always be limited since they rely on availability of pooled human plasma: both short and long term implications

* now the commonest cause of death amongst haemophiliacs
** heat resistant to 120°C, conventional inactivation steps by heating inefficient

cryoprecipitate. From this came antihaemophiliac factors of greater concentrations and stability so that during the 1970s more effective programmes utilising home therapy became established. In this way affected individuals were able to treat themselves the instant a bleeding episode was noted.

Complications occurring with the plasma-derived antihaemophiliac factors remained significant (Table 7.2). The initial problems associated with hepatitis B viral infection from blood products were thought to have been resolved once blood banks introduced screening programmes and viral-inactivating steps were incorporated into the commercial production. Nevertheless, the subsequent recognition of other viruses, e.g. hepatitis C, human immunodeficiency virus (HIV) and parvovirus focused again on the problems of human-derived products. For example, hepatitis C as the cause of non-A non-B hepatitis was initially considered to be a complication which was more likely to occur if plasma had been obtained from paid donors. Subsequently, with the availability of more sophisticated tests for hepatitis C detection, it became apparent that both paid and voluntary donors were sources of this virus. The increasing number of haemophiliacs infected with the human immunodeficiency virus from the late 1970s illustrated further limitations of plasma products, even those that had undergone 'viral inactivation steps'. These processes included various combinations of heating and/or organic solvent exposure. This increased the production cost but gave no guarantee that all viruses (both known and unknown) would be neutralised. For example, the parvovirus can withstand temperatures to 120°C.

Liver disease as a cause of morbidity and mortality in haemophilia is well documented. An association between hepatitis B virus, hepatitis C virus and liver disease is established. Immuno-

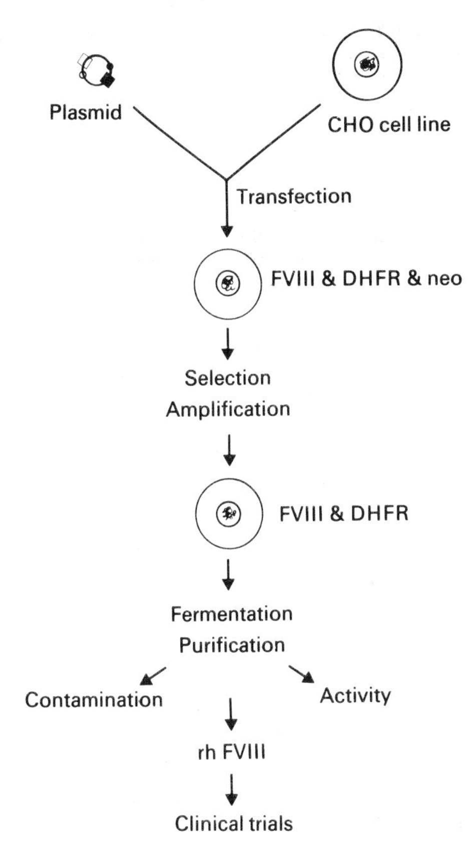

Fig. 7.2 Steps involved in producing recombinant human factor VIII.
The recombinant plasmid has both cDNA for factor VIII (■) and a dihydrofolic acid reductase gene (DHFR = □) as well as their respective promoters (●, ○). This plasmid is transfected into a mammalian cell line such as CHO (Chinese hamster ovary) to enable post-translational changes to occur. Another plasmid containing a gene for the antibiotic neomycin (neo) is co-transfected. The end result is the CHO genome which has integrated into it factor VIII, DHFR and neo. Specific selection for CHO cells which have the neo gene is possible by culturing in a medium containing G418. Amplification for the factor VIII-containing CHO cells is then obtained by culturing the CHO cells in medium with increasing concentrations of methotexate. This will select CHO cells which have the DHFR gene (and so also the factor VIII) since DHFR-deficient CHO cells will not grow in the presence of methotrexate. The CHO cells are then fermented in large, commercial volumes. Protein isolated from these cells is next purified by immunoaffinity chromatography with monoclonal antibodies to factor VIII. The end product is checked for contaminants (mouse, CHO proteins, DNA) and its functional activity assessed. Recombinant human factor VIII is then available for clinical trials.

suppression involving both B and T cell lymphocyte function and independent of infection with the human immunodeficiency virus was also found in haemophiliacs. A mechanism proposed for immunosuppression implicated the many contaminating plasma proteins present in the factor VIII or factor IX products.

An additional and serious complication in haemophilia occurs in approximately 15% of patients who develop factor VIII or factor IX inhibitors (antibodies) as a consequence of exposure to new antigens (neoantigens) to which they have not developed immunological tolerance. These individuals are placed at a much greater risk of dying from an uncontrollable bleeding episode since conventional factor replacement becomes ineffective.

Finally, a short and long term, practical consideration with plasma-derived products is their availability which will always be dependent on obtaining a regular supply of pooled human plasma. This could never be guaranteed.

Modern therapeutic developments in haemophilia

The availability of monoclonal antibodies to factor VIII led to the production in the 1980s of a substance with higher potency and greater purity. The reduction in contaminating proteins was considered to be an added bonus if this effect was significant in the immunosuppression observed in haemophiliacs.

In 1987, the first patient was treated with a recombinant DNA-derived human factor VIII. The steps involved in preparing this product are summarised in Figure 7.2. The use of mammalian cell lines such as CHO (Chinese hamster ovary) enabled complex post-translational steps e.g. glycosylation, to be undertaken. Recombinant factor VIII produced in this way was shown to have fewer protein contaminants although some were invariably present. The long-term effects of

Table 7.3 Purity of various factor VIII preparations

Source	Activity (units/mg protein)
Plasma	0.01
Cryoprecipitate	0.1
Concentrates 1970s	~0.5
Concentrates 1980s	~1.5
Immunopurified	~2000
Recombinant	~2000

chronic exposure to these substances remain to be determined. The activity of the recombinant product was equivalent to monoclonal antibody purified factor VIII (Table 7.3).

The first multicentred human trials of recombinant human factor VIII were started in 1988 and reports issued 5 years later indicated very promising laboratory and clinical responses. The potential risk from immunisation by exposure to neoantigens produced by the recombinant DNA process remained theoretical but did not appear to be a major problem in the above trials. Whether recombinant products will be more or less likely to induce the development of inhibitors will take some years to assess.

Factor VIII inhibitor treatment

Plasma-derived activated coagulation factors (mixtures of activated factors X and VII and tissue factor) were produced to overcome the block on factor VIII activation brought about by the development of antibodies. However, these were expensive, had all the potential complications associated with factor VIII and in addition were reported to produce thrombotic complications. Activated factor VII on its own appeared to be a useful way to bypass the factor VIII block but isolation of this substance from plasma was commercially impractical. This problem was resolved with the development of a recombinant human factor VII product which is presently being tested in patients with factor VIII antibodies. Preliminary results are promising.

Future

The feasibility of recombinant DNA-derived factor VIII and factor VII coagulation products is now being assessed. The ability to have a regular and controllable supply will allow better planning and more efficient and early treatment of bleeding problems. The potential long-term immunological and infective sequelae of recombinant-derived products remain to be determined. However, from experience already gained with insulin, growth hormone and the hepatitis B vaccine, these are unlikely to be major problems. Ultimately, it is likely that plasma-derived substances for the treatment of all coagulation disorders will be replaced by drugs produced by recombinant DNA.

Hormones and growth factors

Erythropoietin

Anaemia is an important complicating factor in chronic renal failure and has a significant effect on morbidity. Components in the anaemia of renal failure include: decreased production of the hormone erythropoietin, iron deficiency and other factors such as chronic infection, which can impair bone marrow function. Erythropoietin is the primary regulator of red cell production. The majority of this hormone is produced in the kidney and its level is regulated by tissue oxygen tension present in the kidney. Erythropoietin then feeds back to the bone marrow where it acts on the committed erythroid progenitors and precursors. Thus, in the presence of hypoxia, the level of erythropoietin increases and the red blood cell mass expands. Once hypoxia is corrected, erythropoietin production in the kidney is suppressed.

Erythropoietin was discovered over 80 years ago. It has also been known for some time that reduced erythropoietin production plays a major role in the anaemia of chronic renal failure. However, it was not possible to take this any further since the amount of erythropoietin which could be isolated from kidneys or the urine was insignificant. Thus, erythropoietin as a therapeutic agent was unavailable until the gene was cloned in 1985. Recombinant erythropoietin is now produced by expression of its cDNA in a mammalian cell line system similar to that described for haemophilia. The necessity for a cell line expression system reflects the requirement for glycosylation of this hormone which would not be possible with the more rudimentary bacterial expression system.

Clinical trials have now confirmed that recombinant erythropoietin given in renal failure is capable of raising the haemoglobin level and so can remove the necessity for blood transfusions which are required in those patients who have severe anaemia. Thus, the potential complications associated with blood transfusions (infection, immunosensitisation to HLA antigens, iron and circulatory overload) can be avoided with the recombinant hormone which is given subcutaneously or intravenously.

Other indications for recombinant erythropoietin are presently being assessed. These include the anaemia associated with cancer or chronic inflammatory disorders such as rheumatoid arthritis. Whether recombinant erythropoietin will have a beneficial role to play in these circumstances remains to be determined. To date no significant side-effects have been reported with recombinant human erythropoietin.

Growth hormone

Human growth hormone, a protein of 191 amino acids, is essential for growth. Because this hormone is species-specific its biological source must be human. Following the successful treatment of a pituitary dwarf in 1958 with human growth hormone, a number of programmes were established to isolate this substance from pituitaries obtained from cadavers. However, the programmes were ceased in the mid-1980s when a number of recipients died from Creutzfeldt–Jakob disease, a fatal slow virus infection of the central nervous system. A direct association between the pituitary extract and the slow virus infection was soon established leading to withdrawal of the human-derived growth hormone.

This has now been replaced with a recombinant product following the cloning and expression of the human growth hormone gene in 1979. Because the protein does not require sophisticated post-translational modifications to its structure, recombinant growth hormone is prepared using a relatively simple bacterial expression system. However, there are two potential disadvantages from a bacterial expression system. They are: the necessity for extensive purification to remove impurities of bacterial origin, particularly endotoxins and the presence of an additional methionine amino acid at the start of the protein. The latter occurs because the eukaryotic start codon (ATG) is translated in the prokaryotic system into a methionine (see Ch. 2).

Clinical trials during the mid-1980s have demonstrated the efficacy of the recombinant growth hormone and it has remained in continuous use since, without significant side-effects becoming apparent. The additional methionine does not, as originally feared, lead to a major increase in antigenicity of the product.

Haematopoietic growth factors

Bone marrow haematopoietic ('blood forming') cells are derived from the proliferation and differentiation of progenitor cells that form specific lineages after they interact with growth factors. The pluripotential stem cell is the ultimate source of the lymphoid and myeloid systems. The latter differentiates into cells of the eosinophil, neutrophil, monocyte, basophil, erythroid and platelet lineages. As the progenitors mature into committed stem cells, they become responsive to more specific growth factors although two or more factors are often required for optimal effect.

Knowledge that haematopoietic growth factors existed has been available since the 1960s. However, the minute amounts able to be isolated made characterisation of these substances extremely difficult. In the past few years this has dramatically changed with the cloning of many genes which produce these growth factors. Apart from erythropoietin (discussed above) there are two groups of haematopoietic growth factors: the colony stimulating factors (CSFs) and the interleukins (ILs).

The colony stimulating factors comprise four glycoproteins which are capable of stimulating the formation and activity of white blood cells (granulocytes or monocytes). The prefix indicates the target cell, i.e. G-CSF (granulocyte); M-CSF (monocyte or macrophage). The remaining two CSFs (GM-CSF and IL3, interleukin 3) are less lineage-restricted since they affect both granulocytes and monocytes. GM-CSF also stimulates the precursor cells in the erythroid (red blood cell) and megakaryocyte (platelet) lineages. Many cell types including fibroblasts and endothelial cells produce one or more of the CSFs. Base-line pro-

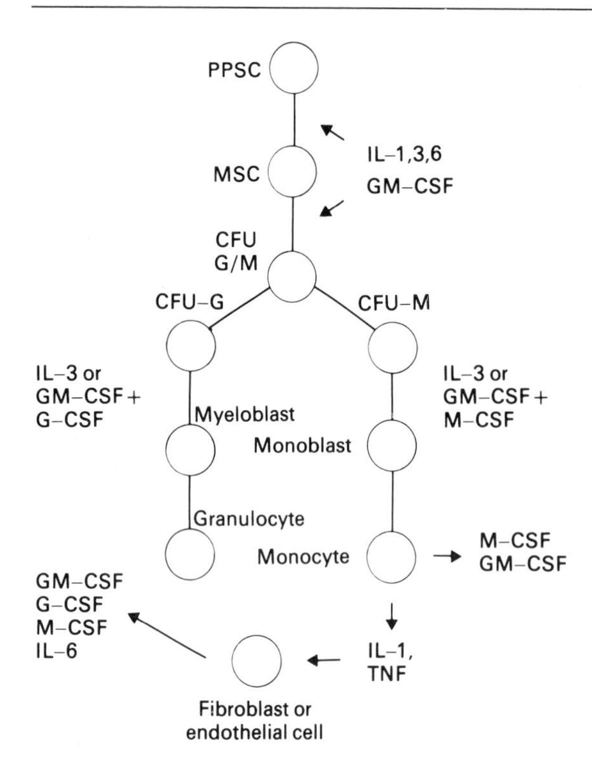

Fig. 7.3 The neutrophil and monocyte differentiation pathways and their associated growth factors.
The ultimate source of the lymphoid and myeloid systems is the pluripotential stem cell. Differentiation occurs in response to growth factors with cells responding to more specific factors as they become committed to a cell line. Neutrophil pathway = CFU-G; monocyte pathway = CFU-M; PPSC = pluripotential stem cell; MSC = myeloid stem cell; TNF = tumour necrosis factor; IL = interleukin; GM = granulocyte/monocyte; M = monocyte (macrophage).

duction is low but rises rapidly following exogenous stimuli such as bacterial endotoxins. In turn, the CSFs provide the proliferative signals for cellular differentiation. Other interleukins (IL-1, IL-4, IL-5, IL-6) are also produced by a variety of cells and have some haematopoietic growth factor activities as well as making important contributions to the immune and inflammatory responses (Fig. 7.3).

The production of recombinant-derived human growth factors e.g. rh G-CSF, rh GM-CSF, rh IL-3 (rh = recombinant human), is providing important information on their physiological functions. The availability of large quantities, which became possible with recombinant DNA technology, has enabled the potential therapeutic roles of growth factors to be assessed in clinical situations associated with low white blood cells or low platelet counts, e.g. following bone marrow transplantation and after chemotherapy for cancer or leukaemia. G-CSF and GM-CSF alone or in combination have shown therapeutic benefits by reducing the duration of leukopenia (low white blood cell count). Thus, there is less risk of infection-related complications which in turn will lead to a decrease in morbidity, mortality and overall in-patient hospital costs since patients can be discharged from hospital earlier. Alternatively, more aggressive chemotherapy will become possible thereby increasing the chance that residual cancer cells will be killed. Zidovudine, used in the treatment of AIDS, has bone marrow suppression as one of its side-effects. An additional use of the recombinant CSFs would be in situations such as AIDs where pre-existing immunodeficiency is exacerbated by treatment with a drug such as zidovudine. In this circumstance, the ability to reduce the degree or duration of marrow depression would have a beneficial effect on overall morbidity and mortality.

In the long term, stimulating the production of progenitor cells might allow peripheral blood to be used instead of bone marrow for transplantation. One such example involves studies which are under way to assess the efficacy of *cord blood* for transplantation. The use of cord blood cells in this situation avoids the morbidity associated with repeated bone marrow sampling particularly if the donor is an infant or child. The risk for graft versus host disease may also be less with cord blood. Preliminary data would suggest that treatment of the recipient with GM-CSF may enhance the rate for engraftment which is slower than that found when the source of tissue is bone marrow. Immunomodulatory and antileukaemic effects of the growth factors are also being investigated at present.

Despite all the apparent advantages of the recombinant growth factors a number of questions remain unanswered. What are the optimal drugs to use, e.g. single factors or combinations and if so which combinations? The routes of administration and regimens require to be identified. When are growth factors given in respect to bone marrow transplantation and so on. Similarly,

Box 7.1 Recombinant growth factors: variations with different production vectors

Human recombinant growth factors have been commercially produced using three vector systems: *Esch. coli*, yeast and mammalian cell lines such as CHO (Chinese hamster ovary). The expressed proteins in each case are structurally different, e.g. the bacterial expression system produces a recombinant product which is not glycosylated and the yeast system a partially glycosylated growth factor. The fully glycosylated substance can only be made with the more sophisticated mammalian expression systems. Results from clinical trials to date have suggested that antibodies can form to growth factors which are not fully glycosylated. Thus, the clinical situation requiring potential long-term use of such drugs may need to utilise only the fully glycosylated forms. It is important to note that the utility of recombinant DNA-derived products to be 'tailor made' so that changes in DNA sequence can enhance desirable properties or diminish unwanted ones, needs careful short and long term monitoring to ensure that sequence changes do not produce conformational effects which are antigenic. The recombinant human insulin story is salutory in this respect. Recombinant human insulin was introduced in 1982 and by the late 1980s had replaced animal derived (bovine or porcine) insulins as the usual therapeutic product. A novel and potentially fatal side-effect of the human-derived product is now being investigated. Whilst it remains to be proven, there is some evidence that the human product leads to 'hypoglycaemia unawareness' so that autonomic nervous system changes which warn a diabetic that he/she is at risk for hypoglycaemia (sweating, palpitations) become dampened. This is not an obvious problem with the animal-derived products.

potential long-term side-effects of the recombinant products still need to be assessed (Box 7.1).

Vaccines

The success stories illustrated above and involving recombinant DNA-derived therapeutic substances have, so far, not been reproduced in vaccine production. With the exception of the hepatitis B virus vaccine, which is described in more detail below, the results with recombinant DNA vaccines have been disappointing. This is despite the considerable financial input into the area which has been driven in part by the urgency to find a vaccine for infection with the human immunodeficiency virus. Commonly used and effective vaccines such as poliomyelitis, measles, rubella are, with few exceptions, composed of *live* (infectious) and *attenuated* (non-pathogenic but immunogenic) organisms. As such they have proven to be highly effective, relatively cheap and so affordable by many communities. Other conventional vaccines comprise *killed* microorganisms (e.g. Salk poliomyelitis vaccine) or the vaccines are composed of one or more antigenic components (i.e. *subunit vaccines* such as are found with influenza and recombinant hepatitis B).

Nevertheless, despite the efficacy of the above vaccines, it should be noted that a number of them might never have become available for clinical use with the stringent licensing regulations currently in force. For example, the oral poliomyelitis (Sabin) vaccine can revert on rare occasions to the wild-type (neurotoxic) strain and so produce poliomyelitis (see Ch. 5). Subacute sclerosing panencephalitis is a very rare neurological complication following infection with the measles virus including some vaccine-derived strains. It can be fatal or lead to permanent neurological sequelae such as mental retardation. It is unlikely that these two vaccines would have been marketed with these potential risks in today's litigation-conscious community.

Novel versus conventional vaccines

Different approaches have been used as alternatives to the three types of conventional vaccines described above. *Synthetic peptides* derived from segments of the infectious agent that are considered to be highly immunogenic have been prepared. In theory, this enables better standardisation since the antigens to which the immune system is stimulated are defined. Synthetic peptides would also circumvent the risk that live attenuated vaccines could revert to wild-type and so become infectious. However, the synthetic peptide approach has been disappointing. In part this reflects the complex physical conformation which may be required to promote optimal antigenic stimulation. Segments of the infectious agent involved in this complex may be discontinuous and so not represented in a restricted and linear peptide.

There is also good evidence that immune responsiveness is dependent on genetic factors, perhaps HLA-related. Thus, limited antigenic exposure such as that resulting from a synthetic peptide may not be equally effective in all circumstances, whereas the organism itself in live attenuated or inactivated forms gives a broader antigenic exposure.

An alternative to the synthetic peptides is the use of recombinant *DNA expression systems to generate viral antigens*. The potential advantages of this approach include the availability of an unlimited source of antigen; there is a better chance that the product can be standardised; there is greater flexibility with the type of structure which is produced and in the long term the vaccine should be cheaper and safer. To date, the only successful human recombinant DNA-derived vaccine is that for hepatitis B virus. However, more successes have been obtained with recombinant DNA vaccines in veterinary practice.

Hepatitis B vaccines

In 1982, a hepatitis B vaccine became available. The source for this vaccine was the plasmas of known chronic hepatitis B carriers. In this circumstance, stringent purification and inactivation procedures became mandatory. Thus, the vaccine was expensive and the amount ultimately limited by the availability of infected plasmas. The vaccine was not well received by the public in view of the theoretical risk that other viruses, e.g. human immunodeficiency virus, might be transmitted despite the inactivation processes undertaken. In view of the above problems and the importance of hepatitis B virus as a cause of liver disease (see Chs 5, 6) a recombinant DNA-derived vaccine was released in 1987.

The recombinant hepatitis B virus vaccine is a subunit vaccine directed to the surface antigen (HBsAg) of the virus. The expression vector required for this vaccine has to be relatively sophisticated since the viral surface protein coat is glycosylated. Thus, either yeast-derived or mammalian expression systems are necessary. The former has been used. Purification steps are therefore required to remove potential contaminating yeast proteins.

The vaccine has now been used extensively and confirmed to be effective although not ideal. This is because the product remains relatively expensive and so is not able to be afforded by all communities. Unfortunately, it is often in these communities that a higher carrier rate for hepatitis B is found and that the vaccine would be of most benefit. The second problem relates to the recombinant vaccine being less immunogenic than its plasma-derived counterpart. This is thought to reflect a different glycosylation pattern produced by the yeast host. The long-term consequences of the vaccine's reduced antigenicity remain to be determined.

Future directions

The hepatitis B virus vaccine has shown that recombinant DNA vaccines can work although the two significant problems identified above need to be resolved. Despite the difficulties experienced with the recombinant DNA-derived products they provide more secure long-term sources for vaccines. The potential for innovative developments available through recombinant DNA technology may enable difficulties associated with vaccination for some infections to be resolved. This is well illustrated by reference to AIDS.

Because the human immunodeficiency virus is a retrovirus, its RNA is capable of integrating into host DNA and so it can remain latent until it is activated. Thus, the use of a conventional live attenuated virus as a vaccine poses a potential risk if the integrated (latent) form were able to become activated and produce a mutated (wild-type) strain at a later stage. The other conventional approaches involving inactivated human immunodeficiency virus-1 or subunit components are also unsatisfactory since gp120 (Fig. 5.2), an important antigenic surface envelope protein which enables the virus to attach to its target, the lymphocyte's CD4 cell receptor, is subject to considerable antigenic variation. Finally, the human immunodeficiency virus may be transmitted by infected cells as well as in free virus form. Thus, intracellular virus may escape immune surveillance.

The above seemingly insoluble difficulties associated with the development of a successful vaccine against the human immunodeficiency virus are more likely to be resolved by recombinant DNA strategies presently available or to be developed in the future. For example, a chimaeric recombinant vaccinia virus vaccine which also expresses the human immunodeficiency virus gp160 envelope gene, is presently under clinical trial. gp160 is a glycoprotein precursor which is cleaved to give the envelope structural proteins gp120 and gp41 (Fig. 5.2). Non-human primates inoculated with this vaccine develop both humoral (antibody) and cellular (T-lymphocyte) responses. Its effect in humans is awaited. Other proteins which may demonstrate less antigenic variation associated with the envelope (Env) and Gag proteins (see Fig. 5.2) have been defined and could be expressed individually or as a mixture by recombinant DNA means.

Chimaeric vaccines, in which the genome of a live attenuated virus (e.g. vaccinia, adenovirus or poliovirus) is genetically engineered to express other viral antigens, offer potentially interesting future developments. In this way a number of antigens are produced simultaneously thereby reducing the overall costs for vaccination programmes. Immunostimulation from the primary viral component (e.g. vaccinia) may be helpful in provoking an additional response to the secondary antigen (e.g. human immunodeficiency virus). At the DNA level, it is not difficult to modify a viral genome to enable it to contain additional genes and so express a greater range of potentially antigenic proteins. What remains to be determined is the efficacy of chimaeric vaccines, e.g. what effect will previous exposure or immunisation to vaccinia have if this is co-expressed with the human immunodeficiency virus?

Infectious agents which are difficult or dangerous to produce by conventional culture techniques, e.g. rabies virus, could also be better developed through recombinant DNA means. Genetic manipulation, which is possible by recombinant DNA strategies, would be useful to reduce the likelihood of reversion to wild-type strains, e.g. poliomyelitis, or to increase the antigenicity of a particular component derived from the infecting organism.

Biotechnology developments

Blood substitutes

A priority in the biotechnology industry is the development of a blood substitute. This substance will need to possess both the properties of oxygen carriage and be able to function as a plasma expander. The urgency to obtain such a substance comes from the increasing concern about viral and other contaminants in blood products and an anticipated diminishing supply of blood as there is more selective donor screening and donors themselves become less willing to donate. An artificial blood substitute is anticipated to have an annual market of US $1 billion and would be used in cases of trauma, surgery and medical conditions associated with anaemia.

A plasma-derived synthetic haemoglobin has been manufactured. Haemoglobin is extracted from out-dated blood donations or from bovine blood. It next has to be modified chemically to overcome two problems which arise once haemoglobin is separated from its red blood cell environment. These are an increase in oxygen affinity (i.e. oxygen is more difficult to extract from the haemoglobin) and an instability of the $\alpha_2\beta_2$ tetramer (see Ch. 1). These modified haemoglobins are known as 'haemoglobin-based oxygen carriers'. Modifications which will overcome the oxygen affinity and instability problems include conjugation of haemoglobin to substances such as dextran or polyethylene glycol and polymerisation of these compounds. The drawbacks to the

Table 7.4 Some recombinant DNA-derived human therapeutic agents

	Recombinant product
Drugs	Erythropoietin
	Insulin
	Growth hormone
	Coagulation factors (VIII, VII)
	Plasminogen activator
Vaccines	Hepatitis B
Cytokines/growth factors	GM-CSF
	G-CSF
	Interleukins
	Interferons

above approaches include the requirement for blood donor-derived products (demand will thus be dependent on donations) and the potential for viral or other infections to be transmitted. The immunogenicity of bovine-derived products remains to be determined.

Recombinant human haemoglobins have been produced using both *Esch. coli* and yeast expression vectors. In both cases, the products have similar structure and function to native haemoglobin. However, limitations such as low yield, high oxygen affinity and instability have made these recombinant substances commercially unattractive. The problem of low yield may be resolved in the long term with the use of transgenic animals as expression vectors. This was briefly mentioned earlier in describing transgenic sheep which produce recombinant human factor IX in their milk (Table 7.1). Similarly, transgenic pigs, which produce a considerable amount of human haemoglobin in their own blood, have been made.

The potential to genetically engineer the α or β globin genes prior to their expression has also been utilised in an attempt to overcome other limitations described above. For example, in a recent report (Looker et al 1992) a genetically engineered recombinant haemoglobin was produced. The feature of this synthetic substance was that it had both stability and low oxygen affinity. Thus, it was possible that oxygen could be released to surrounding tissues. The stability was obtained by fusion of the two α globin subunits in haemoglobin by expression of a tandem duplication of the α globin genes. The reduced oxygen affinity was accomplished by incorporation of an asparagine to lysine mutation at codon 108 in the β globin gene. This produces a variant haemoglobin (equivalent to the naturally occurring mutant called Hb Presbyterian) which has reduced oxygen affinity.

There is some way to go before the cell-free haemoglobin products are shown to have adequate physiological function and to be free of side-effects. Nevertheless, the example of how synthetic recombinant DNA-derived haemoglobin may be produced illustrates the utility of the molecular techniques in the synthesis and then commercial production of a range of therapeutic agents (Table 7.4).

Novel expression systems

The initial high costs in preparing genes, vectors and evaluating the mandatory in vitro and in vivo parameters of safety and efficacy should be offset by the long-term and reliable production of recombinant products. More efficient expression systems would reduce costs even further. Large transgenic animals, e.g. sheep, goats and pigs, are being investigated with this in mind. Recombinant drugs which are expressed in the milk or blood of these animals would ensure a source of high yield was available. Examples being evaluated at present include coagulation factor IX and α_1 antitrypsin (sheep milk), tissue plasminogen activator (goat milk) and haemoglobin (pig blood). The long-term developments in the transgenic 'pharmyard' are very promising.

IMMUNOTHERAPY

Monoclonal antibodies

Structure and function of antibodies

Antibodies are proteins made by higher vertebrates. These proteins form a defensive system against foreign tissues, cells and organisms. Antibodies comprise two heavy and two light chains (Fig. 7.4). The part which binds to the recognition sequence on the target (called the antigen) is designated the Fab portion. The remainder of the antibody is the Fc segment which defines a number of the antibody's properties. The ability of each antibody to recognise a single unique antigen is put to use in various diagnostic strategies. More recently, the potential for antibodies as therapeutic agents is being explored.

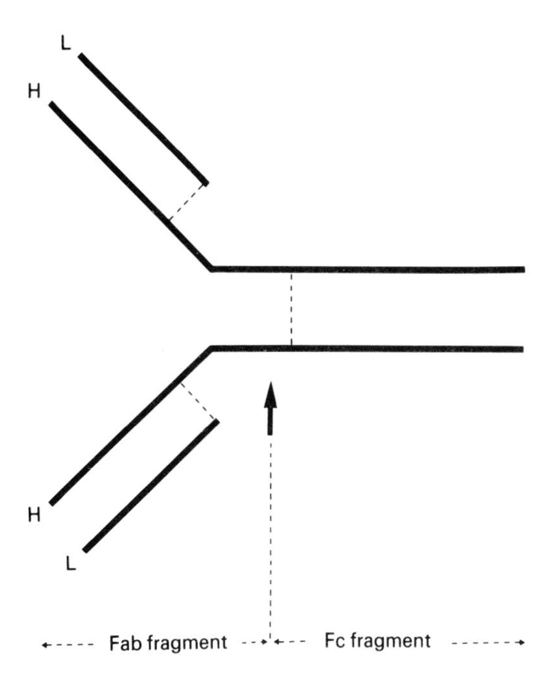

Fig. 7.4 Structure of an antibody (immunoglobulin).
Components are two heavy chains (H) and two light chains (L) which are held together by disulphide bonds (----). Papain breaks an immunoglobulin molecule into two components (Fab and Fc). These are abbreviations for **f**ragment **a**nti**b**ody and **f**ragment **c**rystalline. The Fab portion of the antibody binds to antigens (foreign materials). The Fc portion is important since it contains the receptors which allow the antibody to bind to macrophages and B lymphocytes. The Fc portion also determines the class of immunoglobulin, i.e. IgG, IgM, IgD, IgA or IgE.

Antibodies are conventionally produced by immunising animals, such a rabbits or goats with an antigen. The antiserum produced contains a mixture of antibodies in terms of both their Fab

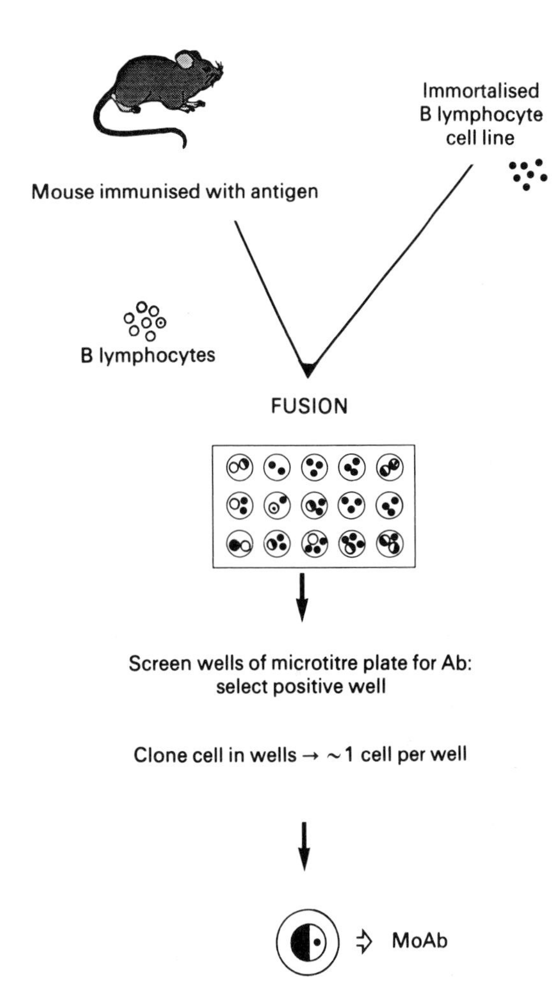

Fig. 7.5 Production of monoclonal antibodies (MoAb).
An immunised mouse will produce a polyclonal antibody response. Of the many lymphocytes involved in this response (○) there will be one which produces a specific antibody (○ with a dot inside). To isolate the latter, the mouse's antibody producing B lymphocytes are fused with an immortalised B cell line (●). In the wells after fusion may be found many combinations: mouse B lymphocytes; immortalised B cells and various fused cells called hybridomas (indicated by half-filled circles). Enrichment for hybridomas is possible by using special medium to grow these cells. Hybridomas in the wells are then screened for the antibody being sought. A positive well is found but this may contain a number of hybridomas only one of which is likely to be the correct one. A single (monoclonal) hybridoma is isolated by serial dilutions until there is only one hybridoma per well. This cell line then becomes an unlimited source of the monoclonal antibody.

and Fc components. The potential problem of heterogeneity resulting from the above immunisation strategy was overcome in 1975 with the development of monoclonal antibodies (Fig. 7.5). Thus, antibodies of a single antigenic specificity and type could be produced in an unlimited amount and with defined activities. Monoclonal antibodies have now made a major impact as diagnostic tools in clinical medicine. In the long term, the therapeutic potential of monoclonal antibodies will also be significant.

Monoclonal antibodies in clinical medicine

Diagnostic applications of the monoclonal antibodies include radioimmunoassays, enzyme immunosorbent assays, flow cytometry and in situ hybridisation for histological immunotyping of tissue sections. When linked to radionuclides, monoclonal antibodies provide useful diagnostic agents for cancer diagnosis, the detection of secondary deposits and monitoring progress of treatment.

In therapeutic terms, monoclonal antibodies are used as an in vitro means to purge bone marrow of residual neoplastic cells prior to bone marrow transplantation. Alternatively, monoclonal antibodies to tumour specific antigens can be linked to effector molecules such as radionuclides, drugs or toxins. In theory, this combination enables in vivo targeting of specific therapy to a localised region thereby minimising treatment side-effects.

Three problems have prevented monoclonal antibodies from having effective in vivo activity particularly for therapeutic purposes. These are: the immunogenicity of monoclonals, the ability of these antibodies to penetrate target sites and the satisfactory linking of effector compounds to the monoclonals. Immunogenicity of monoclonals simply reflects the murine source of the antibody. Thus, a mouse-derived monoclonal antibody will have a short survival time in a human since it is of foreign origin and so induces an immunological response against both its Fc and Fab segments. The murine-specific Fc component of the monoclonal antibody is in fact of limited value because it can only weakly recruit human effector elements which are particularly useful in the antibody response. For example, human IgG subclass 1 demonstrates better antitumour activity than mouse and other human IgG subclasses since it can more readily induce cell mediated killing. Solutions to the above problems are now being sought with genetically engineered monoclonal antibodies.

Genetically engineered monoclonal antibodies

'Humanised' monoclonal antibodies

The antigenicity of monoclonal antibodies can be reduced by taking the mouse gene for the antibody and replacing portions of its heavy and light chains with gene segments which are human in origin. Thus, the antigen recognition portion remains murine but the remainder (including the Fc component) is human. The chimaeric monoclonal formed is less antigenic but still retains some antigenicity which is directed to the Fab component. An antibody to the latter component is called an anti-idiotype antibody.

More sophisticated refinements at the DNA level can be undertaken to replace murine components involved in anti-idiotype antibody formation. At the same time the Fc portion coding for human IgG subclass 1 can be incorporated into the 'humanised' monoclonal antibody to enhance its effector activity. Clinical trials using 'humanised' monoclonal antibodies are now being conducted. Results would suggest that the genetically engineered antibodies are less immunogenic and have a longer survival time.

Improving tissue penetration

Since an antibody is a relatively large molecule it cannot easily cross the cell's membrane. Chemical cleavage of antibodies by pepsin or papain enables separation of the Fab and Fc portions thereby reducing the size of the antibody and so enhancing tissue penetration. However, chemical proteolysis is technically difficult and produces an end-product with considerable batch variability.

Genetic engineering is again proving useful to

overcome such problems since it is possible to alter the DNA sequence to place a premature stop codon between the Fab and Fc components thereby making a similar structure to that produced by chemical cleavage. The advantage of the rDNA-derived agent over the chemically derived substance is a more precise and uniform product which does not require extensive purification beyond that which would be necessary for any recombinant derived drug.

Enhancing the effector function

A number of agents can enhance the monoclonal antibody's ability to kill cells. They include cytotoxic drugs and the toxin known as ricin. To bind these compounds to the monoclonal antibodies by current technology requires the antibody to be chemically modified. This is difficult to accomplish without interfering with the affinity of the monoclonal antibody for its target. Furthermore, chemical modification steps are not easily standardised and so batch variability occurs.

Biotechnology companies are now approaching this problem with strategies based on genetic engineering. For example, binding sites for the effector molecules can be created as part of the primary structure of the 'humanised' monoclonal antibody. More futuristic approaches might entail the fusion of the antibody and effector genes so that both are expressed as the one substance. The potential application of the monoclonal antibody and its alterations through genetic engineering offers some exciting developments both in the diagnostic and therapeutic areas.

MONITORING DRUG RESPONSES

Drug resistance

Resistance to chemotherapy remains a major problem in the treatment of infectious diseases and cancer. In the latter, resistance can be *intrinsic*, i.e. the cancer is unresponsive to any form of chemotherapy, or *acquired*. In the acquired circumstance, initial responsiveness ultimately gives way to a drug-resistant tumour (Fig. 7.6). To overcome or delay resistance, combinations of drugs rather than single agents are frequently used in chemotherapy regimens. Nevertheless, resistance will develop to a wide range of drugs.

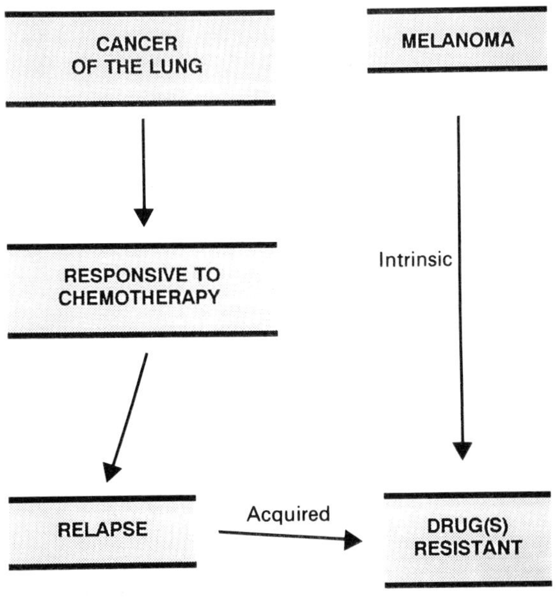

Fig. 7.6 Drug resistance in cancer can be acquired or intrinsic.

There are many factors which can influence drug resistance, e.g. route of administration, concentration attained at target site and so on. At the level of the tumour itself, it is becoming apparent that a number of cellular and molecular (DNA) modifications are associated with drug resistance. Changes at the molecular level include: (1) mutations which alter protein binding affinity, (2) increased gene expression via amplification or enhanced transcription and (3) increased efflux or decreased uptake of the drug. These are illustrated by reference to the drug methotrexate and the substance P-glycoprotein.

Methotrexate

Folic acid, a key vitamin in the synthesis of DNA, must be maintained in its fully reduced tetrahydrofolate form to be active. Inhibition of the enzyme required for this (dihydrofolate reductase, DHFR) will interfere with cell growth. Methotrexate is an antimetabolite drug whose anticancer effect occurs through inhibition of dihydrofolate reductase. Mention was made in Chapter 5 of the malaria parasite which can overcome the effect of antifolate drugs such as pyrimethamine, by disrupting the binding between the antifolate drugs and dihydrofolate reductase. At the DNA level, this occurs on the basis of point mutations at critical sites in the gene coding for the dihydrofolate reductase molecule.

A different defect has been seen in tumour cells which acquire resistance to methotrexate. Elevated dihydrofolate reductase activity, as evident from increased expression of the appropriate mRNA species, has been shown to occur secondary to gene amplification. Thus, the effect of the antifolate drug methotrexate is negated by an increase in its substrate dihydrofolate reductase. There are two ways in which this can occur at the DNA level. The original tumour could have comprised two populations of cells: drug sensitive and drug resistant. During treatment, the latter has a growth advantage and ultimately becomes the predominant population. Alternatively, a single drug-sensitive population has undergone a series of mutations one of which produces a cell with increased dihydrofolate reductase activity. This clone is thereafter positively selected for by continued drug treatment.

P-glycoprotein

The observations that resistance to cytotoxic drugs could involve a wide range of seemingly unrelated compounds and that a single defect was likely to be associated led researchers to seek a gene encoding for multiple drug resistance ('MDR'). One potential site for drug resistance to occur was the cell's membrane since it became apparent that drugs were being excluded from drug-resistant cell lines. Comparisons of plasma membranes isolated from different drug-sensitive and drug-resistant tumour cell lines identified a unique glycoprotein in the latter. This substance was called P-glycoprotein in view of its apparent function as a permeability barrier to drugs in association with multiple resistance. P-glycoprotein was detectable in a number of tumour cell lines resistant to a wide range of cytoxic agents. The initial identification of P-glycoprotein was by monoclonal antibodies. The gene coding for this substance was next cloned and characterised. A number of important observations subsequently emerged.

The DNA and its associated amino acid sequence suggested that P-glycoprotein is a transmembrane ATP-dependent active transporter which pumps hydrophobic compounds out of cells. Normal cells have this gene and its associated P-glycoprotein but drug-resistant cells demonstrate increased expression of the gene as seen from Northern blot analysis for its 4.5 kb mRNA. The reason for the increased expression became evident in Southern blot studies which indicated that gene amplification had occurred with multiple copies, e.g. $\times 60$, of the gene present in resistant cell lines. In vitro evidence for the above came from transfection studies in which drug-sensitive cells were able to be converted to the resistant phenotype following acquisition of additional copies of the P-glycoprotein gene.

The normal function of the gene remains to be determined. One hypothesis is that it enables the cell to extrude unwanted substances. Alternatively, P-glycoprotein is involved in some transport mechanisms required for the cell's normal functions. With respect to the latter hypothesis, it is relevant to note that the structure of P-glycoprotein bears many similarities to the CFTR gene (cystic fibrosis transmembrane regulator or cystic fibrosis gene, see Ch. 3). Verapamil, as described previously in relation to malaria, (Ch. 5) is able to reverse the drug-resistance associated with P-glycoprotein. This is thought to reflect inhibition of drug transport by the P-glycoprotein.

Experimental work comparing P-glycoprotein expression in tumour cell lines that have intrinsic drug resistance or acquire resistance following treatment has shown a good direct correlation between the level of the mRNA specific for the gene and the sensitivity/resistance status of the underlying cell. Thus, a means may become available to assess drug susceptibility in cancer cells. Potential strategies to overcome this type of multiple drug resistance are next required.

Drug sensitivity

Debrisoquine, an anti-hypertensive agent, illustrates how DNA technology can assist in identifying those individuals who are more likely to have side-effects to this drug because they have a genetic inability to metabolise it efficiently. The 'poor metaboliser phenotype' is inherited as an autosomal recessive trait and is found in approximately 5–10% of Europeans. Identification of at-risk individuals can be difficult. The conventional approach requires a test dose with assessment in the urine of the drug to metabolite ratio over a period of several hours. This in itself can lead to drug-related side-effects and interpretation of results can be difficult if there are other drugs being taken or the patient is ill.

Debrisoquine metabolism occurs via a liver specific cytochromal enzyme P450IID6. Inability to metabolise debrisoquine efficiently reflects reduced to absent activity of this enzyme. The gene coding for P450IID6 has been localised to chromosome 22. By use of DNA amplification through the polymerase chain reaction it has been possible to characterise the gene in patients with poor metabolising phenotypes and identify either point mutations or rarely a deletion in DNA. This opens up the possibility that drug metabolism status can now be assessed by study of DNA from peripheral blood white blood cells. Both heterozygote and homozygote-affected individuals are identifiable and a prediction can be made about their ability to handle debrisoquine and a number of other drugs which appear to be metabolised in a similar manner. The problems of drug-exposure and interpretation of the conventional urine assays can thus be avoided.

GENE THERAPY

Gene transfer

Gene therapy can be defined as the transfer of genetic material into the cells of an organism in order to treat disease. In practical terms gene therapy refers to *somatic cell gene therapy* which means the target cell does not form part of the germline so that transmission to future generations cannot occur. Germline gene therapy (an example of which would be the transgenic animals described in Ch. 2) has been proscribed for various reasons which are discussed further in Chapter 9. In the first instance, gene therapy is being directed towards the genetic disorders. In the long-term, gene therapy in some form is likely to have a role in more complex situations such as cancer and the infectious diseases. Genetic disorders for which gene therapy might be indicated include:

- Coagulopathies, e.g. haemophilias A, B
- Haemoglobinopathies, e.g. β thalassaemia, sickle cell disease
- Immunodeficiencies, e.g. adenosine deaminase deficiency
- Storage disorders, e.g. glucocerebrosidase deficiency (Gaucher disease)
- Urea cycle disorders, e.g. ornithine transcarbamylase deficiency
- Other metabolic disorders, e.g. hypoxanthine guanine phosphoribosyl transferase deficiency (Lesch–Nyhan syndrome).

A number of criteria have been proposed to identify the types of genetic disorders for which gene therapy would be indicated. These include: (a) a life-threatening condition for which there is no effective treatment, (b) the cause of the defect is a single gene and the involved gene has been cloned, (c) regulation of the gene need not be precise and (d) the technical problems associated with delivery and expression of the gene have been resolved (see below).

Monitoring of potential gene therapy protocols by various government and institutional biosafety committees has been intense. It was not until September 1989 that the USA National Institutes of Health approved the first trial involving transfer of DNA into patients with melanoma, a malignant skin cancer. In September 1990, the first transfer of a genetically engineered cell was undertaken in a 4-year-old child with the potentially fatal genetic disorder, adenosine deaminase deficiency. The long-term effects, particularly of the latter trial, are awaited.

Methodologies for gene transfer

There are a number of delivery systems for ex vivo gene transfer (Table 7.5). The ultimate aim is to get DNA into a cell. The cell and nuclear membranes can be made more permeable to DNA following co-precipitation of DNA with calcium phosphate or an electric shock (called electroporation). Using micropipettes it is possible to inject DNA into the nucleus of a cell. More novel approaches to facilitate movement of DNA into a cell include: the injection of DNA directly into muscle cells; insertion of DNA into liposomes i.e. synthetic spherical vesicles which have a lipid bilayer and so are able to cross the cell membrane; the coating of DNA with proteins.

The current preferred method for gene transfer in the context of gene therapy, involves the use of *retroviruses* (Box 7.2, Fig. 6.2). Conventional RNA and DNA viruses are unsuitable as vectors because nucleic acid is rapidly degraded if it cannot integrate into the DNA of the host genome. A further advantage of the retroviral vectors relates to their high efficiency of gene transfer which in theory can approach 100%. Physical methods such as calcium phosphate co-precipitation, electroporation and microinjection are relatively inefficient when it comes to cells taking up DNA. More importantly, DNA integ-

Box 7.2 Use of retroviruses to transfer genes

Wild-type retroviruses can convert their RNA into double-stranded DNA. The latter is then able to become integrated into the host's genome. Viral proteins encoded by the *gag, pol* and env genes make up approximately 80% of the retroviral genome. These RNA segments can be deleted and replaced by a gene of interest, e.g. adenosine deaminase. Now the recombinant retrovirus will not form infectious particles because it has lost the ability to produce its structural proteins. This is a good thing for gene therapy. Persistent infection by the genetically engineered retrovirus would not be permissible since it might lead to neoplastic changes, the wrong cells expressing the gene or even the germ cells becoming infected and so expressing the gene. However, unless the retrovirus can infect in a *controlled* way, it will not be useful as a means of transferring DNA. Inserted DNA cannot infect target cells unless a source of the structural proteins is available. This can be provided by use of 'packaging cells'. These contain a 'helper' retrovirus which also has been genetically manipulated so its structural proteins are produced but these cannot form an infectious virion. However, the retroviral vector with its adenosine deaminase gene insert can utilise the structural proteins produced by the helper virus in the 'packaging cells' to form an infectious virion which is capable of one round of infection. This would be enough to get the genetically engineered retroviral RNA into the target cells and thence the latter's DNA. A further consideration in using retroviruses to transfer DNA is that the target cell must be dividing before the retrovirus can integrate into the cell's genome.

Table 7.5 Delivery systems for gene transfer.
Methods are devised to get DNA into the cell and located so that gene expression can occur

	Delivery method
Physical methods	Improving membrane permeability to DNA: calcium phosphate coprecipitation of DNA electroporation (electric shock) Microinjection: into the cell nucleus Novel methods: insertion into liposomes; coating DNA with proteins; injection into muscle cells
Viral methods	Retrovirus, adenovirus, herpes simplex virus

rated into the host genome by physical means is usually present as multiple copies. Following retroviral induced insertion, the gene of interest is present as one copy in a single, *random* site in the host's genome. Thus, progeny of the infected cells are more likely to retain the gene.

There are a number of potential problems with the use of retroviruses as vectors for gene transfer. These include the concern that the retroviruses will revert to replication competent organisms which would give them the theoretical risk of inducing cancer. DNA insert size is limited which can be a problem if a large gene is involved. Since retroviral vectors are produced from living cells there is the worry that contaminants derived from these cells will be present.

Target cells for gene therapy

Another consideration in gene therapy is the *target cell.* An important pre-requisite for the target cell is for it to be dividing so that the retrovirus can integrate into the host genome. The target cell should also be appropriate to the type of expression required. For example, a neurologic disorder may derive no benefit from the transfer of genes into haematopoietic cells. Finally, the target cell needs to be long-lived to prolong the effects of gene therapy. For this condition, the use of pluripotential stem cells would be ideal since integration of a gene into this type of stem cell should produce a cure. Because of the potential availability of stem cells and the relatively advanced state of bone marrow transplantation, considerable work to date has focused on the haematopoietic stem cells as targets for gene transfer.

The bone marrow pluripotential stem cell is a rare cell which to date has only been satisfactorily isolated and characterised in the mouse. Gene transfer into human bone marrow-derived stem cells has been possible because of the infectious capability of the retroviruses. Nevertheless, expression observed in these instances has been low and of short duration. Thus, gene therapy in this circumstance would not be appropriate in disorders such as the β thalassaemias for which significant gene expression would be required to produce an adequate supply of protein. This may be overcome in the near future with recent developments in molecular technology. These include: (1) the potential to stimulate division of the pluripotential stem cells with the recombinant human growth factors thereby making these cells move out of the G_0 phase of the cell cycle and so become more accessible to infection with a retrovirus (Box 6.1); (2) the availability of DNA sequences which can significantly up-regulate (i.e. increase) gene expression. Enhancer elements recently located in the β globin gene locus are able to increase expression of transfected genes. The incorporation of these enhancer elements in retroviral constructs is now being attempted (see

Table 7.6 Target cells for gene transfer.
Target cells need to be of a suitable type for the expression required, dividing and long-lived

Cell	Utility
Haematopoietic stem cells	Useful sources are bone marrow and umbilical cord blood. The possibility for gene transfer into the pluripotential stem cell means a life-long cure is feasible. Haematological and immunological defects are the types of disorders which could be corrected.
Fibroblasts, keratinocytes (skin)	Easy to access and grow in culture. Can produce biologically active compounds, e.g. coagulation factor IX. Main problem is short-term effect which may be due to graft rejection.
Hepatocytes	Cultures can be obtained from liver tissue. Cells transduced with retrovirus can be reimplanted in the liver via the spleen. Novel methods to get DNA into cells involve coating DNA with a protein receptor which is recognised by hepatocytes. Stability of expression in animal systems satisfactory, up to 6 months.
Muscle cells	Injection of DNA in plasmid form into muscle cells enables expression of the DNA without it necessarily incorporating into host genome. The expression is relatively prolonged over a few months. Promising approach which requires further assessment.
Lymphocytes	The major advantages of the lymphocyte are its role in immunity, its relatively long life and its ease of access in the blood. This cell has been the target for gene transfer in melanoma and adenosine deaminase deficiency.

Ch. 1 for a further description of the enhancer elements).

Other cells being assessed as potential target cells for gene therapy are summarised in Table 7.6. These cells display a number of features which make them useful targets for the transfer of genetic information. However, none promises a cure even if their effects are long-lived, because they are not stem cells, i.e. their life span is finite as they do not have an unlimited potential for self-renewal and they do not differentiate into various types of secondary cells.

Tumour infiltrating lymphocytes and melanoma

Tumour infiltrating lymphocytes are lymphoid cells which invade solid tumours. They can be grown in culture and have tumour killing potential. For example, removal of these cells from a melanoma and growth in the presence of an interleukin (IL2) enables the tumour infiltrating lymphocytes to be cultured and then reinfused back into patients. The lymphocytes target to the melanoma where they can induce regression of the tumour. The first gene transfer study involved the genetic engineering of the tumour infiltrating lymphocytes obtained from patients with advanced melanoma. The antibiotic resistance gene neomycin was added to the lymphocytes through a retroviral vector. The transduced lymphocytes were then reinfused into the patients and their survival duration and potential toxic effects noted. The results were satisfactory in that the neomycin-containing tumour infiltrating lymphocytes persisted in the circulation; they were able to target to the melanoma cells and there were no apparent side-effects directly attributable to the retroviral vector.

The success of this limited form of gene transfer enabled the next step to proceed. In this case, the gene for tumour necrosis factor (TNF) was inserted into the tumour infiltrating lymphocytes using a retroviral vector. Tumour necrosis factor is a potent agent which can cause regression of tumours but has limited use because of its toxic side-effects. Thus, the aim was to deliver tumour necrosis factor directly to melanoma cells and so reduce systemic toxicity. Since January 1991, a number of patients have received tumour necrosis factor. The therapeutic benefits of tumour necrosis factor delivered by genetically engineered lymphocytes remain to be determined. Nevertheless, the results of both studies demonstrated that the retroviral-mediated gene transfer strategy was feasible.

Adenosine deaminase deficiency and addition gene therapy

In 1990, a 4-year-old child with the potentially fatal autosomal recessive disorder, adenosine deaminase deficiency, received an infusion of her own lymphocytes which had been genetically altered by a retrovirus containing a normal adenosine deaminase gene. Adenosine deaminase deficiency was chosen for a number of reasons. It constituted an important cause of the severe combined immunodeficiency syndromes in children. Death within the first 2 years of life was common. Medical treatment was suboptimal. This included a recently released drug called PEG-ADA which comprised the natural product (ADA) coupled to polyethylene glycol (PEG) to increase its half-life. PEG-ADA was very expensive and follow-up time had not been sufficient to confirm its long-term efficacy. Similarly, bone marrow transplantation was a therapeutic option which produced cure in most cases following a successful transplant. However, less than one third of patients had an appropriately matched sibling for transplantation. The above 4-year-old girl had been treated with PEG-ADA but had responded inadequately to this form of therapy. In these circumstances, approval was given to attempt gene therapy.

Features which made adenosine deaminase deficiency a good candidate for gene therapy included: (1) the target or affected cells were lymphocytes and so accessible through the blood, (2) T lymphocytes have a relatively long life span, (3) the gene had been cloned and was only 3.2 kb in size and (4) it was expected that a moderate level of gene expression would be sufficient to reduce mortality in this condition (Fig. 7.7).

At first, stem cells isolated from the bone

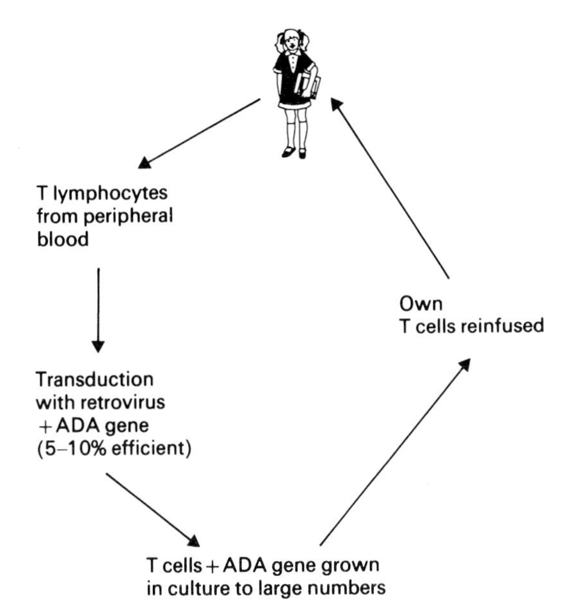

Fig. 7.7 Protocol for adenosine deaminase gene therapy. T lymphocytes are removed from the patient's circulation and infected with a retrovirus containing the wild-type ADA gene. The patient is then re-infused with her own lymphocytes after sufficient numbers of the transduced cells are grown in culture.

marrow of the 4-year-old girl were proposed as the targets for gene transfer. Subsequently it was found that mature T lymphocytes which could be isolated from peripheral blood were more practical alternatives to the elusive stem cells (estimated to be between 1 in every 10 000–100 000 cells). Multiple infusions of genetically engineered autologous T lymphocytes were given at 1–2 monthly intervals. These were followed by improvement in lymphocyte function as well as perceptible numbers of genetically engineered lymphocytes in the child's blood. Since the target cell in this case was no longer the pluripotential stem cell, a cure was unlikely. The child is presently receiving maintenance infusions at 3–4 monthly intervals. Long-term follow-up is necessary to assess the advantages as well as potential side-effects. It remains to be determined how frequently lymphocyte transfusions will be required. Modifications to the protocol have been proposed to enable more primitive lymphocytes to be isolated by monoclonal antibodies and then these are returned to the patient following genetic manipulation. This change may overcome some of the immune deficiencies which are still apparent following gene transfer with mature T lymphocytes. At present three children with ADA deficiency (two in the USA and one in Italy) are being treated by gene therapy.

Gene manipulation

HbF – a model for gene manipulation

As described in Chapter 1, there is progressive switching of globin genes during development. The fetal genes (Gγ and Aγ) are replaced by adult β and δ globin genes at approximately 6 months of age. A delay in switching from fetal to adult globin genes has been seen in infants of diabetic mothers and in association with the sudden infant death syndrome. The reason for this remains unknown. A second group of disorders leading to incomplete HbF switch are genetic in origin, e.g hereditary persistence of fetal haemoglobin (HPFH). This has been described in Chapter 3 and the molecular defects reviewed. Thus, the HbF (fetal) to HbA (adult) switch can be altered. This has clinical relevance to the β thalassaemias and sickle cell disease.

Patients with the severe homozygous form of β thalassaemia do not manifest clinical problems until after 6 months of age when their normal fetal genes are replaced by the non-functioning adult β gene. Since HbF is physiologically normal, the potential to postpone or induce an incomplete switch of the fetal genes is appealing since this could correct the clinical problems and complications of β thalassaemia. Similarly, HbS (sickle cell haemoglobin) is less severe if the concentration of HbS present in the red blood cells can be lowered. One way to reduce the HbS level is to increase the HbF content.

Pharmacologic manipulation of HbF

A number of chemotherapeutic agents are able to increase the level of HbF. In humans, 5-azacytadine and hydroxyurea produce a consistent increase in HbF although there are clear genetic effects also operating since the changes in HbF levels differ considerably from individual to individual. The obvious drawback to treatment with these agents is their potential side-effects particularly marrow toxicity and the long-term worry that leukaemia or another form of cancer will be induced. The mode of action of these drugs remains unclear. The initial observation that 5-azacytadine demethylated DNA and so potentially exerted its effect through gene reactivation is no longer considered to be its mechanism of action. Since fetal haemoglobin in adults is derived from a population of cells called F cells, it is proposed that the cytotoxics exert their effect through a change in cellular kinetics brought about by pulsed doses of cycle-specific drugs. Thus, rapid erythroid regeneration after each drug cycle enables predominantly the formation of F cells.

Clinical trials are presently under way to monitor the effects of hydroxyurea in patients with severe forms of sickle cell disease. Preliminary data show some useful responses although these may represent individuals who also have a co-existent hereditary persistence of HbF type of defect and so are able to increase their HbF from the normal level of $<1\%$ to the expected therapeutic range of approximately 20% HbF. More specific agents which can interfere with the HbF to HbA switch or reactivate HbF once it has switched off will provide an important therapeutic benefit for the haemoglobinopathies which are found in many parts of the world.

Future developments

Replacement gene therapy

The adenosine deaminase deficiency example given above illustrates gene therapy in which a normal gene is added to the patient's deficient cells. Hence, it represents *addition* rather than *replacement* therapy. One potential problem with

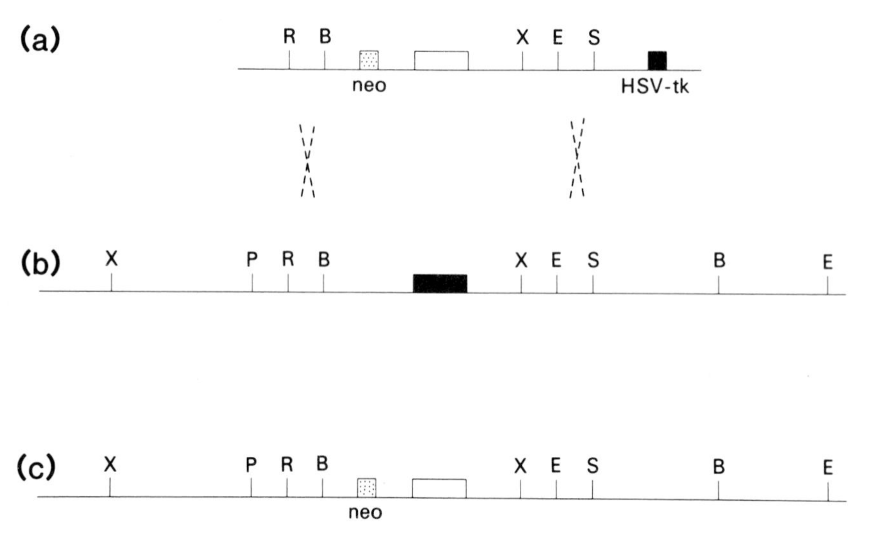

Fig. 7.8 Homologous recombination to insert a gene in the correct location.
Genes added in gene therapy can be targeted to the correct positions and replace the defective gene thus avoiding many of the problems inherent in adding a gene randomly to the genome. Homologous recombination between an incoming DNA segment (**a**) and target DNA (**b**). The normal gene is depicted as □ and the mutant as ■. The incoming DNA segment has two genes which will be used to enrich for cells which have undergone homologous recombination. The selectable markers are neo (neomycin resistance) and HSV-tk (herpes simplex virus thymidine kinase gene). If there is homologous recombination at the corresponding DNA sequences marked by (X) the structure depicted in (**c**) can result, i.e. the mutant gene is replaced by the normal gene which also brings with it neomycin resistance but not the HSV-tk gene. On the other hand if there is random integration of segment (a), the whole of (a) will be acquired since random integration occurs through the ends of linearised DNA, i.e. (a) will simply link to the end of (b) and so there will be both mutant gene, normal gene plus the neomycin and HSV-tk genes in tandem array. Cells which contain the neomycin gene can be selected for by growing them in the presence of a drug such as G418. This will select cells which have both homologous and random integration. The two options can be differentiated by using another drug (gancyclovir) which is cytotoxic to cells which contain the HSV-tk gene. The end result is selection for cells which have undergone homologous recombination. The letters R,B,X,E,S indicate recognition sites for different restriction endonuclease. Similarity in the restriction enzyme patterns for (a) and (b) show that the two DNA areas are the same, i.e. homologous recombination is possible at this locus.

addition therapy is that the inserted genes are randomly integrated into the genome. This can lead to: inefficient expression, inappropriate expression or interference with the function of nearby genes. Apart from the ethical issues, the inability to direct where DNA will be transferred into the genome is a significant factor in the prohibition of germline gene therapy. Thus, a more satisfactory approach to gene therapy would be replacement of a defective gene.

One strategy which enables the inserted gene to be targeted to its correct position is called *homologous recombination*. This has been demonstrated to occur in yeast in which recombination between incoming DNA sequences and their homologous regions in the yeast genome enable the incoming DNA to be directed to its appropriate locus (Fig. 7.8). In vitro studies using mammalian cells have shown that it is also possible to target by homologous recombination, although the frequency of this occurring is low, e.g. 1 in 10^6. The difference between yeast and mammalian cells is thought to reflect the smaller genome of the former enabling unique sequences to be more readily identifiable by incoming DNA. A higher proportion of homologous recombination (e.g. 1 in 100) can be achieved if it is possible to select for cells in which this has occurred.

Two developments have made homologous recombination an achievable goal in mammals. First, the polymerase chain reaction enables many cells to be screened for the appropriate recombination event. This is achieved by constructing DNA amplification primers which are able to detect the creation of a novel junction formed between target and incoming DNA. DNA inserted elsewhere in the genome is not detected because the junction fragment would not be present. Since homologous recombination will be rare it requires a technique with the sensitivity of the polymerase chain reaction to detect it.

The second development has been the availability of mouse embryonic stem cells (usually abbreviated to ES cells). Embryonic stem cells are pluripotential cells which can be established from an early embryo. A vector containing the gene of interest is transferred into these cells by physical means, e.g. microinjection. In the great majority of cases random insertion of DNA will occur. However, in a very few cells, the gene of interest will pair with its corresponding DNA sequence. Transferred and targeted DNA can then be exchanged by homologous recombination. Embryonic stem cells which have undergone targeting are selected by the polymerase chain reaction technique and then injected into the blastocoel cavity of a fertilised mouse embryo. The embryonic stem cells become incorporated into the latter which is allowed to develop in a foster mother. The chimaeric animal resulting will express the transferred gene in its appropriate location in the genome.

Work is now under way to isolate and utilise embryonic stem cells in domestic animals. In terms of human gene therapy, the approach using embryonic stem cells is an example of germline therapy and so the prohibitions mentioned earlier apply. However, on this occasion the gene to be inserted is now in its correct position and so is less likely to interfere with the function of other genes or be expressed inappropriately. Once homologous recombination is shown to be consistently successful and without complications the moratorium on germline gene therapy may be reviewed. At present, many technical problems related to homologous recombination remain to be resolved. However, the potential to target genes to their correct locus in the genome, particularly if this were to involve somatic cells, is very appealing.

Novel gene transfer systems

Cystic fibrosis

An abnormality in the chloride transport channel in cystic fibrosis leads to failure of chloride to exit from the cell and so water is forced back into the cell. This produces an accumulation of thick, dry mucus in the lung, pancreas and other organs which is the hallmark of this disorder. A major cause of morbidity and mortality in cystic fibrosis occurs secondary to the chronic respiratory infections which ultimately lead to respiratory failure. Thus, local 'gene therapy' at the level of the respiratory epithelium would be beneficial. This would be a viable option since it is possible in vitro to correct the chloride channel defect by transfer of the cystic fibrosis transmembrane regulator (CFTR) gene into CFTR-deficient cell lines.

The conventional gene therapy approach which has been illustrated in Figure 7.7, i.e. take cells, genetically manipulate ex vivo and then return them to the appropriate environment, is more difficult to achieve in the respiratory epithelium. Moreover, the epithelium divides slowly and so a retroviral vector would be less satisfactory. In this circumstance, an alternative vector is required. One virus which would have appeal in the context of respiratory epithelium is the adenovirus. This virus is ubiquitous in humans and will target to respiratory epithelium. Adenoviruses are non-pathogenic integrating DNA viruses that are replication deficient. The adenoviral genome will accept a relatively large DNA insert such as the 6–7 kb CFTR cDNA and it produces a stable recombinant. Host cell division does not appear to be a prerequisite for integration to occur.

Despite its obvious advantages, a number of questions related to the adenovirus as a vector for gene transfer remain unresolved. It is not clear how (and if) integration between virus and host occurs and so beneficial effects may not be permanent. Nevertheless, this may not be a major problem since it would be relatively simple to undergo repeated exposures, e.g. a nebulising spray. The long-term safety of an attenuated adenoviral vector and the immune response to the cystic fibrosis protein delivered in this way remains to be seen. Overall, this novel way of introducing the cystic fibrosis gene product to a specific region is highly attractive and is being pursued. Future clinical trials will ultimately determine the efficacy of this approach for the treatment of cystic fibrosis.

Familial hypercholesterolaemia

This is an autosomal dominant disorder associated with a deficiency of the low density lipoprotein (LDL) receptor. Hypercholesterolaemia results in affected homozygotes dying pre-

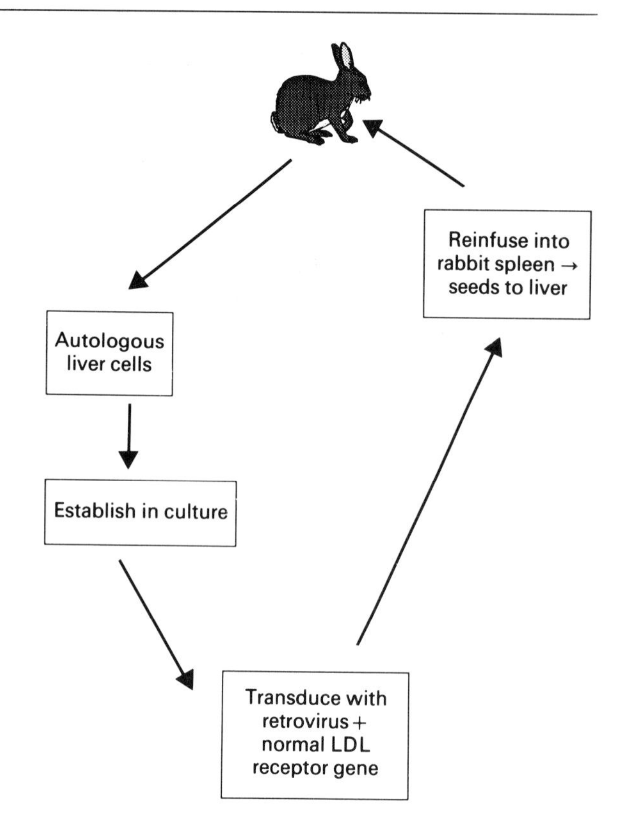

Fig. 7.9 Gene therapy protocol to treat LDL-receptor-mediated hypercholesterolaemia in the rabbit.
This is a model system for familial hypercholesterolaemia, a disorder associated with a deficiency in the low density lipoprotein receptor.

maturely from coronary artery disease in their childhood. Heterozygotes are also at increased risk from coronary artery disease. Drug options for treatment of familial hypercholesterolaemia are limited. Liver transplantation (an extreme form of treatment) produces a cure indicating that the liver plays a key role in the function of the low density lipoprotein receptor.

A rabbit animal model for familial hypercholesterolaemia is available. A successful gene therapy strategy in this animal has led to a reduction in the serum cholesterol which persisted over 6 months (Fig. 7.9). A human clinical trial protocol has now been proposed based on preliminary successful results from primate studies. A small amount of liver tissue (e.g. 35 g) is removed from the patient. The tissue is grown in culture and a normal LDL-receptor gene transduced into liver cells with a retroviral vector. The genetically engineered liver cells are then reinfused into the patient's portal circulation.

A promising in vivo gene therapy strategy also exists using the liver as a target cell. DNA to be transferred is ligated into a plasmid and then coated with a protein which has a receptor recognised by the liver cells (hepatocytes). The compound is taken up by hepatocytes. Most DNA would be degraded but a proportion reaches the nucleus and expresses the LDL receptor product. Irrespective of the approach used for gene therapy, the end result is unlikely to be curative but a means to reduce the serum cholesterol level for a period of time. The long-term efficacy of this remains to be determined. Similarly, the appearance of a 'foreign substance' in affected homozygotes, who have never produced the LDL-receptor, has the potential for an adverse immunological response in the form of antibodies.

Antisense RNA, antisense oligonucleotides

Perturbation of DNA function is possible with antisense RNA (see Ch. 1, Fig. 7.10). The potential for antisense technology has been demonstrated in agriculture, e.g. genetically engineered tomatoes that are mush-resistant because the gene producing polygalacturonase, which breaks down the cell wall, has been inhibited by antisense RNA. A number of plants have now been genetically engineered in similar fashion to give them resistance to particular viruses. In the human, antisense strategies would be useful in diseases for which there is inappropriate expression of a gene, e.g. oncogenes and the development of cancer as well as foreign genes, e.g. an invading pathogen. An antisense approach would not be effective in genetic defects associated with a lack of expression.

The synthesis of antisense oligonucleotides, rather than the use of RNA, provides an alternative strategy with which to manipulate gene function. These oligonucleotides have a DNA sequence complementary to target DNA or RNA (see Ch. 1). Antisense oligonucleotides bind to their targets and inhibit transcription or translation. An oligonucleotide 20–30 base pairs in size

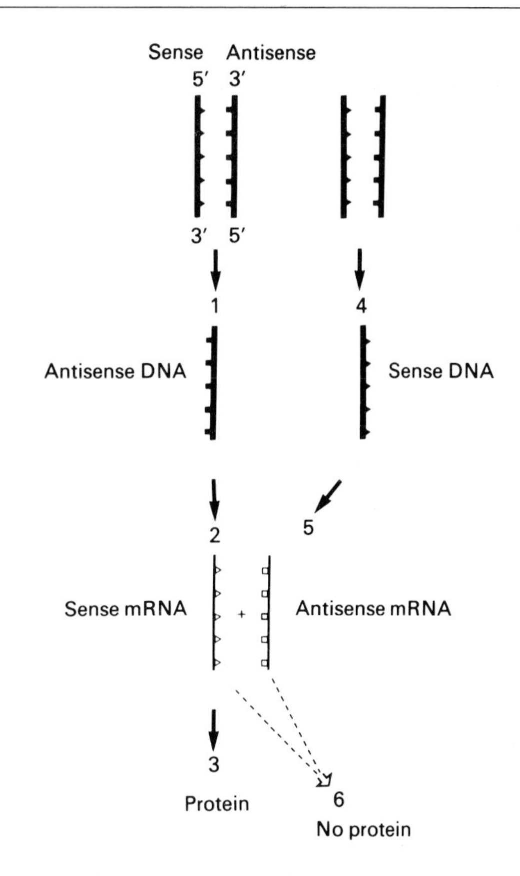

Fig. 7.10 Antisense technology as a form of gene therapy. Antisense RNA can inhibit the expression of a gene and would be effective therefore in diseases where unwanted gene expression occurs. The underlying mechanism(s) which produce the antisense effect are not fully known. One explanation involves a binding of the antisense RNA (or an antisense oligonucleotide) to sense mRNA and so inhibition of the latter during translation (protein synthesis). The normal transcription → translation pathway is illustrated on the left of the diagram (**1**,**2**,**3**) with the antisense DNA strand providing the template for transcription by RNA polymerase to make (sense) mRNA. On the right (**4**,**5**,**6**), the DNA sense strand has been copied. This could be a normal response by an organism to foreign DNA or a genetically engineered gene which has been 'flipped' around so that the sense sequence becomes the template for mRNA synthesis. Alternatively, oligonucleotides with the antisense sequence are introduced into the cell. The end result is the same with the antisense mRNA/oligonucleotide binding to the (sense) mRNA and inhibiting its activity.

will usually ensure attachment occurs to a *unique* sequence within the genome. In vitro, antisense oligonucleotides are able to decrease the tumourigenicity of cell lines. Antisense oligonucleotides have been synthesised to a number of potential oncogenic sequences and have specifically inhibited their expression in cell lines, e.g. the bcr-*abl* hybrid transcript which is unique to the leukaemic cells in chronic granulocytic leukaemia.

Inhibition of mRNA specific for the human immunodeficiency virus through an antisense oligonucleotide strategy would be a novel way to attempt control of AIDS and other viral infections. In the case of AIDS, the lymphocyte, which is the target for the human immunodeficiency virus, could be genetically engineered to produce an antisense mRNA inhibitory to viral replication. The identification of mRNAs which are pathogen-specific, e.g. the common 35 nucleotide mRNA leader sequence in all trypanosomes or the prokaryote-specific rRNAs of bacteria, would be useful targets for therapy via antisense technology.

An extension of antisense technology is *DNA triple helix formation*. This relies on the ability to accommodate a third strand of DNA within part of the DNA duplex to form a triple helix. Binding of the third base occurs to the already formed A/T or G/C pairs. Triple helix formation was first demonstrated in the 1950s. Confirmation of its existence in double-stranded DNA came in the late 1980s. Subsequently, in vitro studies were able to show repression of transcription by the proto-oncogene *myc*. Triple helix formation can inhibit DNA replication or DNA/protein interactions thereby affecting transcription. Potential targets for triple helix formation include viral infections and neoplastic disorders. The former are particularly relevant since DNA or RNA specific oligonucleotides are able to be synthesised. The complex abnormalities involved in the pathogenesis of cancer and the possibility of single-base DNA differences between oncogenes and proto-oncogenes makes cancer a less attractive target for antisense technology in the first instance.

Antisense technology has in part been driven by developments in automated DNA synthesisers. These have enabled the synthesis of large amounts of relatively cheap and good quality oligonucleotides. Successful chemical modification of the oligonucleotides reduces their potential for endogenous breakdown by cellular and nuclear nucleases. Considerable technical problems still remain to be resolved. For example, antisense molecules are not catalytic. Binding to target, which is reversible, could turn out to be ineffective. Delivery of antisense compounds also needs to become more efficient. Membrane-based vehicles, e.g. liposomes, similar to those described in gene therapy, are able to internalise antisense oligonucleotides into cells. Once in the cell the oligonucleotide can act within the cytoplasm or gain access to the nucleus if DNA is the target. The pharmacokinetic properties and potential toxicity of oligonucleotides are still to be fully determined. Studies to date have utilised non-physiological conditions since saturating quantities of oligonucleotides have been required to achieve adequate intracellular and nuclear concentrations. Nevertheless, the novelty of this technology and the successful in vitro results obtained to date make it likely that in vivo testing will proceed at some future date.

Ribozymes

A potential form of gene therapy would involve the expression of genes which produce ribozymes. These are RNA species with enzyme-like activity (see Ch. 1). Ribozymes would have similar functions to those described for antisense oligonucleotides, i.e. DNA or RNA species whether they are from tumours or infectious agents could be specifically inhibited. A potential advantage of ribozymes over antisense molecules lies in the former's catalytic activity so that following binding there is cleavage of target RNA. Specificity of the ribozyme rests with the hybridising (antisense) arms located on either side of the molecule's catalytic activity domain. The target triplet recognition site (GUC, with uracil the critical component) ensures that there are many potential sequences within RNA that can be targeted for ribozyme catalysis. For example, in vitro studies have demonstrated that expression of the p24 and Gag proteins of the human immunodeficiency virus can be suppressed.

Technological constraints for ribozymes include their design which makes construction of oligonucleotides more difficult. They are also susceptible to degradation by RNAases. Replacement of some RNA components of the ribozyme with DNA sequences will reduce the latter prob-

Table 7.7 Current status of gene therapy in some genetic diseases

Disease	Defective product	Target cell(s)	Status
Severe, combined immuno-deficiency disorder (SCID)	Adenosine deaminase deficiency (20–30% of SCID)	Bone marrow stem cells or T lymphocytes	Started: long-term effects being assessed
Cystic fibrosis	Cystic fibrosis trans-membrane regulator (CFTR)	Respiratory epithelium	Animal studies promising. Potential for aerosol delivery of CFTR to respiratory epithelium
Familial hyper-cholesterol-aemia	Low density lipoprotein receptor	Liver	Animal stage completed; human trials to start
Duchenne muscular dystrophy	Dystrophin	Skeletal muscle	Animal studies promising, efficiency of transfection and long-term stability to be defined
β Thalassaemia	β Globin	Haemato-poietic stem cells	Could start if problem of low expression corrected
Haemophilia	Factor VIII or factor IX	Fibroblasts, liver cells	A trial involving factor IX transduced into autologous skin fibroblasts is presently under way in China

lem. More efficient methods for delivery of ribozymes into cells will also need to be developed. A combination of ribozyme and antisense technology is a promising future development. In this strategy, ribozymes are incorporated into antisense oligonucleotides thereby providing the latter with catalytic activity. These catalytic antisense molecules have twice the efficiency of the conventional substances. Just as described for the antisense oligonucleotides, considerable technological developments are still required before the novel therapeutic approaches just described will have successful clinical outcomes.

Gene therapy in genetic diseases

In a short time frame, over a dozen protocols for gene therapy are undergoing clinical trials in three continents. Just as many are about to begin. These include the diseases mentioned previously, as well as proposals involving AIDS, both the acute and chronic granulocytic leukaemias and advanced cancers. A summary of the current status of gene therapy in a number of genetic disorders is given in Table 7.7.

FURTHER READING

Adair J R, Whittle N R, Owens R J 1990 Designer antibodies. In: Carney D, Sikora K (eds) Genes and cancer. J Wiley, New York, p 151–161

Anderson W F 1992 Human gene therapy. Science 256: 808–813

Bloom A L 1991 The evolution and future of haemophilia therapy. Transfusion Medicine 1: 5–12

Capecchi M R 1989 The new mouse genetics: altering the genome by gene targeting. Trends in Genetics 5: 70–76

Cohen J S 1992 Oligonucleotide therapeutics. Trends in Biotechnology 10: 87–91

Gutierrez A A, Lemoine N R, Sikora K 1992 Gene therapy for cancer. Lancet i: 715–720

Heim M, Meyer U A 1990 Genotyping of poor metabolisers of debrisoquine by allele-specific PCR amplification. Lancet ii: 529–532

Kartner N, Ling V 1989 Multidrug resistance in cancer. Scientific American 260: 26–33

Kaufman R J 1991 Developing rDNA products for treatment of haemophilia A. Trends in Biotechnology 9: 353–359

Looker D, Abbott-Brown D, Cozart P et al 1992 A human recombinant haemoglobin designed for use as a blood substitute. Nature 356: 258–260

Lusher J M, Arkin S, Abildgaard C F et al 1993 Recombinant factor VIII for the treatment of previously untreated patients with hemophilia A. New England Journal of Medicine 328: 453–459

Metcalf D 1989 Haemopoietic growth factors 1. Lancet i: 825–827

Miller A D 1990 Progress toward human gene therapy. Blood 76: 271–278

Ogden J E 1992 Recombinant haemoglobin in the development of red blood cell substitutes. Trends in Biotechnology 10: 91–96

Recombinant factor VIII 1991 Proceedings of the first international symposium on recombinant factor VIII. Seminars in Haematology 28(2): (suppl 1)

Schild G C, Minor P D 1990 Modern vaccines – human immunodeficiency virus and AIDS: challenges and progress. Lancet i: 1081–1084

Tannock G 1991 What hope for recombinant vaccines? Today's Life Science 3: 12–15

Wickstrom E 1992 Strategies for administering targeted therapeutic oligodeoxynucleotides. Trends in Biotechnology 10: 281–287

8

FORENSIC MEDICINE

INTRODUCTION

Genetic differences identifiable by protein polymorphisms have been used in forensic laboratories since the late 1960s. Initially, protein markers were based on the ABO blood groups. Subsequently, other blood groups, serum proteins, red blood cell enzymes and more recently histocompatibility (HLA) antigens have been typed. One disadvantage of protein polymorphisms has been the limited degree of variability associated with these markers. Thus, the identification of commonly occurring protein polymorphisms between two samples is of lesser value if the probability is sufficiently high that these could represent chance events. Therefore, the emphasis in the legal sense has been on *exclusion* rather than positive identification when two samples have been compared.

Other problems inherent in protein analysis relate to the amount of tissue required for testing and the relative ease with which proteins degrade. These considerations are particularly relevant to the scene of a crime where the 'ideal' laboratory conditions will not be found and tissue available for analysis will more often than not be limited in amount.

In 1978, the first human *DNA polymorphism* related to the β globin gene was used to identify genetic disease. In 1980, it was reported that small variations in DNA detected using restriction endonucleases, the restriction fragment length polymorphisms (RFLPs) (see Chs 2, 3), were dispersed throughout the entire human genome, thereby showing the potential for a more sophisticated approach to tissue comparisons or even the *identification* of individuals. More complex and so potentially more informative DNA polymorphisms were described in 1985. These were called 'minisatellites' and it soon became possible in British and American courts of law for DNA evidence to be used in criminal and civil cases. The first such trial occurred in Bristol, England in November 1987. DNA evidence in this particular case was crucial in providing the link between a case of burglary and rape.

By the late 1980s, over 1000 cases in the USA had involved DNA evidence. Today, UK, USA, European and other courts have allowed, to varying degrees, the admission of DNA data as evidence in criminal trials and paternity disputes. The important appeal of DNA lay in its intrinsic variability so that exclusion was no longer the major intent. Thus, it became possible to aim for a unique DNA profile for each individual similar to dermatoglyphic fingerprints.

It was only in 1989 during a pretrial hearing for a double murder case involving the *State of New York versus Castro* that DNA evidence was first seriously questioned. This led to the demonstration of suboptimal laboratory practices as well as doubtful interpretation of the statistical significance of DNA polymorphic data. Evidence based on DNA studies in this case was thereby deemed inadmissible. Subsequently, a number of other cases have had to be withdrawn by the prosecution because DNA data comprised an important component of the evidence. Cases already decided will no doubt be appealed.

Thus, in a very short time, DNA technology has had a major impact on the judicial system which is extraordinary given the slow pace with which the system usually moves. Nevertheless, the rapid utilisation of DNA technology has produced significant problems. These reflect differences in the interpretation of DNA polymorphic data particularly in relation to minority ethnic groups, the types of laboratory protocols and the standard of quality assurance practised in some of the laboratories. Such problems are being slowly resolved. Legislation enacted in many communities now requires the highest code of practice for laboratories involved in forensic DNA technology.

For the purpose of this chapter, the application of DNA technology in the criminal case as well as the establishment of familial relationships will be considered under the one category of forensic medicine.

REPETITIVE DNA

DNA, which comprises the 3.3×10^9, base pairs of the human haploid genome has a number of functions. Approximately 70% codes for genes or is involved in a number of gene-related activities such as regulation of expression. For example, DNA provides the signals for its own replication as well as those required for chromosomal replication, division and segregation. The remaining 30% of the eukaryote genome is composed of repetitive DNA sequences which appear to have no function. The term 'junk' DNA has been used to describe these areas although this would seem inappropriate since the distribution of such DNA is non-random in places and there remains some inter-species homology. A potential role of repetitive DNA as 'hot spots' for recombination has been proposed. This is an appealing hypothesis since the repeat sequences have no apparent coding (exon) function and so there would be less evolutionary pressure for conservation. A greater degree of mutational activity would thus be possible at these loci. In an evolutionary sense, this would be useful, e.g. for the development of new genes.

Repetitive DNA can be divided into two major classes. The tandemly repetitive sequences (known as *satellite* DNA) and the *interspersed* repeats. The term 'satellite' has been used to describe DNA sequences which comprise short head-to-tail tandem repeats incorporating specific motifs. These make up one third of DNA repeats (i.e. 10% of the total genome) and are exemplified by the microsatellites, minisatellites and macrosatellites. A classification for repetitive DNA is given in Table 8.1.

Table 8.1 Types of repetitive DNA in the human genome.
These have no known function and are divided into two classes: the satellite DNA which has tandemly repetitive sequences and the interspersed repeats

Designation	Size range	Examples	Features
Satellite DNA			
Micro-satellites	<1 kb	$(AC)_n$ repeats	$(XXX)_n$, $(XXXX)_n$ are potentially more useful
Mini-satellites	1–30 kb	Probes called 33.6, 33.15 3'α HVR	'multilocus' (common core)$_n$ Multilocus/single locus Multiple repeats in tandem: VNTRs
Macro-satellites	Can be megabases in size	Alpha satellites	Mostly in centromeres/telomeres
Interspersed repeats	≈300 bp >500 bp to 10 kb	Alu repeats Kpn or L1 repeats	Interspersed repeats which are not necessarily repetitive internally or in tandem array

n = number of repeats; X = nucleotide base

Microsatellites

Microsatellites comprise small DNA polymorphisms usually <1 kb in size. The best described are the dinucleotide repeats $(AC)_n$ where n (the number of repeats present) varies from 10–60. Because of the tandem array of repeat units, these polymorphisms are examples of the type called **v**ariable **n**umber of **t**andem **r**epeats (VNTRs). Individually, the microsatellites are considered to be *single-locus VNTRs* because each can be made to identify one segment of the genome. It is estimated that the human genome contains approximately 50 000 of the $(AC)_n$ repeats. Thus, the value of these polymorphisms lies in their widespread distribution throughout DNA which makes them ideal for *genome mapping*. As DNA polymorphisms, they are highly informative in family studies to identify wild-type versus mutant alleles or in paternity testing (see p. 184). Microsatellites, because of their potential hypervariability, are more informative than the biallelic RFLP system (see Chs 2, 3).

One technical consideration with the $(AC)_n$ repeats is the necessity to use the polymerase chain reaction since size differences between alleles are small and it is essential to utilise oligo-

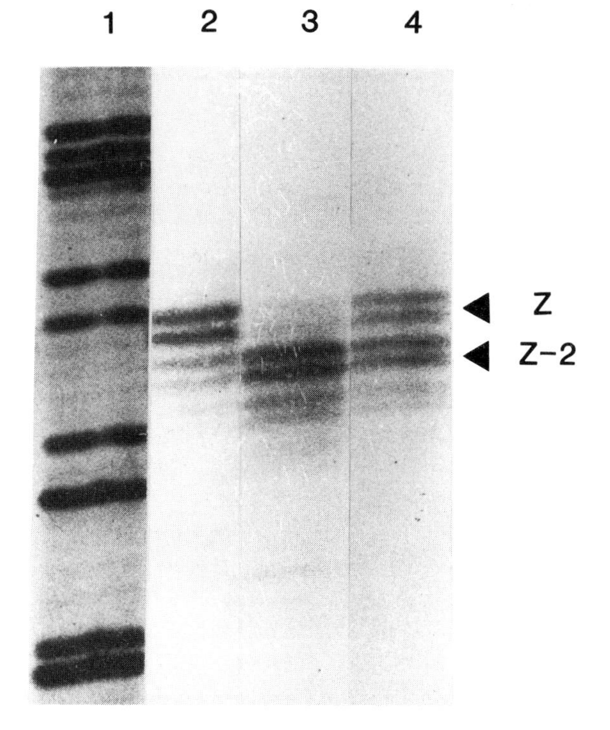

Fig. 8.1 Detection of microsatellite DNA.
$(AC)_n$ microsatellite repeat patterns from locus D15S10. Track 1 = DNA molecular weight marker; track 2 = top $(AC)_n$ allele designated as 'z'; track 3 = lower $(AC)_n$ allele differing from track 2 by 2 base pairs (designated 'z-2') and track 4 = a heterozygote since both z and z-2 alleles are present. $(AC)_n$ repeat patterns can be difficult to interpret, e.g. as shown here each allele comprises two bands. This occurs because the method used to identify the repeat has led to the (AC) strand and its complementary (TG) strand *both* being radiolabelled and so each shows up on autoradiography.

nucleotides to target a specific region in the genome so that one microsatellite locus alone is tested (Fig. 8.1). On the positive side, this means that an automated and more rapid testing procedure is available to detect the microsatellites. However, this needs to be balanced with the problem of DNA amplification in the forensic situation (discussed further on p. 190). More recently, microsatellites comprising a three or four base pair core (e.g. $(AGC)_n$ or $(AATG)_n$) have been described. Interpretation of gel patterns which contain amplification products for these is easier since the difference between alleles is now greater i.e. from two bases to three or four. Amplification by the polymerase chain reaction is also more reliable compared to that found with the dinucleotide $(AC)_n$ repeats.

Minisatellites

Of greater value in the forensic laboratory are the minisatellite repeats. Here the common core sequence which will be repeated is *longer* than that found with the microsatellites. This produces restriction fragments which are in the kilobase range compared to the microsatellite alleles which extend from 20–120 nucleotide bases in size. Thus, minisatellites give a much wider size range for their polymorphic DNA fragments. Because of this, it is possible to use either DNA Southern blotting or the polymerase chain reaction to identify minisatellites.

Minisatellites are either *multilocus* (repeated throughout the genome in many loci) or *single locus*, where the position of the minisatellite in the genome can be localised to one place. The chance of finding differences between two alleles using minisatellites is very high (up to 99% in some cases) and it is also possible to use the complex polymorphic patterns arising from multiple loci to construct a unique DNA profile or 'fingerprint' for an individual.

Multilocus minisatellite VNTRs

The first minisatellites to be used in the courts of law were described by Jeffreys in the United Kingdom. These were DNA probes designated 33.6 and 33.15. The former constituted a core sequence of motif $(AGGGCTGGAGG)_3$ repeated 18 times and the latter was a 16 base pair motif, AGAGGTGGGCAGGTGG, which was repeated 29 times. The loci detected by these two probes were dispersed throughout the genome (hence their name, multilocus VNTRs). A composite of these multiple loci produced intricate band patterns depending on how many repeats of the 'core sequence' were present. For example, in one study probe 33.6 gave on Southern blotting of DNA digested with the restriction endonuclease *HinfI* (a frequently cutting restriction enzyme since it recognises a four base pair sequence) an estimated 43 loci from the paternal DNA and 27 from the maternal DNA. The restriction fragments in the multilocus 'DNA fingerprint' clustered in the 2–4 kb range. In contrast, the single-locus

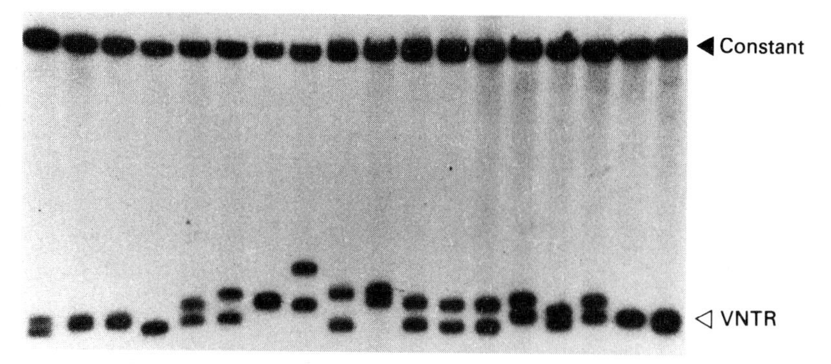

Fig. 8.2 Single locus minisatellite VNTR (variable number of tandem repeats) patterns.
Each of these VNTRs can have a wide range of band sizes for each of the two alleles which increases the chance of finding different patterns in different individuals. A Southern blot analysis of a VNTR type polymorphism derived from the immunoglobulin heavy chain gene is illustrated. There is an upper band which is the same for all individuals (i.e. it is a constant band) and the lower bands are polymorphic i.e. variable in size. The latter display considerable diversity amongst a random normal population. Since these VNTRs are single locus changes, a maximum of two polymorphic alleles is possible.

VNTR patterns identified by the microsatellites produce a maximum of two DNA fragments which reflects the two alleles present. The biallelic pattern of the latter makes interpretation easier although the amount of information provided is correspondingly reduced.

Disadvantages of the multilocus minisatellite probes described above included: (1) the technical expertise necessary to get good DNA patterns; (2) the relatively large amount of DNA required for the Southern blotting studies (e.g. multilocus minisatellites may require as much as 25 times the quantity of DNA used to detect a single locus VNTR) and (3) the difficulties which can arise in the interpretation of the large number of DNA fragments present. In particular, individual variations in the agarose gel tracks may alter the mobility of DNA as it is electrophoresed (called *band shifts*). This could be controlled, in the scientific research laboratory, by procedures such as running multiple gels and varying the position of the samples being tested. However, the forensic laboratory, with limited amounts of DNA which are often degraded or contaminated with other sources of DNA or extraneous material capable of altering the mobility of DNA, is in a less fortunate position. Thus, it is not surprising that, apart from the relatively 'clean' samples and unlimited quantity of DNA available in family studies such as paternity disputes, the crime-related applications using the multilocus minisatellites were confined to very few experienced laboratories and special circumstances.

Single-locus minisatellite VNTRs

A second type of minisatellite VNTR identifies a single locus in the genome. The many examples of these types of polymorphisms, which were first characterised in 1986–87, are found throughout the genome, e.g. in association with the α globin gene locus on chromosome 16 (called 3′ α HVR where HVR is an abbreviation for hypervariable region) or the immunoglobulin heavy chain gene locus on chromosome 14 (Fig. 8.2). The polymorphic fragments for the single-locus minisatellite VNTR are usually of larger size than the multilocus VNTR fragments because there is a greater number of repeat units in tandem. Localisation to the one region of the genome is possible by utilising a DNA probe and stringent hybridisation and washing conditions when DNA mapping is undertaken (Fig. 8.3).

Some applications of the minisatellite VNTR

A number of commercial and government laboratories involved in DNA testing for legal purposes have utilised the single-locus minisatellite VNTRs described above. Each of these VNTRs is highly informative producing two alleles but with a *wide range of band sizes per allele*. Therefore, the chance of finding different patterns between individuals is considerably higher than that possible

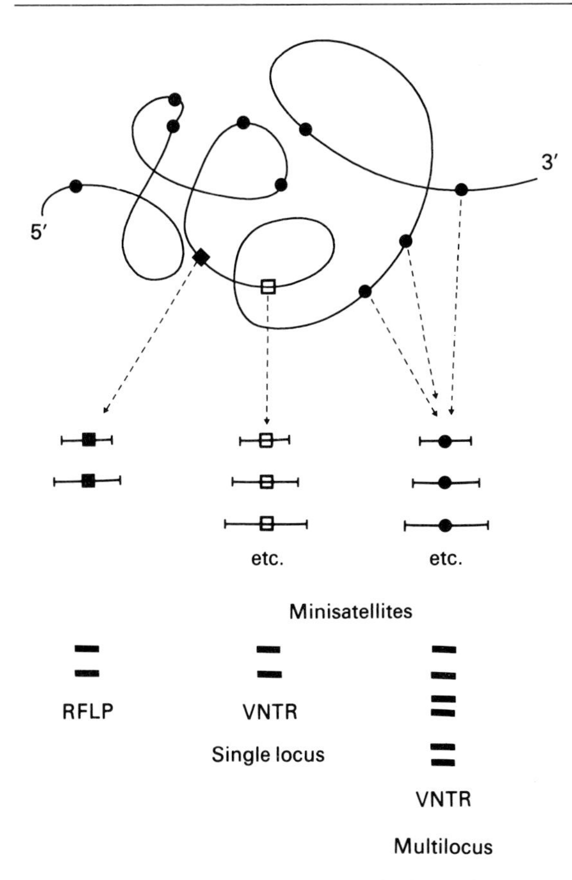

Fig. 8.3 A comparison of the various types of polymorphisms available for DNA testing.
■ Indicates an RFLP which is present at a single locus and will produce two polymorphic bands—large and small—which are of fixed size. Thus, the number of combinations generated by two alleles in all individuals is limited to: large/large; small/small and large/small. □ Two polymorphic bands are also obtained for the single locus VNTR minisatellite but these polymorphisms are more informative because there is greater variability between the sizes obtainable for each of the two bands and so there is more chance that individuals will have different profiles (see Fig. 8.2). *Combinations* of a number of different single locus VNTRs produce an even more characteristic set of markers for an individual. ● is the most informative of all because the multilocus VNTR pattern is a composite of many VNTRs which are scattered throughout the genome. A complex DNA profile ('fingerprint') results (see also Figs 8.4 and 8.5 which show actual autoradiographic patterns for a multilocus VNTR).

with an RFLP since variability with the latter will be limited to one of three options (large/large, small/small and large/small) (Fig. 8.3). A combination of 4–6 single-locus VNTR markers gives an overall DNA profile which is very polymorphic and so is considered to be unique to a person. The probability of a chance match between individuals was estimated in some cases to be 1 in 10^6. The latter claim may in fact be less accurate than originally proposed and is discussed further below (p. 188). Southern blotting analysis was required for these VNTRs. However, the limited amount of DNA available from a crime scene was less of a problem since filters could be hybridised with one VNTR probe and then rehybridised with a second, third and additional DNA probes.

Other repetitive DNA

Located near centromeres and telomeres are a third form of hypervariable satellite DNA: the macrosatellites. In contrast to the microsatellites and minisatellites, the macrosatellites can be very large (e.g. megabases in size) and so pulsed field gel electrophoresis may be necessary for their identification and characterisation. In practical terms, DNA polymorphisms associated with the macrosatellites are not used in forensic practice since DNA is often degraded to some extent. Thus, high molecular weight DNA essential for pulsed field gel electrophoresis would be unavailable in these circumstances.

By comparison to the 'satellite' DNA repeats, *interspersed repeats* are more frequently occurring. These are not usually found in tandem array and are not necessarily located as multiple repeats. Therefore, they have little role to play in comparisons between individual DNA specimens. Two repetitive elements in this class are the 'Alu' repeats and the 'Kpn' repeats. The Alu repeats are so named because they frequently have a cleavage site for the restriction endonuclease *AluI*. Alu repeats are estimated to occur every 5–10 kb of DNA. Kpn repeats (named after the restriction endonuclease *KpnI* and also called L1 repeats) occur in approximately 1–2% of the genome and are larger in size than the Alu repeats. The distribution of Kpn repeats is consistent with mobile elements such as the transposons (DNA sequences which can replicate and insert a copy at a new location in the genome). Alu repeats have a forensic application since they are human (higher primate) specific and so are useful in determining a specimen's origin as human or non-human.

DNA COMPARATIVE STUDIES

Family studies

In addition to assessing family relationships for genetic disorders, the use of DNA studies in legal assessments of relationships has become increasingly common.

Paternity disputes

Protein typing for HLA and other polymorphisms still remains a valuable approach for paternity testing. In this situation, fresh blood can be obtained from the various parties and analyses are conducted under optimal laboratory conditions. The polymorphic nature of the HLA (**h**uman **l**eukocyte **a**ntigens) markers makes them very useful in paternity studies. For example, commercial kits which utilise DNA amplification and allele-specific oligonucleotides (see Ch. 2), enable typing for the six alleles associated with the HLA-DQα gene. The six alleles are designated A1.1, A1.2, A1.3, A2, A3 and A4. Combinations of these alleles produce 21 genotypes. One estimate suggests that there is only a 7% chance that two individuals selected at random will share the same HLA-DQα type. Whilst the variability in a six allelic system is inadequate for positive tissue identification, it provides a rapid and relatively simple DNA marker which is helpful as evidence that a tissue sample did not belong to a suspect or more frequently for the exclusion of an individual in the case of disputed paternity. A combination of the HLA types and DNA polymorphisms would go one step further to allow an estimate to be made whether the individual in dispute is in fact the biological father. This would rely on the markers obtained, their frequency in the population and the likelihood that the combination detected could occur by chance alone.

The multilocus minisatellites are even better than the HLA markers in this respect, since they produce a greater number of variable alleles which would allow both exclusion and more definitive identification of the biological father. Again, the material for analysis could be prepared

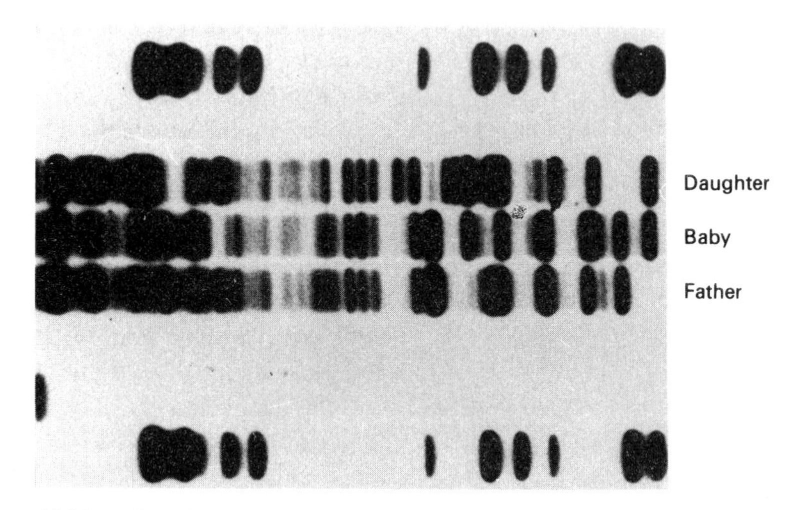

Fig. 8.4 A case of incest which is confirmed by DNA fingerprints using a multilocus minisatellite probe.
The three DNA tracks in the centre depict the accused's daughter who is also the mother of the baby; the baby and the daughter's father who is also the putative grandfather. The father, his daughter and the baby share a number of common bands. In a first degree relationship, it would be expected to see on average a bandshare in the order of 62% which is what is found between father and daughter in this case. However, the bandshare between father and baby is 78% which shows that not only is he a first degree relative of the baby but also closely related to the daughter. All paternal bands in the baby can be assigned to the putative father proving conclusively that he is both the father and grandfather (case and photo provided by courtesy of Mr T Whiting, Cellmark Diagnostics, Abingdon UK).

Box 8.1 DNA typing with multilocus minisatellites to confirm family links

A black African male (called X) who was born in the United Kingdom left that country to live with his father in Ghana. On return to the United Kingdom to rejoin his mother and three siblings he was refused residency since there was suspicion that a substitution in the form of a nephew or an unrelated male may have occurred. The father was not available to be tested and in fact there was the possibility of non-paternity involving 'X'. Protein markers confirmed that the woman (called 'M') who claimed to be the mother of 'X' was related to him. However, it could not be excluded that she was in fact his aunt. Minisatellite probes described previously (33.6 and 33.15) were able to define 61 distinct fragments in DNA from X. These were all found in the putative mother or the father (the latter's pattern was inferred from DNA profiles obtained from the three available siblings). This confirmed that X was related to the family since DNA fingerprints are seldom shared amongst unrelated individuals. 25 fragments in X were shown to have come from M and so there could be little doubt that M was the mother of X. On the basis of this evidence X was granted residency (Jeffreys et al 1985).

under optimal conditions and so the potential complicating effects of degradation or contamination on the DNA patterns, which is a vital consideration in the crime scene, would be less relevant. Paternity determination in the case of incest is particularly difficult to resolve if conventional protein methods are used since the suspect and the related victim are bound to share a number of common types. In these circumstances, the more highly polymorphic DNA markers become extremely valuable (Fig. 8.4).

Relationships within families

An extension of the paternity situation can be found when it is necessary to confirm relationships between individual members of a family. This is illustrated in a case involving an immigration dispute with one individual being refused residence in the United Kingdom. A satisfactory conclusion to the case was possible when DNA typing with the multilocus minisatellites confirmed the individual's identity and so his right to live in the country (Box 8.1).

The parent of origin who carries a defective gene(s) associated with the phenomenon of imprinting (see Ch. 3) can now be accurately identified using microsatellite or minisatellite probes. Monozygotic or dizygotic twins are distinguishable on the basis of their minisatellite patterns.

HLA markers

Tissue typing

An important set of markers for tissue typing and comparative studies of populations are the antigens which make up the HLA (human leukocyte antigen) complex. This is also known as the MHC (**m**ajor **h**istocompatibility **c**omplex). To date, HLA typing has been predominantly based on serological markers although very subtle differences between antigens shared by individuals may only be detectable by what is called a mixed lymphocyte reaction. HLA typing is required in most types of tissue transplantation since the closer the graft is to the donor in the immunological sense, the less likely is graft rejection to occur. Comprehensive tissue-typing can be time-consuming and despite histocompatibility at various HLA loci, examples of rejection still occur, which may in part reflect hidden antigenic determinants.

The genes which comprise the HLA complex on the short arm of chromosome 6 are divided into three classes which occupy a physical distance of approximately 3000 kb. Class I genes are called HLA-A, HLA-B and HLA-C. Next to these are found the class III genes which include various components of the complement cascade, a heat shock protein, tumour necrosis factor and a number of other genes. Class II genes are called HLA-DP, HLA-DQ and HLA-DR. The role of the class I and II genes in the immune response involves the presentation of antigens to the T lymphocyte. This is important both in recognition of self ('tolerance') and foreign antigens (e.g. the transplant situation). Class III genes are also implicated in the immune response, e.g. as components which mediate inflammation. The functions of many genes in the MHC remain unknown. Characterisation of these as well as the class I–III genes will further our knowledge of this important locus.

On a more practical basis, DNA typing will avoid the time-consuming assays which are otherwise necessary to obtain a comprehensive HLA profile at the protein level. The various HLA genes are extremely polyrnorphic which facilitates DNA analysis. Kits which allow identification of HLA determinants by oligonucleotide probes are now commercially available and in routine use in many of the tissue typing laboratories.

Disease associations

There are over 40 human diseases which appear to involve the HLA complex For example, class I genes are associated with haemochromatosis and ankylosing spondylitis. Class II genes with autoimmune disorders such as systemic lupus erythematosus, multiple sclerosis and many others. The class III locus is associated with psoriasis. The relationship between HLA and these disorders will become better understood once the locus is fully characterised.

TISSUE IDENTIFICATION

Evidentiary samples

At the scene of a crime there may be stains (e.g. blood, semen on the victim's clothing) or other tissues (e.g. skin, hair under the victim's fingernails) which require identification and characterisation. The first question which might need to be answered is whether these samples are human in origin. As indicated above, the Alu repeat sequences can resolve this problem. The next step is to obtain DNA from evidentiary samples to enable DNA profiles, in the form of polymorphic patterns, to be constructed. Later, these will be compared to DNA patterns from the suspect(s). In the criminal scenario, it is important to realise that there are likely to be other sources of DNA, e.g. the victim him/herself or the environment. The potentially complicating issues facing the forensic scientist are well illustrated by the type of evidentiary samples obtained following a case of rape. In this situation, DNA can come from multiple sources: (1) the victim in the form of blood, tissues (vaginal, anal or oral in origin) and bacteria; (2) there may have been multiple assailants; (3) semen may also be present from voluntary intercourse which had taken place prior to the assault and (4) animals or bacteria from the crime scene.

DNA testing is still feasible in the complex circumstances just described. DNA originating from microorganisms or other animals does not usually cross-hybridise with human-specific DNA probes although feint bands attributed to bacterial or non-human products have been reported by some laboratories. DNA obtained from sperm is more 'robust' when it comes to isolation procedures. Therefore, laboratory protocols can be designed to utilise this property and enhance the isolation of sperm DNA at the expense of DNA from other tissues. Thus, the level of the contaminants can be reduced. The dilemma arising from multiple human sources for DNA can be approached by comparisons. Blood or tissue from the victim is obtained to identify his/her DNA band patterns which are then 'subtracted' from the overall profile. The contribution from an innocent third party (e.g. sexual partner) can be treated in the same way. DNA patterns from evidentiary samples can then be compared to those obtained from potential assailant(s). From these comparisons it might be possible to identify specifically the assailant(s) as the source of the DNA stain or semen. Alternatively, blood from the victim may have spilled onto an assailant's clothing. DNA isolated from the blood spot will sub-

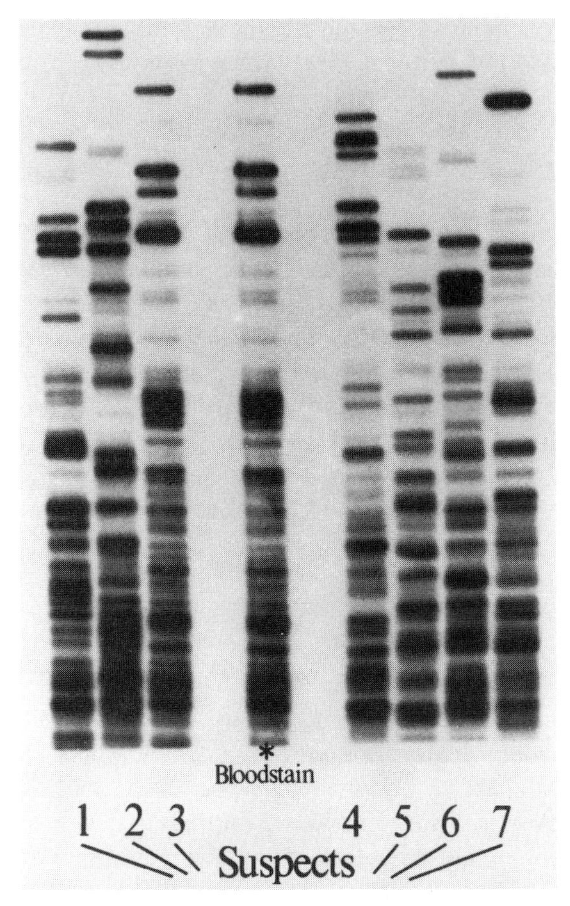

Fig. 8.5 A scene of crime DNA fingerprint comparing a blood stain with DNA from seven suspects.
Suspect number three gives a positive match since bands shared between suspect and the blood stain are identical (case and photo provided by courtesy of Mr T Whiting, Cellmark Diagnostics, Abingdon UK).

sequently provide important evidence connecting the victim and the individual wearing the clothing (Fig. 8.5).

Practical applications

Evidence based on DNA testing will be used by the prosecution to confirm a link between the victim and the accused. On the other hand, DNA evidence could turn out to be more beneficial to the defence if DNA patterns were able to exclude a match. An accused who is on trial because of evidence obtained from an eye-witness may request DNA testing as the only means by which his/her innocence can be proven. In one US case, the accused was actually acquitted because the prosecutor failed to test for a DNA match! DNA fingerprinting will save time in police investigations since suspects can be quickly excluded if their DNA profiles are different. Two experienced forensic laboratories (the US Federal Bureau of Investigation and the British Home Office) have found an exclusion rate of approximately 25% in cases studied to date.

DNA profiles have a role to play in the identification of human remains. For example, the availability of parental DNA samples might allow identification of a body when conventional means (physical appearances, dermatoglyphic fingerprints, dental charts) have been unsuccessful. Thus, dissimilar DNA profiles will exclude a relationship and similarities in DNA polymorphisms will confirm that the deceased is likely to be or is an offspring. The availability of DNA fingerprints for military personnel would be of practical assistance in the combat situation when identification of a dismembered body may be required.

Special considerations

To avoid prejudicing juries with scientific data, pretrial hearings are usually undertaken to assess the admissibility of new scientific evidence which in this case is the DNA profile/fingerprint. At these hearings, the judge evaluates the evidence before it is presented to the jury. Expert witnesses from both sides are called and the evidence considered in the light of various opinions. One important feature of DNA fingerprints is the way in which the data are imparted to the judge and jurors who are unlikely to have much knowledge of the subject which in itself is a complex topic. For example, a claim by the prosecution that a DNA match between the accused and blood obtained from the victim's clothing represents a 1 in 10^6 chance of a random event is very persuasive evidence. However, an equally crucial component to this evidence is the requirement to explain to the judge and later the jury how the test was done and what are its potential drawbacks including the method used to assess the statistical probability of a random match. In this respect it is essential that the expert witnesses have both a scientific and practical knowledge of the subject. In turn, the lawyers need to understand the advantages and disadvantages of the technology in sufficient depth to enable them to impart this information to the court.

It should also be noted that the same DNA methods which are routinely used in the research or diagnostic laboratories can be technically more demanding in the forensic laboratory. In the latter circumstance, DNA is frequently tested under suboptimal conditions. For example, the stability of DNA will depend to some extent on the way it has been maintained or stored. This cannot be controlled at the crime scene. The older the sample or the longer it has been exposed to the atmosphere (particularly ultraviolet light, moisture or high temperature), the more degraded the DNA becomes. Artefacts or technical problems resulting from contamination or degradation have in fact led to over 40% of evidentiary samples in the US and UK courts being considered inconclusive on the basis of DNA tests.

Forensic laboratories

The potential applications of DNA testing led to the setting up in laboratories, in some cases prematurely, of DNA diagnostic services for forensic purposes. Tests usually employed were the single-locus VNTR probes since these were technically less demanding than the multilocus probes and were considered to be less likely to produce artefacts on the basis of the material being examined. To approach the degree of variability possible with one multilocus VNTR, laboratories obtained composite DNA profiles which were based on polymorphic patterns from four or more of the single-locus VNTR probes.

A number of problems have now become apparent in relation to DNA tests in forensic cases. Laboratory standards have at times been inadequate with band shifts being ignored or controlled for by 'correction factors'. Band shifts are a particular problem in forensic practice since they not only represent artefacts from the procedures themselves (e.g. agarose gel variation described above) but they also result from contaminated or degraded material. Correction factors are suboptimal controls for band shifting and this source of error on the migration of DNA fragments cannot be ignored.

A second difficulty reflected the inability to measure accurately the sizes of restriction fragments, particularly large ones. Related to this was the problem of distinguishing closely co-migrating band fragments. The US Federal Bureau of Investigation attempted to overcome this by defining a number of standard size ranges so that alleles falling within each range were treated as the one marker. Furthermore, to resolve sizing and mobility problems, error ranges from 2.0–5.0% were included in assessing the significance of band patterns. However, these precautions are arbitrary measures which may reduce to some extent the error inherent in the technique but do not correct it.

Population comparisons

A third issue which still remains, even if technical problems are overcome, is the validity of population data against which the DNA polymorphisms from evidentiary samples are being compared. In the *State versus Castro* example, the laboratory reported that a DNA match between a blood stain found on the accused and blood from the victim had a 1 in 10^8 probability of occurring by chance alone. However, the comparisons to derive the chance association may not have been valid unless it had been made against an ethnic group which was relevant to the accused (Hispanic in the above case).

The major advantage of DNA testing is its ability to show identical patterns between two samples being compared and then to take this one step further and demonstrate that the patterns observed are unique, i.e. the probability of them being present in another individual is insignificant. In this respect, it is essential to know the normal distribution of the various alleles in the population. This assumes random mating which may not be occurring in ethnic or minority groups within a community. A further assumption is that the alleles for the four or so VNTR markers used by the forensic laboratory segregate independently of each other, i.e. there is no linkage disequilibrium between the alleles. The presence of linkage disequilibrium can be detected by sampling a large population but does this approach exclude linkage disequilibrium in minority ethnic groups? Would comparative studies be affected adversely in the case of the latter occurring? These are some of the questions which remain to be resolved. The answers will not be easy since population geneticists themselves do not express uniform opinions on the validity of the population comparisons being made at present. Racial admixture, which is becoming increasingly more common in many of the large cosmopolitan cities, will provide an additional complexity in the statistical analysis of probabilities when there is a match. In the meantime, to reduce this type of error, laboratories have accumulated their own databases which are considered to be relevant to the local population. For example, in one US city, a database comprising polymorphism statistics for black, Caucasian, Hispanic and Oriental populations is maintained.

FUTURE DEVELOPMENTS

Quality control

The problems created by incorrect applications of molecular techniques in forensic medicine provide a useful illustration of how utilisation of this technology must be appropriate and scientifically based. Following on from the *State versus Castro* case there are now more stringent legal requirements to ensure that laboratories involved in DNA typing are able to maintain the highest standards which include regular quality assurance. This can be difficult to organise since many of the commercial forensic laboratories utilise their own specific single-locus VNTR probes which are protected from unrestricted use or dissemination of information by patents. Therefore, direct interlaboratory comparisons become a practical problem. In the first instance, however, quality assurance programmes based on the provision of unknown samples will give laboratories and the Courts of law an indication of the laboratory's performance in DNA typing. This is being demanded by the legal profession which is becoming more aware of the technology's advantages and disadvantages. The public's perception of science is also changing. A juror's opinion obtained early on in the DNA era that 'you can't argue with science' is no longer accepted blindly but will need to be confirmed by the highest scientific standards.

Another reason for complexity in quality assurance reflects the types of evidentiary samples provided. For example, some specimens are fresh, others have been exposed to the atmosphere for variable periods of time. The different methods used to store the evidentiary sample once it has been collected might also be another variable in the DNA's electrophoretic migration. Finally, there is the problem of contaminating DNA and the overall effect this will have on the test results. An experienced forensics laboratory will need to know how the above variables might influence the DNA testing procedures and if necessary have the appropriate control samples or data should these be required. There are additional issues pertaining to the polymerase chain reaction which must also be considered in defining a quality assurance programme (see following section).

An important consideration for the defence (or the prosecution) is the frequent inability to retest the evidentiary material to confirm the findings and/or enable the test to be performed in a 'neutral' laboratory. Stains or tissues samples will usually yield a small amount of DNA. An aliquot could be saved for retesting if DNA amplification were to be used. However, the amount of DNA necessary for minisatellite assessment will, in most cases, deplete the sample. DNA evidence obtained from a single laboratory is less than ideal since confirmation will not be possible. In these circumstances, there is even greater need for good quality assurance to guarantee the appropriate laboratory standards.

DNA amplification

Three properties of DNA amplification by the polymerase chain reaction make it an ideal test in the forensic situation. First, minute amounts of evidentiary material left behind at the scene of the crime will provide enough template for DNA analysis. Second, DNA which has been degraded can still be amplified since only a small segment of DNA is required for primers to bind in the polymerase chain reaction. Finally, it would be possible to retest the sample at another laboratory. Balancing the above are a number of potential problems which will need to be resolved before DNA amplification becomes a routine procedure in the forensic situation. These include:

1. Does exposure to the environment with its consequent DNA-damaging effects lead to erroneous results following amplification? Preliminary data would suggest that this does not occur. In other words, if amplification is possible after environmental exposure, then the end product is free of artefacts resulting from damaged DNA.
2. The effect of contaminating sources of DNA on the polymerase chain reaction has already been mentioned in relation to genetic disorders and the detection of pathogens. Contamination occurs in the ideal laboratory despite strict care and the highest standards. Sources of contamination include the laboratory scientists, equipment or more frequently other amplified products. Contamination becomes an even more significant issue in the poorly controlled crime scene. The sources of DNA mentioned in the rape scenario above would need to be expanded considerably if DNA amplification is used, e.g. all who have had contact with the victim or the clothing from which DNA is subsequently extracted may *in theory* leave behind some tissue which can serve as the template for amplification.
3. The problem of errors occurring through misincorporation by the *Taq* polymerase enzyme or differential amplification of DNA sequences has been mentioned in Chapter 2. On present information, this is not seen as a drawback to the use of the polymerase chain reaction in forensic practice. However, the eccentricities of false positives and false negatives attributable to the polymerase chain reaction need to be considered both in practical terms and in designing appropriate quality assurance measures.

Because of the above considerations, DNA amplification in its present form is not routinely used in the criminal case.

Digital DNA typing

An innovative development in typing minisatellite DNA patterns has recently been described which has the potential to overcome the problems of band shifts, the separation of co-migrating DNA fragments and the difficulties associated with sizing the minisatellite-specific bands. As indicated previously, the multilocus minisatellite probes detect specific DNA motifs which are repeated a variable number of times in different loci throughout the genome. The sum of these repeat patterns makes up the fingerprint. Would it be possible to construct a fingerprint from an individual minisatellite locus? For example, DNA could be amplified from one specific locus. The amplified product would actually contain the coding information from two alleles which would enhance its potential variability. Since only one specific locus is involved, variations other than the measurement of length differences would be characterised.

Minisatellite variant repeat maps (also called digital DNA typing) do not rely on the number of repeat copies present in individual loci but on the *variation present at the one locus*. This variation reflects two classes of repeats: one which creates a restriction endonuclease recognition site for the enzyme *Hae*III (recognition site for one strand of DNA is 5′-GGCC-3′) and the second which lacks this site (e.g. the DNA sequence could now read 5′-GACC-3′). These two classes of repeats were designated by their original discoverers (Jeffreys and colleagues) as 'a' (=*Hae*III digests) or 't'

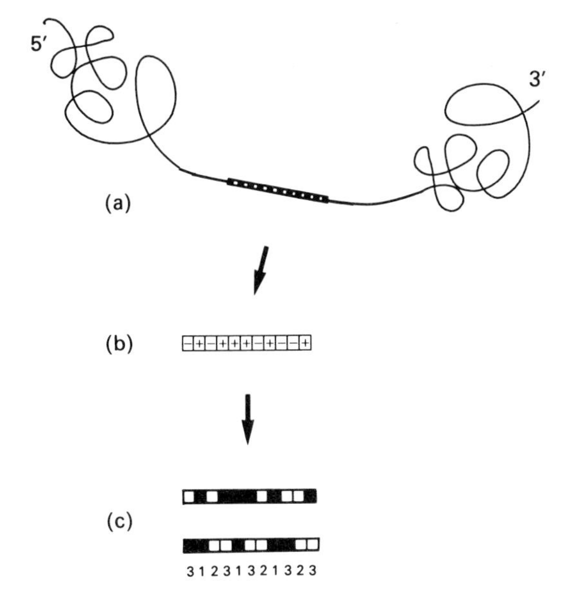

Fig. 8.6 Minisatellite variant repeats producing a ternary bar code for locus identification.
Also known as digital DNA testing this is based on the variation present at one locus. (**a**) Shows a minisatellite locus with 11 repeat units in total genomic DNA. (**b**) The individual locus can be amplified by the polymerase chain reaction. The 11 repeat units within the locus can be characterised on the basis of whether they contain (+) or do not have (−) a restriction enzyme recognition site for *Hae*III. (**c**) Patterns described in (b) become more complex if both alleles are studied. A ternary code can be derived on the basis of both alleles having a *Hae*III site (code = 1); neither allele having the restriction site (code = 2) and one of two alleles with the *Hae*III site (code = 3). □ = restriction site in the repeat is absent; ■ = restriction site in the repeat is present. The complexity of the ternary code becomes unlimited as more minisatellite loci are characterised. The bar code does not involve sizing of DNA fragments.

(=*Hae*III does not digest). One allele at this locus might have a sequence such as 'aaatta' etc. and the second allele could be 'attaat' etc. When the two alleles are measured together following locus specific amplification by the polymerase chain reaction, a ternary bar code could be constructed on the basis of whether there is 'a' or 't' or both 'a' + 't'. The bar code would read 1 = a, 2 = t, 3 = a + t (Fig. 8.6).

Minisatellite variant repeat maps do not involve an actual sizing of individual bands. Thus, this bar code approach has considerable appeal in forensic typing. More complex codes can be obtained by characterising other minisatellite loci. The contribution of individual alleles could be determined by undertaking a family study to distinguish the two parental codes. DNA amplification by the polymerase chain reaction will be required and so even degraded forensic samples could be used.

The issue of contamination and its effect on the bar code remains to be resolved. Mixtures of DNA will not be a problem. These could be reconciled by determining the victim's DNA bar code and then 'subtracting' this from the mixture profile to identify the suspect-specific DNA bar code. The power of this technique is illustrated by the finding that 326 of 347 alleles tested by Jeffreys and colleagues in one study were different. The theoretical calculation made from this is that 10^{15} different allelic states could be distinguished! This would make population-specific data (as is required for fragment size comparison) less of a limiting factor in overall interpretation.

Digital DNA typing will need to be evaluated carefully in the context of the forensic scenario. Because fragment sizing is no longer necessary it will avoid many of the criticisms and faults of the current DNA typing procedures. Interlaboratory comparisons will also be possible which would enable a more objective quality assurance programme to become available. Just as dermatoglyphic fingerprints revolutionised forensic practice it is likely that DNA barcode profiles will be included in computer databases to provide evidence of an individual's presence at the scene of a crime. The DNA information does not indicate that an individual is guilty, it is simply *evidence*, the value of which will be enhanced by the care and the way in which it is obtained.

CONCLUSIONS

The law and science can at times appear to have differing standards. For example, a legal justification for utilisation of DNA testing, despite the possibility of errors resulting from the technology,

Table 8.2 Some advantages and disadvantages of DNA technology in the identification of evidentiary samples

Advantages	Disadvantages
Variability from minisatellite fragment size characterisation makes both exclusion *and* positive identification possible	Minisatellite fragment size estimations are difficult* (co-migrating bands, size calculation, population-specific frequencies, band shifts caused by analysis conditions)
Small amounts of evidentiary samples will yield adequate DNA for typing, especially if the polymerase chain reaction is used	Degraded DNA will be unsatisfactory for Southern blot analysis but still useful for the polymerase chain reaction once the problem of contamination can be resolved.

* the recently described minisatellite variant repeat mapping may solve the methodological problem

is based on the fact that such errors are statistically unlikely to convict an innocent person but if they occur the probable end-result will be the exclusion of a guilty party. Whilst this is a comforting thought, particularly if capital punishment is involved, it is not scientifically sound.

Developments in molecular medicine are leading to important changes in the forensic sciences. Just as has been described in the previous chapters, the utilisation of recombinant DNA technology will need to be applied appropriately and with foresight to prevent problems some of which are avoidable. Dermatoglyphic fingerprints are an integral component of the legal system. Whether an individual's fingerprints are in fact unique to that person has not been determined by the rigorous scientific standards which are demanded today. There remains a lot to be learnt and developed in relation to DNA fingerprinting which will play an important role in the forensic laboratory of the future (Table 8.2). The ethical and privacy issues emerging from this technology are also considerable and will be discussed further in Chapter 9.

FURTHER READING

Farley M A, Harrington J J (eds) 1991 Forensic DNA technology. Lewis, Michigan, USA

Jeffreys A J, Brookfield J F Y, Semeonoff R 1985 Positive identification of an immigration test-case using human DNA fingerprints. Nature 317: 818–819

Jeffreys A J, MacLeod A, Tamaki K, Neil D L, Monckton D G 1991 Minisatellite repeat coding as a digital approach to DNA typing. Nature 354: 204–209

Lander E S 1991 Research on DNA typing catching up with courtroom application. American Journal of Human Genetics 48: 819–823

Neufeld P J, Colman N 1990 When science takes the witness stand. Scientific American 262: 46–53

ETHICAL AND SOCIAL ISSUES

INTRODUCTION

Ethics can be defined as: (1) a system of moral principles by which human actions and proposals may be judged good or bad, right or wrong; (2) rules of conduct recognised in respect of a particular class of human actions (Macquarie dictionary).

The fundamental principles of good medical care are also the basic principles of ethics: *beneficence* or 'do good' and *non-maleficence* or 'do no harm'. These have become very significant particularly with the rapid advances in medical research which have occurred since the Second World War. In 1964, the Declaration of Helsinki made recommendations which were meant to guide physicians in their conduct of biomedical research on human beings. These were intended to improve diagnostic, therapeutic and prophylactic procedures which of necessity would sometimes be combined with professional care. The original Declaration of Helsinki has since been reviewed and amended by a number of World Medical Assemblies, the last being in 1989.

Moral values which might be utilised to clarify ethical issues are many. Of particular relevance to medicine (including molecular medicine) are: beneficence, autonomy, confidentiality and justice. *Beneficence* refers to the obligation to do good as well as the avoidance or removal of harm. The ethical principle of *autonomy* arises from respect for the individual and the recognition that the person has the right to truthful information about his/her clinical condition, his/her options for treatment and the opportunity to participate in the treatment decision. If the process is experimental, the subject needs to understand this and consent freely. Professional *confidentiality* and *privacy* are important ethical principles related to autonomy. Confidentiality and privacy, as will be illustrated, can lead to difficult dilemmas in relation to the types of strategies required for DNA diagnostic testing. The principle of *justice* comes into consideration in the distribution of resources. Included in this is equal access to health services and the avoidance of unnecessary or wasteful procedures. In terms of research, the latter principle would require useful, high quality research and the application of new knowledge in a way which is most valuable clinically but not necessarily the most profitable.

Because genes and their genetic information are integral to the human being, developments in molecular technology have attracted the attention of the public and have been monitored by a number of statutory bodies and ethics committees. Input into these committees from laypersons has enabled more broadly based assessments of protocols and options. To date, there have been few instances of unethical behaviour by physicians and scientists in the area of genetic engineering. Careful monitoring needs to be continued to ensure that this is maintained. On a broader front, communities have followed established and accepted principles for the conduct of genetic engineering. However, there are examples which illustrate how unethical practices can develop in medical and scientific research. There is no reason why the above would not apply to genetic engineering if the advantages to be gained were perceived to be worthwhile. The risk of this happening would be less if the community were both well-informed and involved in decision making.

Training in ethics for health care professionals and scientists does not comprise a significant component of many curricula. This is unfortunate since situations which have both ethical and social implications are likely to develop, particularly in the area of molecular medicine. For example, an expanding pool of knowledge about human DNA is already placing some pressure on available resources. The ethical and social implications of this will only increase as the Human Genome Project (see Ch. 10) gains momentum. Unfortunately, the Human Genome Project as an example of scientific endeavour is being matched in many communities by shrinking resources which will be further depleted to meet the needs of high technology research.

The innovative developments which are possible through genetic engineering need to be balanced by their potential ethical and social implications. Professional knowledge and expertise carries with it the responsibility for the careful weighing of *risk and benefit* for the individual. Therefore, while the degree of benefit may not always be predictable, at least there should be no harm to emerge. The predictable risk should never outweigh a possible benefit. A number of issues which have potential ethical and social implications are summarised in this chapter. They illustrate the present situation as well as making predictions about the future. As has been shown consistently in the previous chapters, the advances in the field of molecular medicine will be substantial. Accompanying these developments will be questions of ethics and the effects of these changes on society as a whole.

MOLECULAR MEDICINE AND CLINICAL PRACTICE

DNA and diagnostic testing

There are several features of DNA which are relevant to diagnostic testing. They include: human tissue for study is easy to obtain since DNA is identical in all cells and at all times during development. The range of diseases which can be detected will be increasing in the foreseeable future as more DNA probes are described. Finally, DNA diagnosis does not require a particular gene to be identified before its presence can be sought. Therefore, DNA has a certain versatility which makes it ideal for genetic testing. On the other hand, this can lead to unnecessary or inadequate testing which produces more harm than good.

Presymptomatic DNA testing

Huntington disease provides a useful paradigm of what is potentially good and bad in molecular medicine. Until DNA tests became available in 1986, the individual who was at-risk for this disorder needed to wait until about his/her 4th decade before the first clinical features became apparent. The gene for Huntington disease was first localised to the short arm of chromosome 4 although the gene itself was not isolated for another decade. Early evidence suggested that there were very few mutations leading to Huntington disease. It had even been proposed that all cases had arisen from the one founder.

Until the gene for Huntington disease had been isolated, an indirect approach to diagnosis was required. This, and the apparent lack of heterogeneity favoured the use of a DNA linkage analysis strategy to detect the mutant gene within the context of a family study (see Box 3.6). From this, the known a priori risk of 50% for an offspring of an affected individual could be lowered to 1–5% or raised to 95–99% depending on whether the individual had inherited a detectable DNA polymorphism marker co-segregating with the normal or mutant clinical phenotypes respectively. The risk estimates given include a conservative 1–5% error rate which reflects the recombination potential for the DNA markers which are in linkage with the putative Huntington disease gene. The 'good' or positive side to DNA testing in Huntington disease is that the known risk for an individual can be reassessed at any time in his/her life. With this knowledge, the individual can make informed decisions about reproduction, life-style and other key issues. Prenatal diagnosis is possible should a couple wish to have children who will be at low risk.

The negative aspect of the DNA technology reflects the way in which testing for Huntington disease is undertaken. For linkage analysis, it is essential to study a number of family members. Key individuals are particularly important since it is these who will allow linkage between a clinical phenotype and a DNA marker to be made. In the context of Huntington disease, a key individual would be one who has uneqivocal evidence for the disorder. Less helpful would be a person who is sufficiently old to enable a confident diagnosis of 'not affected' if the clinical examination is normal (see Ch. 3).

There are a number of ethical issues which can arise in the circumstances of a family study, particularly one which may need to be extended to include distant relatives. Confidentiality will invariably become a concern. For example, even the first step which involves construction of a pedigree is undertaken without the consent or knowledge of family members. Disclosure of such a pedigree may lead to key individuals being coerced by others into supplying information or giving blood. This could even reach the courts of law with an uncooperative relative being threatened by legal action. At present, the example given above produces no major dilemmas since the unwilling participant would not be compelled by the courts to give blood or have his/her privacy infringed by the enforced disclosure of personal details. Nevertheless, it should be kept in mind that some communities have already legislated for compulsory blood testing in certain circumstances, e.g. the determination of blood alcohol levels after road accidents. Whilst the two situations bear little resemblance to each other, it would not be inconceivable in the future, particularly after DNA testing had become more firmly established as a routine procedure, that an individual might influence a judge or court to create a precedent on the grounds that good for the majority (the individual and his/her family) should outweigh what might appear to be trivial concerns by one other person.

What about the key individual in a family who is demented with Huntington disease? He/she would be unable to give informed consent for study. This situation can be overcome by permission on his/her behalf from a spouse or court appointed legal guardian. However, the latter may be placed in a predicament since authority for an invasive procedure to be conducted on his/her ward may only be possible if it is in the ward's best interests. Does giving blood to benefit a relative fall within this guideline?

The ethical dilemma which arises as a consequence of too much information becoming apparent through DNA testing can also become an issue. An example is non-paternity. This may be put aside and not disclosed in some circumstances to comply with the principle of non-maleficence. However, what if the non-paternity then excludes the consultand or other family members from being at-risk? Autonomy and beneficence in respect to the consultand must be balanced by confidentiality and privacy of a family member. Will the ethical dilemma be made easier if the latter is deceased but his/her spouse is alive?

Monozygotic twins at-risk for Huntington disease have been used to illustrate another potential predicament. This would occur if one of the pair wanted to have predictive testing and the other declined. However, because they are monozygous, a result from one would automatically apply to the other. The first individual has the right of autonomy, the second is entitled to his/her privacy. Involving both, is the principle of beneficence. In one particular Huntington disease testing programme, this potential dilemma has been considered and resolved by linking the testing to both or none, i.e. beneficence in this circumstance overrides autonomy.

Does the researcher who is studying the molecular basis of genetic disorders have specific responsibilities? As indicated previously, early studies of Huntington disease suggested a limited, perhaps even a single, mutation for this disorder. There is now doubt that this may be correct. A definitive answer will come now that the actual gene(s) has been cloned. Fortunately, there is still good evidence that *all* Huntington disease is related to chromosome 4. In comparison is adult polycystic kidney disease which is likely to involve at least two loci with the second only becoming apparent nearly 3 years after DNA testing for this disorder had started (see Ch. 3). The difficult question therefore arises: when does research end and routine clinical service begin? This will be asked more frequently as other DNA probes become available. Pressure from peers or consumer groups will make it more difficult to 'postpone' changing what is still in effect a research study into the more readily available routine diagnostic service. In the above circumstances, the researcher has the onerous task of ensuring that data are of sufficiently high standard that appropriate long-term decisions will become possible.

The Huntington disease example is but one involving the presymptomatic testing of individuals at-risk for a genetic disorder. A number of the ethical issues will be solved following identification of the Huntington disease gene when direct detection for mutation(s) will be possible. In the meantime, to reduce potential harmful aspects of DNA testing in this disorder, the International Huntington Association (lay group) and the World Federation of Neurology (medical group) issued a formal ethical issues policy statement in 1989. DNA markers for other genetic disorders will become available on a regular basis for many years to come and similar ethical issues to those described for Huntington disease will continue to occur.

Prenatal diagnosis

The increasing scope for presymptomatic testing is being matched by the number of prenatal diagnoses which are being undertaken. Prior to the availability of chorion villus sampling, prenatal detection of genetic diseases was only possible during the 2nd trimester of pregnancy. Delayed results meant a difficult termination of pregnancy should this be requested. Today, 1st trimester diagnosis by chorion villus sampling makes termination of a pregnancy less traumatic and increases the number of disorders detectable since DNA can be isolated with relative ease from chorionic tissue. The above, as well as the increasing availability and commercialisation of DNA diagnostic kits, has meant that the decision whether or not to undergo prenatal diagnosis and so perhaps terminate a pregnancy is being faced by an increasing number of couples. The US Congress has recently expressed concern that the proliferation of laboratories offering DNA genetic testing is placing, in some circumstances, unnecessary pressure on individuals to make decisions about their genetic status.

The ethical and social implications of terminating a pregnancy have not been discussed but these will also influence a couple's final decision. In this respect it is sobering to read what individuals who have the severe form of β thalassaemia think about prenatal diagnosis. In one study, all stated that they wished to have children of their own. The great majority indicated their willingness to undergo prenatal diagnosis for β thalassaemia should their spouse also be a carrier. Less than half indicated that, taking into consideration the constraints placed on their life, they were still happy to have been born. The remainder expressed a wish that their parents should have undergone prenatal diagnosis and termination of pregnancy if necessary. Thus, there is no easy solution from the patients' perspective. The dilemmas facing the couple can be even more difficult since there is often no first-hand experience of the genetic disorder being tested.

Counselling

The emphasis these days is on non-directive counselling. Thus, individuals seeking advice because of a family history of genetic disease or a couple in the prenatal diagnosis situation, will need to make decisions based on the information they are given. However, the rapid advances in molecular genetics can complicate counselling. The following illustrates how information gained from molecular genetics can make it more difficult to answer a question which is frequently asked, i.e. how severely affected will an offspring become if he/she has a particular genetic disorder? In β thalassaemia, the molecular basis for the milder, non-transfusion form called 'thalassaemia intermedia' has been determined in some cases. However, other factors, which are as yet undefined, are also involved. Has knowledge based on molecular analysis reached the stage where co-inheritance of β thalassaemia with one such modulating change (e.g. an increase in fetal haemoglobin or co-existing α thalassaemia) is sufficiently well understood to enable a confident prediction of severity? Similarly, therapeutic options, particularly those related to molecular medicine, are rapidly changing. For example, gene therapy in cystic fibrosis and the availability of an oral iron chelating agent in thalassaemia will become future additions to treatment. These should improve the outlook for both disorders and would need to be considered by the consultand(s).

A greater demand is also being placed on the counsellor. There will be many who are unable to provide a comprehensive counselling session which is based on a wide range of issues including the clinical consequences, recent advances in molecular medicine and potential future developments. Despite this, the number of genetic counsellors has increased considerably with the emphasis, in some cases, relating to the way in which the information is imparted rather than the actual content. The complexities associated with DNA testing, in particular, require a team approach, since it is unlikely that a single individual will have the breadth and depth of knowledge required to understand *all* genetic disorders and then convey this information to the consultand in a way which will be comprehensible to the latter.

Population screening

The utility of DNA testing strategies, particularly the potential of the polymerase chain reaction to test many samples quickly and relatively cheaply, has meant that widespread screening of the population becomes a practical consideration. As more genes are sequenced, the number of diseases identifiable by the polymerase chain reaction will increase. There are many types of screening programmes which can be implemented (Box 9.1).

Two contrasting examples of *selective screening* programmes targeted to at-risk populations and implemented before the DNA era, illustrate the potential advantages and disadvantages of this type of testing. *Tay Sachs disease* is a fatal neurodegenerative disorder of childhood. It is inherited as an autosomal recessive trait. Since the early 1970s, individuals at-risk for Tay Sachs have been screened and counselled. The incidence of this disorder has been reduced without the problems associated with population screening illustrated by *sickle cell disease*. This is also transmitted as an autosomal recessive disorder. It leads to considerable morbidity and mortality although the ultimate outcome is not entirely genetic in origin as environmental factors may play a part. The US-based sickle cell screening programme, which was also started in the early

Box 9.1 Population screening strategies

There are a number of different screening strategies possible. *Selective screening* where individuals at increased risk for a genetic disorder are targeted. DNA testing is very useful in this circumstance since a mutation known to be present in a family can be sought. Alternatively, DNA polymorphic markers enable a linkage approach, within the constraints of a family, to be followed.

Mass screening describes the testing of entire population groups usually looking for a common disorder. An example of this is cervical cancer checks. To be effective, a large percentage of women over the age of 45–50 must participate and the test needs to be sufficiently sensitive and specific to make the programme cost-effective.

Epidemiological surveys often utilise mass screening-type approaches to ascertain the prevalence and incidence of a particular disorder. Important ethical considerations in setting up screening programmes include the ability to do good by early detection, i.e. the disorder is potentially serious and there is effective treatment available. Privacy and confidentiality issues will arise since the information gathered could be acquired by third parties.

Occupational screening, genetic DNA screening in the workplace, can have two different aims. The first identifies individuals at greater risk for industrially-related complications. For example, the combination of α_1-antitrypsin deficiency in a worker and a dusty workplace would be more likely to lead to a chronic lung disease. The second utilises DNA technology to detect DNA damage which is related to the workplace. This would provide objective data in terms of cause and effect. Although screening at the workplace has merit, a potential criticism would be that emphasis is being placed on exclusion of the at-risk worker rather than attempting to make the workplace a safer environment.

1970s, was targeted to the at-risk black population. The initial version of this programme produced more harm than good. Results led to a lowering of the self-image, overprotection by parents and discrimination. The discrimination came from employers, insurance companies and potential spouses. What were some of the differences between these two programmes? One important consideration in the success of Tay Sachs screening was that the target group comprised individuals of Jewish origin who had, in contrast to the black population, better *educational* opportunities. Thus, the necessity for counselling and public education to explain the significance of mass screening and the results obtained are important ethical considerations. Subsequent changes to the sickle cell screening programme, which removed the legal compulsion to screen and improved counselling/education facilities, enabled more successful testing to be pursued.

The modern screening dilemma involves cystic fibrosis. DNA amplification enables the common ΔF508 defect to be detected in peripheral blood or more accessible tissues such as the buccal cells present in a 10 ml mouth wash specimen. As summarised in Chapter 3 (Medical genetics) there are over 200 mutations which produce cystic fibrosis although the ΔF508 is the most common amongst individuals of northern European origin. Other mutations are much less frequent (Table 9.1). Debate continues as to the value of *mass population* screening in contrast to testing individuals or families at-risk i.e. *selective screening*. The ethical and social issues raised include the use of a test which will not detect all affected

Table 9.1 Ethnic distribution of the ΔF508 mutation of cystic fibrosis (Tsui, Buchwald 1991).
There are over 200 mutations which produce cystic fibrosis. Mutations other than ΔF508 occur less frequently and will not be detected in a screening programme for ΔF508

Population	ΔF508 mutation in cystic fibrosis chromosomes (%)
Northern European	70–75
Italian	41
Greek	56
Turkish	30
Yugoslav	54
Ashkenazi Jewish	40

individuals. For example, if only the ΔF508 mutation is sought, false negative results in couples from a population with a frequency for the ΔF508 mutation of 70% will be $(1-(0.7 \times 0.7)) = 0.51$, i.e. approximately half the couples will not be informative by this approach. The detection of the less common mutations (some of which are only present at a 1–2% frequency in the population) will add to the workload but not increase substantially the information to come from the screening programme. Even if laboratory facilities are available, major efforts directed towards genetic counselling and public education will be required to ensure that those tested fully understand the implications of the results. The financial resources to carry out a mass screening programme will be enormous. In view of this and the inability to detect all mutations with present technology, most national genetic societies have recommended limited or selective screening of groups who are at higher risk than the general population.

Additional problems which need to be resolved before embarking on a widespread cystic fibrosis screening programme include the uncertainty of the mutations in respect to disease severity. Thus, counselling will be difficult and incomplete. Potential therapeutic developments (including gene therapy) may alter the affected individual's quality of life and survival and need to be considered in the overall equation. Data storage will have to be both accessible and at the same time secure to ensure confidentiality.

Both employers and insurance companies are becoming increasingly more conscious of the potential for DNA screening. There are a number of reasons given to justify this including the comment that failure to identify those at-risk will unfairly discriminate against others in the workplace or policyholders of an insurance company. To the individual, this type of screening has the advantage of detecting potentially reversible problems early or allowing him/her a longer disease-free period by preventing exposure to industrial toxins which are particularly noxious to an individual with his/her genetic constitution. On the other hand, the disadvantages are that discrimination, loss of privacy, loss of health benefits and even loss of employment can follow.

Testing children

One important advantage of DNA testing is that it can be undertaken at any age. However, this can lead to an ethical dilemma which is centred around the investigation of otherwise healthy children. In these circumstances informed consent is given by their parents. Why are children being tested? An acceptable indication is a childhood disorder for which treatment or prognosis can be improved if diagnosis is made early. Dubious indications include the options for planning of future educational, career or even reproductive decisions *by the child*. In these circumstances, it is difficult to know what component reflects the parental wishes, what directly relates to the child and whether it would be more appropriate to test at a later age when the child himself/herself would be in a better position to make an informed choice. In the end, it is the parents who will make the request for DNA testing. Whether this is to relieve anxiety on their part or a legitimate medical indication can be difficult to assess. However, relief of parental anxiety must be balanced by the problems of DNA testing mentioned already and the potential harm to a child whose disclosed genetic status may lead to an unnecessary change in the way he/she is allowed to develop. To avoid making difficult decisions in these circumstances, DNA programmes, for example that for Huntington disease, will not test individuals unless they themselves are able to give informed consent, i.e. they have attained the age of 16–18 years. Whilst guidelines are helpful in complex cases, some flexibility needs to be maintained to deal with those situations where DNA testing will be beneficial to the child.

Tissue storage

Human-derived tissues are stored in a number of ways. Storage may relate to 'routine laboratory practice', for example, the keeping of residual tissue sections, blood samples or even blood spots present on Guthrie cards obtained during a community neonatal screening programme. On the other hand, storage may be more formal as a DNA bank where immortalised cell lines enable tissues to be kept and then retrieved for an indefinite period of time (see Ch. 3). Tissue storage has produced additional ethical and legal considerations since DNA technology has shown that even archival samples are able to be tested. A trivial but recent example of this was the suggestion that the long-standing dilemma whether the US President Abraham Lincoln had Marfan syndrome could be resolved through a DNA amplification study of tissue fragments and blood spots taken from clothing worn at the time of his assassination. This became possible when the DNA sequence related to fibrillin, the likely candidate gene for Marfan syndrome on chromosome 15, was published. Opinions about the ethical aspects of such a study were divided. One group considered that there were no legal and ethical issues involved. Another questioned the ethics of this proposal.

Formal guidelines have been set down to define the rights of the individual who has had DNA banked. How the material is used, the anticipated condition of the material and mode of storage are clearly stated. Ownership of the material remains with the depositor. Nevertheless, dilemmas arise when material has been stored for one purpose but then another DNA test becomes available and the stored material (the depositor may be deceased at this time) would be helpful in defining the genetic status of other family members.

Ownership of biological material, e.g. frozen embryos, may also come into dispute when the involved partners have separated and there is conflict as to what should be done with the stored sample. Another practical example would be the Guthrie card. In many communities, the state has legislated that Guthrie spots will be taken and these are used to screen for a number of medical conditions. Could blood taken from an infant for the purpose of neonatal screening be used to test for other genetic diseases? The good that this can bring to the individual and/or the community needs to be balanced with the person's right to privacy. Another question is whether the Guthrie card could be subpoenaed by a court of law to resolve a legal matter?

Genetic registers and databases

The keeping of registers is an integral part of a genetics service. Registers come in various shapes and sizes. The compilation of a pedigree is one form of a register. As indicated previously, many individuals identified on the pedigree are not aware of its existence nor have given permission for their inclusion. In fact, to acquire informed consent in these circumstances is both difficult and in itself an intrusion of an individual's privacy. A further extension of the genetic register is the availability, in a central database, of a list of names or identities of individuals who have a particular type of genetic disorder. The significance of this in providing information for health planning or to assist other family members is balanced by the potential for unauthorised disclosure of such data. The privacy issue is particularly significant when third parties, e.g. employers, insurance companies or the courts of law may gain access to this information. In these circumstances, the ethical principles of autonomy and confidentiality outweigh the considerations that third parties may derive benefit or harm from such information. However, this can be interpreted in different ways and a court of law may consider the reverse holds in a particular situation.

DNA fingerprints provide a unique profile for an individual. It is likely that in future they will be used in place of the dermatoglyphic patterns which are now an acceptable routine procedure in criminal cases. However, unlike dermatoglyphics (which would not be universally available for all individuals in a community), there will be alternative sources of DNA for most, if not all, members of a community. These would have been obtained for a specific purpose, e.g. Guthrie spots for disease testing, but could now be used (or misused) for another reason.

Quality assurance

The rapid advances in molecular medicine and the negative effects these can have on laboratory practices have been illustrated by suboptimal standards practised in some forensic laboratories (Ch. 8). An important ethical issue (and a legal issue) in laboratory and clinical practice includes the obligation to ensure high standards. Quality assurance programmes have been implemented in many clinical and laboratory situations to monitor and maintain standards. On the other hand, the very rapid developments in molecular technology, particularly in respect to diagnostic laboratories, have managed in a number of cases to overtake the implementation of formal quality assurance programmes. Deficiencies in this respect may not reflect a primary reluctance by the laboratory to participate in such a programme but may result from the intense pressure to start a new diagnostic test because a probe or DNA sequence is available. Thus, expansion to increase the quantity of testing may take priority over the quality of the results. In the circumstance of increasing knowledge, as will be observed in molecular medicine for some time, a balance is essential to ensure that data provided by a laboratory are of the highest standard possible given the resources available.

Future issues

All the ethical and social implications of DNA testing raised above have centred on single-gene disorders. In these circumstances, diagnosis is relatively straightforward and the resulting phenotypes more or less predictable. However, in the not too distant future, the more complex *multifactorial and somatic cell disorders* will become recognisable at the DNA level. Thus, coronary artery disease, neuroaffective disorders, dementias and cancers will have their underlying DNA components defined. On the positive side, there is the potential to understand more fully the DNA to DNA and the DNA to environment interactions which are necessary to produce the various clinical disorders. This will lead to the implementation of more effective preventative and therapeutic measures. The negative components include the increasing gap which will emerge between what is known about many diseases and what can be done to treat them. There will also be the potential for misuse of this information by the individual, the state, industry or other parties.

Taking multifactorial genes one step further will be the DNA characterisation of genes involved in the individual's fundamental make-up including physical features, behavioural and other components such as intelligence. This knowledge is presently a long way in the future. There is also an informal agreement that 'normal traits' should in no way be the subject of genetic diagnosis. It would be interesting to see if this decision holds once the genetic components for these characteristics start to be defined. Even if scientific curiosity can be kept in check, the pressure from the public to measure these normal traits will be immense. A type of precedent, the IQ test, already exists.

Recombinant DNA and therapeutics

Gene therapy

The issue of germline versus somatic cell therapy has already been raised in Chapter 7. The former is unacceptable for whatever clinical indications. This reflects the inevitable consequence of germline manipulation in that any changes which result will be transmitted through the germ cells to offspring in subsequent generations. On the other hand, somatic cell gene therapy is considered similar to other accepted medical procedures, e.g. manipulations required for bone marrow or organ transplantation. Ultimately, it is the *risk:benefit ratio* which determines whether somatic cell therapy is medically and ethically acceptable in individual cases. Thus, a potentially life-threatening disorder for which there is no effective treatment or what is available is associated with significant complications, e.g. bone marrow transplantation, is the first prerequisite. Once the somatic cell therapy approach can be justified as technically and therapeutically acceptable, it is allowed to proceed after protocols are reviewed by the appropriate monitoring bodies and a follow-up process is set into place. The various steps mentioned above have already taken place in the example of adenosine deaminase deficiency. Informed consent, since the affected individual was a 4-year-old child, was provided by her parents (see Ch. 7).

Recombinant DNA drugs

Mention was made in Chapter 7 of the potential problem of 'hypoglycaemia unawareness' associated with human recombinant insulin but not the animal-derived insulins. Irrespective of whether this is proven to be correct, the marketing of the recombinant human insulin product is now being reviewed. Although introduced in 1982, its use 7 years later exceeded that of the animal products which on retrospection seems hasty particularly when the latter were and still are therapeutically effective. Even now, it is not clear if porcine insulin is more immunogenic than recombinant human insulin. Contributing factors to the change from animal to recombinant insulins were the marketing campaigns and a belief by physicians that animal insulins were being withdrawn. In some countries the latter has in fact occurred or the use of animal-derived insulins is restricted to specific indications.

An additional factor which is having a greater influence in drug regulation is the consumer. This has already become evident in some countries in respect to drugs which have not been fully evaluated for the treatment of cancer or AIDS. In some circumstances, governments have modified or 'fast tracked' drug evaluation procedures following external pressures. Whilst bureaucratic delays are totally unacceptable when it comes to evaluating the efficacy of a therapeutic agent, it is a concern that the long-term assessment of a drug's potential may be adversely affected by ad hoc decisions which have resulted from consumer pressure. This is of relevance to molecular medicine from which novel synthetic compounds will increasingly become available. A case in point might be the recombinant DNA-derived coagulant factors or haemoglobin-derivatives. The appropriate protocols to evaluate beneficial as well as toxic effects are essential to ensure the long-term safe usage of these products.

MOLECULAR MEDICINE AND RESEARCH

Pursuit of knowledge

The pursuit of knowledge for the sake of knowledge is becoming more difficult in today's goal-driven research which is frequently encouraged to have a marketable end-product. Whilst shrinking resources are in part the reason for this type of rationalisation, such a trend has the long-term potential to stifle creative or innovative work. Governments provide a variable proportion of the research funds and so demand some input into the types of work being undertaken. Thus, in subtle ways the state is able to influence the direction that research will take. Whilst this is a reasonable request on the part of government, it suppresses the creative individuals who have made significant contributions in the past but will have difficulty thriving in a regulated environment.

Similarly, there would be more than one Nobel prize winner who would not be competitive on current research funding allocations which are based on a strict process of peer review. On the positive side, peer review ensures that there is some form of justice when it comes to distribution of grants and ensures that the money has been spent 'wisely'. On the negative side, the type of research proposals submitted are frequently designed with short-term goals in sight so that the next round of peer review will be successful.

The molecular medicine era is one of the most exciting in the modern history of medicine. Many developments have occurred at a time when there have been significant shifts in the way in which research is being undertaken. The free-thinking or hypothesis driven research is giving way in some circumstances to goal driven activities which have attracted funding because of perceived commercial benefits. The Human Genome Project illustrates another approach in which a mammoth undertaking will be resolved by technological blitzkrieg. The effects that these shifts in strategies or emphasis will have on society through future developments in medicine will take time to assess.

Patents and ownership of intellectual property

The medical and scientific community is increasingly being encouraged to derive benefits from the patenting of important discoveries. To date these have been therapeutic substances or technologies. More recently, genes and their products have become the subject of patents. Thus, DNA diagnostic kits utilising a particular DNA sequence for amplification by the polymerase chain reaction will involve royalty payments to cover both the patented DNA sequence and the polymerase chain reaction technique. Initially, the patenting of genes or DNA sequences raised concern that there would be a reduction in dissemination of information throughout the scientific community. To date, this has not occurred to any significant extent. However, a US based group at the National Institutes for Health recently surprised the scientific community by filing patents for over 2000 human brain-derived genes. These 'genes' were isolated from a brain cDNA library and their uniqueness demonstrated by sequencing a segment of the cDNA and showing on DNA database searches that the sequences were not present in the databases. Thus, they represented unique DNA segments which, since they come from a cDNA library, were likely to be genes of hitherto unknown function.

The controversy arising from the cDNA clones described above reflects the philosophy behind a patent, i.e. a novel idea or invention which has some utility. Since the above cDNA clones have no known function their utility is difficult to assess. Various governments and groups of scientists on both sides of the Atlantic were drawn into this controversy which was eventually defused when the patents application was rejected. However, the example highlights a potential ethical issue about scientific practice and how knowledge will be pursued and shared in the future.

MOLECULAR MEDICINE AND GOVERNMENT

Allocation of resources

The rapid advances in medical knowledge are starting to produce ethical dilemmas even in the affluent societies. Information about the human genome is expanding rapidly in terms of the single gene disorders but is still in its infancy for the more complex multifactorial conditions. The issues of what are priorities will need to be addressed. Health planners are demanding more accountability and resource allocation based on value for money. In this context the moral principle of justice applies to ensure that access is equitable. On the other hand, to achieve value for money and efficiency, it is essential that the other principles of autonomy, confidentiality and privacy are not discarded or manipulated.

Consumer pressure and lobbying will play an important role, with politicians being influenced considerably by the perceived gains or losses from making such decisions. It is essential that the public are exposed to complete information which they can understand. Therefore, educational processes, particularly at the school level, are a priority so that the full medical, ethical and social implications of new developments (in this case relevant to genetic engineering) can be appreciated by the majority and not the vocal minority groups. Informed community input involving ordinary human wisdom is required to play a role in the direction or application of research and clinical medicine.

Embryo research

The ethical and social issues concerning termination of pregnancy have, with recent technological developments, become relevant to manipulation or research into the embryo or pre-embryo. The pre-embryo is defined as the stage of development from fertilisation until the product of conception becomes implanted in the uterus (approximately 14 days after fertilisation). In this circumstance, the key issue is what constitutes an individual and when during development does an individual become a distinct entity. There are two commonly held views in this respect. An individual becomes a discrete entity at the time of fertilisation. In this case, embryo research would be forbidden since it would have the potential to interfere with the individual's right to life. Another opinion is that an individual cannot exist until about 14 days after conception by which time the primitive streak is being formed. Prior to that time it is possible for the embryo to split and so produce identical twins, i.e. it is not a distinct entity. Based on the above and a number of other arguments, research involving the embryo is allowed in a number of countries until 14 days post-conception. There are additional safeguards imposed, e.g. research is only permitted in the case of spare embryos obtained at in vitro fertilisation which would otherwise be destroyed; there is a prohibition on transfer of embryos to the uterus of any species once they have been genetically manipulated.

Monitoring committees

A number of ethical and biosafety committees are in place to monitor activities in genetic engineering. They include the local or institutional committees, the statutory State-based committees and various ad hoc committees which are formed to review and examine topical or sensitive matters. Membership of these committees is usually designed to ensure that there is an appropriate mix of professionals, consumers, ethicists and other interested parties. The work done by these committees, which is often voluntary, has been an important factor in the smooth progress of molecular medicine to date.

FURTHER READING

Fost N 1992 Ethical issues in genetics. Pediatric clinics of North America 39: 79–89

Harper P S, Clarke A 1990 Should we test children for 'adult' genetic diseases? Lancet i: 1205–1206
Huggins M, Bloch M, Kanani S et al 1990 Ethical and legal dilemmas arising during predictive testing for adult-onset disease: the experience of Huntington disease. American Journal of Human Genetics 47: 4–12
Recommendations of European Medical Research Councils 1988 Gene therapy in man. Lancet i: 1271–1272
Richards J R, Bobrow M 1991 Ethical issues in clinical genetics. Journal of the Royal College of Physicians of London 25: 284–288
Schiliro G, Romeo M A, Mollica F 1988 Prenatal diagnosis of thalassemia: the viewpoint of patients. Prenatal Diagnosis 8: 231–233
Statement of the American Society of Human Genetics 1992 Cystic fibrosis carrier screening. American Journal of Human Genetics 51: 1443–1444
Tsui L-C, Buchwald M 1991 Biochemical and molecular genetics of cystic fibrosis. In: Harris H, Hirschhorn K (eds) Advances in human genetics. Plenum, New York, p 153–266
Walters W A (ed) 1991 Human reproduction: current and future ethical issues. In: Clinical obstetrics and gynaecology. Baillière Tindall, London, vol 3

10

MOLECULAR MEDICINE AND THE FUTURE

Human Genome Project

Introduction

It is estimated that the human genome consists of approximately 50 000 to 100 000 genes located on the 23 pairs of chromosomes. To date only 2000–3000 of these genes have been mapped to specific chromosomes. A list of some genetic disorders for which DNA markers are available and their chromosomal locations is given in Table 10.1. The potential to map and sequence the 3.3×10^9 base pairs which make up the human haploid genome was first discussed in 1986–7. Benefits from such a project would be in many areas including clinical medicine, biological research and biotechnology. Following extensive scientific and public discussion the US Congress launched the Human Genome Project in 1988. It is estimated that it will cost 3 billion dollars and the proposed completion date is the year 2005. The project officially started in October 1990 and will be conducted in many international laboratories. It is described as making available the source book for biomedical science in the 21st century. Coordination and central planning will be undertaken by the US National Institutes of Health and the US Department of Energy.

Table 10.1 An abridged list of genetic disorders for which there are DNA markers available
Of the 50 000–100 000 genes only 2000–3000 have been mapped to specific chromosomes

Chromosome	Disease (chromosome arm)
1	Congenital adrenal hyperplasia (p), Gaucher disease (q), Charcot–Marie–Tooth type 1B (q)
2	Ehlers–Danlos syndrome IV (q), Waardenburg syndrome (q)
3	von Hippel–Lindau (p)
4	Huntington disease (p)
5	Familial polyposis coli (q)
6	Haemochromatosis (p)
7	Glucokinase (p)*, cystic fibrosis (q)
8	Langer–Giedion syndrome (q)
9	Friedreich ataxia (q), hereditary fructose intolerance (q)
10	Multiple endocrine neoplasia 2A,2B (q)
11	β Thalassaemia (p), multiple endocrine neoplasia 1(q), Wilms' tumour (p), Beckwith–Wiedemann syndrome (p)
12	von Willebrand disease (p), phenylketonuria (q)
13	Retinoblastoma (q), Wilson disease (q)
14	Familial hypertrophic cardiomyopathy (q), α_1-antitrypsin deficiency (q)
15	Prader–Willi and Angelman syndromes (q), Marfan syndrome (q)
16	α Thalassaemia (p), adult polycystic kidney disease (p)
17	Neurofibromatosis 1 (q), Miller–Dieker syndrome (p), Charcot–Marie–Tooth 1A (p)
18	Erythropoietic protoporphyria (q)
19	Myotonic dystrophy (q), malignant hyperthermia (q)
20	Fanconi anaemia (q)
21	Early onset familial Alzheimer disease (p-q), amylotrophic lateral sclerosis 1 (q), Down syndrome locus (q)
22	Neurofibromatosis 2 (q), diGeorge syndrome (q)
X	Duchenne muscular dystrophy (p), retinitis pigmentosa (p), Norrie disease (p), haemophilia A, B (q), agammaglobulinaemia and severe combined immunodeficiency (q), fragile X (q)
Y	XY sex reversal (SRY gene) (p)

p = short arm and q = long arm of the chromosome. Additional disease loci are not excluded in some of the above examples
* Linked to maturity onset type diabetes of the young

Components of the Project

The Human Genome Project has seven distinct programmes which are summarised in Table 10.2. The first program involves the construction of comprehensive genetic and physical maps of the human genome. *Genetic maps* have been described earlier and are obtained by using polymorphic DNA markers within the context of family studies. Polymorphisms include the **r**estriction **f**ragment **l**ength **p**olymorphisms (RFLPs), the **v**ariable **n**umber of **t**andem **r**epeat polymorphisms (VNTRs) and the microsatellites. Examples of microsatellites are the $(AC)_n$ repeats. The number of $(AC)_n$ repeats available for genome mapping has now reached over 400 although these polymorphisms were only described a few years ago. Details of the microsatellites are being systematically stored in the Genome Database (GDB) which is sponsored by the Howard Hughes Medical Institute and the Johns Hopkins University (Baltimore). Linkage between a DNA marker and a genetic disorder or trait will provide a starting point from which the associated gene can be isolated and characterised.

Table 10.2 Seven components of the Human Genome Project and the goals for the years 1991–1995.
The Project is planned to cover the period 1990–2005 with goals divided into 5-year groups

Programme	Goals for the first 5 years
1. Mapping and sequencing the human genome	a) Construct genetic maps with DNA markers at 1 cM distances b) Construct physical maps with 2 Mb contigs and 100 kb spaced STSs c) Sequence small defined regions of the genome. Develop existing technology and encourage innovative approaches to sequencing
2. Mapping and sequencing genomes of model organisms	a) Construction of genetic, physical maps of the mouse genome b) Sequencing of the genome of the mouse and selected bacteria, yeast, plant and the fruit fly
3. Informatics: data collection and analysis	a) Develop software and database designs to support large scale collection, storage, distribution and allow ready access to databases b) Develop new methods for interpretation and analysis of genome maps and DNA sequences
4. Ethical, legal and social considerations	Issues of privacy, confidentiality, stigmatisation, discrimination, knowledge of future health/ill health in an individual will be researched in terms of their interactions with the Project
5. Research training	Fellowships created , training courses provided
6. Technology development	Develop current approaches and encourage novel ones
7. Technology transfer	Close liaison and interaction with industry and the private sector. Technology transfer to the medical community

The text describes the long-term aims of the Project. Contigs are overlapping sets of clones, STSs are unique **s**equence **t**agged **s**ites used to identify each DNA marker

The distance between markers on a genetic map is defined in terms of a centiMorgan (1 cM is very approximately equivalent to 1 Mb). At present, genetic maps are available for a number of defined loci throughout the human genome. The aim of the Human Genome Project is to refine this so that the entire genome is covered by DNA markers which are 1 cM apart. Each of the DNA markers generated will need to have an unambiguous identity and for this the concept of **se**quence **t**agged **s**ites (STSs) has been proposed. This means that sequencing of DNA markers will be required. Each marker will then be identified by the part of its sequence that is unique to that particular marker. Thus, data emerging from the work of different groups and the use of different technologies can be easily accessed and compared. The technology to undertake this is available but the amount of work required is considerable.

From a genetic map it will be necessary to construct *physical maps* so that the distance between the DNA markers can be determined in absolute terms (i.e. base pairs) and a restriction map can be constructed for each locus. This is a mammoth undertaking since it will be necessary to have entire regions of the genome characterised in terms of overlapping clones. Generation of clones in the form of libraries is a routine procedure. The difficult task is obtaining sufficiently large overlapping sets of clones (called 'contigs'). At present, contigs stretch over a relatively small distance (perhaps a few hundred kilobases). In the first 5 years, it is proposed to increase the contig distance to 2 Mbs. DNA markers identifed by sequence tagged sites will be located approximately 100 kb apart. Subsequently, the aim of the project is to generate cosmid or yeast artificial chromosome (YAC) contigs which span the entire genome.

Another goal of the first programme will be to improve the current methods for DNA sequencing and encourage the development of novel approaches. Defined regions of the genome will be sequenced but the actual sequencing of the *entire* genome will be deferred until technology enables it to become more efficient and less costly (e.g. one goal would aim to reduce the present US $2–5 estimated for each base pair sequenced to less than $0.5). It is highly likely that when the time comes for the entire genome to be sequenced the methodology used will be of a form not yet described! The development of high speed robotic work stations will be an integral component of this phase.

The second programme involves mapping and sequencing of model organisms. This has two ultimate aims. First, the provision of less complex genomes to facilitate technology development. Second, to enable comparative studies to be made between the human genome and those from non-human sources. This will provide information essential for an understanding of the evolutionary processes, how genes are regulated and the basis of some genetic disorders (as an example, see the discussion on Imprinting in Ch. 3). The model organisms chosen include:

1. Bacteria: *Esch. coli*, *Bacillus subtilis*, two species of mycobacteria
2. Yeast: *Saccharomyces cerevisiae*
3. Simple plant: *Arabidopsis thaliana*
4. Nematode: *Caenorhabditis elegans*
5. The fruit-fly: *Drosophila melanogaster*
6. Mammal: the mouse.

The third programme is termed informatics and will involve methods to store the data which will be generated by the Project, i.e. genome maps and DNA sequences. A considerable amount of software development will be required in terms of data storage, data accessibility and data analyses. Basic questions such as whether there will be one central database or multiple smaller but networked ones remain to be resolved.

The Human Genome Project has started despite considerable criticism and controversy. This has ranged from economical considerations to those who do not approve of broad 'shot-gun' approaches to research. Therefore, an additional (fourth) programme was included to consider the ethical, legal and social implications of the Project. 3% of the total budget has been set aside to research issues such as privacy and confidentiality (e.g. who will have access to an individual's genetic makeup); stigmatisation or discrimination (e.g. what might insurance companies or employers do with the information generated from the Project); what adverse effects might result from knowledge of the genome's sequence and so the potential to predict health or disease in an individual? The importance of the public's education about the project and its implications will also be addressed.

The fifth programme will set out to train at the pre- and post-doctoral levels individuals who will have a good knowledge of genome research methodologies. Skills resulting from this will not only be in the area of molecular biology but will include computer science, physics, chemistry, engineering and mathematics. Interdisciplinary approaches to training and skill-acquisition will be encouraged. Training and short courses in defined areas of the Project will be supported. This programme will in the long term be beneficial to industry and the private sector.

The last two programmes will concentrate on technology development and transfer. These components of the Project alone are considered to justify the money being spent since they will lead to technological developments which will have widespread use in research, industry and the practice of medicine. Development of novel methodologies will be strongly encouraged. Technology transfer to industry, the private sector and the medical community will be rapid so that discoveries from the Project can be developed to their full potential. Direct funding of private companies will be made available to expedite needed developments.

Consequences of the Project

The Human Genome Project will have far-reaching effects. Information obtained about single-gene disorders will in itself generate both good and potentially adverse effects. For example, a lot will be known about genetic diseases but the gap between what is known and what can be done in the therapeutic sense will widen. The cystic fibrosis and Huntington disease examples have illustrated some of the advantages and disadvantages to emerge from DNA technology. However, the consequences of single gene disorders will appear to be modest when information from the Human Genome Project starts to flow over into the multifactorial and somatic cell diseases. Information about human chromosomes will also be gained from the Project. Since a large proportion of morbidity in utero and in the neonatal period may be attributable to chromosomal abnormalities, further knowledge on how chromosomes divide and replicate will be extremely valuable.

Funding required for the Project will be enormous. Priorities in terms of resource allocations are likely to produce widespread ethical, social and political dilemmas. One justification for the US space programme has been the technology developments which have contributed to our daily lives. The rapid emergence of the biotechnology industry during the 1980s would suggest that this aspect of the Project will also make important contributions to its final outcome.

Completion of the last segment of DNA to be sequenced will mark the end of the Human Genome Project but the beginning of much hard work directed to 'decoding'. At this stage it will be necessary to determine the functional significance of many of the genes or DNA sequences, e.g. regulatory elements. This will be particularly challenging in complex genetic traits for which there are multigenic effects.

HUMAN GROWTH AND DEVELOPMENT

Two of the great unknowns in human biology include the way in which the mind works and how the human body develops embryologically. These complex, multifactorial traits are now being approached from a number of directions. A wide range of human neuropsychiatric disorders and mental retardation are being studied by positional cloning. One aim of this strategy is to identify the DNA mechanisms involved in abnormalities and from these understand better the normal functions of the brain. DNA is also the focus in studies of human structural development. In both these areas some interesting results are starting to emerge.

Senile dementias

It has been estimated that over $80 billion annually is spent in the US for the care of individuals with Alzheimer disease, a degenerative brain disorder leading to dementia. The disease was first described in 1907 by Alzheimer who showed the characteristic pathological changes of senile plaques and neurofibrillary tangles. The former is now known to be due to deposition of amyloid β protein, a modified form of a normally occurring neuronal protein called tau. These changes are not only found in Alzheimer disease but are present to a lesser degree in other chronic disorders of the human brain including advanced old age.

Two significant clinical observations were made about Alzheimer disease before positional cloning strategies were initiated. First, it was apparent that in some cases the disease was familial. In this circumstance, onset was often at an earlier age. Second, individuals with Down syndrome (trisomy 21) were particularly predisposed to develop this type of dementia.

In view of the above, it is not surprising that some familial forms of Alzheimer disease are now known to be associated with a gene on chromosome 21. This gene, called β-APP (β amyloid precursor protein) encodes the precursor protein for the amyloid β protein. How β-APP leads to Alzheimer disease remains to be determined. It is possible that this is a dose effect (e.g. Down syndrome where there are presumably three copies of the gene) or alternatively point mutations which have been found in this gene may be involved. Intensive work at the molecular level in Alzheimer disease is producing many results most of which are leading to questions rather than answers. Nevertheless, the positional cloning strategy in this important disorder has identified a subset of individuals whose defect lies within the β amyloid precursor protein on chromosome 21. Other chromosomal locations are being sought.

Ultimately, characterisation of the molecular defects will provide knowledge of the pathogenesis of Alzheimer disease. This will enable a more effective approach to treatment or prevention and hopefully some insight into the workings of the normal brain.

Homeotic genes

Mutations in the body form of the fruit fly *Drosophila melanogaster* which caused a part of the body to be replaced by a structure normally found elsewhere were shown in the early 1980s to involve a number of genes called *homeotic* genes. All vertebrates including the human have four homeotic gene complexes located on different chromosomes. In *Drosophila*, it has been possible to show that the physical arrangement of the homeotic genes in these complexes is identical to the order in which the genes are expressed along the anteroposterior axis of the embryo during development. In mammals, the *homeotic* genes are known as *hox* genes and are also believed to specify cell identity along the anteroposterior axis of the embryo.

Subsequently, a conserved DNA sequence was found in all homeotic genes. This was called the *homeobox*. To understand the role of the homeoboxes in development, *Drosophila* or animals with mutations which occur spontaneously or are created by techniques such as homologous recombination are being studied. The latter approach, i.e. the targeting of specific homeoboxes and then their removal or inactivation by recombination with identical DNA sequences, promises to provide important information (see also disease models, p. 214). Despite the identification of these highly conserved genes, the search for natural mutants involving the human homeoboxes has been less fruitful. Thus, the role that these genes play in human development remains obscure.

Recently, another conserved DNA sequence has been characterised in mice and other species as divergent as worms and humans. This is called the *paired box*. The relevant genes are known as **Pax** (**pa**ired bo**x**). An interesting developmental disorder in the mouse is called *Splotch*. This affects neural crest-derived components leading to the development of spina bifida and dysmorphic features including white spots on the body. It has now been shown that *Splotch* is the result of a 32 base pair deletion in the mouse's *Pax*-3 locus. The corresponding genetic locus in the human is called *HuP*2 and is found on chromosome 2q35-q37. Recently, a mutation involving the *HuP*2 gene has been identified in the human autosomal dominant disorder called Waardenburg syndrome. In this condition there is deafness and pigmentary disturbance. The two tissues involved are both of neural crest origin. Thus, a genetic defect has highlighted the role that a conserved gene plays in development of the neural crest in the human. The *Splotch* animal model also becomes available to study normal neurological development including the abnormality which leads to spina bifida. Analysis of the function of homeotic, Pax and related genes will allow further understanding of how human development is programmed.

PREDICTIONS FOR THE FUTURE

Technology

As indicated previously, the development of novel technological approaches to DNA sequencing is a priority of the Human Genome Project. In the long term, cheaper automation will facilitate DNA sequencing so that sequencing rather than specific mutation analysis will become the method of choice for the diagnosis of genetic defects. Important future developments will also include the increasing availability of DNA diagnostic kits which have non-radiolabelled DNA or RNA probes. These will be user-friendly and so applicable across a wide range of clinical laboratories. The development of newer methods to amplify DNA or the removal of the contamination problem with the present techniques will see greater use of DNA amplification across many areas including clinical practice, research and industry.

Somatic cell genetics

This now joins the single gene disorders, multifactorial disorders and chromosomal abnormalities as one of the four categories for genetic disease. The role to be played by changes in the genetic material of somatic cells is well exemplified by cancer, e.g. the familial adenomatous polyposis example described in Chapter 3 in which a series of mutations affecting somatic cells enable progression from the localised tumour stage to a malignant, metastasising cancer.

The potential for mutation at the somatic cell level has recently been highlighted by a novel molecular defect which produces *anticipation* (increasing severity or earlier age at onset of a genetic disease in successive generations). In the case of the fragile X syndrome (X-linked disorder) and myotonic dystrophy (an autosomal dominant disorder) it has been shown that anticipation is associated with a heritable and unstable DNA nucleotide triplet ($(CCG)_n$ in fragile X; $(AGC)_n$ in myotonic dystrophy) which can increase in size from one generation to another and demonstrates both *somatic instability* as well as *inheritance* through the germ cells. A new mechanism for instability at the somatic cell level has thus been identified.

The detection of specific environmentally-related DNA changes, as illustrated in Chapter 6, will enable a better understanding of how the environment can influence mutagenesis of somatic cells. Changes in somatic cells associated with ageing, autoimmune disease and congenital malformations will become fruitful areas of research.

Multifactorial disorders

The many forms of heart disease, the dementias and adult onset diabetes have both genetic and environmental components in their aetiologies. Knowledge of the former will lead to improved means to detect those who are at-risk. The environmental contributions to the multifactorial disorders will be understood better if their effects on DNA can be identified and characterised. In practical terms this will enable more effective and focused public health preventative measures. Therapeutic options will also become more specific and directed to the interactions between genes and environment.

Disease models

There are a number of animal models for human diseases. These often arise spontaneously and provide a means by which the natural history of a disorder can be followed progressively over a number of generations. Therapeutic options also can be tested prior to human trials being undertaken. Animals, unlike humans, can be manipulated experimentally and genetically. As discussed previously in Chapter 3 (genetics of hypertension) some of the more complex multifactorial diseases in humans may best be resolved by research using a genetic breeding approach coupled to DNA analysis. Over the years, inbred strains, particularly the laboratory mouse, have provided models for cancer, autoimmune diseases, infectious diseases, alcohol intolerance, hypertension and atherosclerosis. Future directions in this area will enable a particular disease to have its animal counterpart through a recombinant DNA approach.

Transgenic mice illustrate a useful way to develop animal models for human disorders. Although the gene of interest cannot be targeted to its correct locus, it is possible to insert genes which become functional. Transgenic mice produced by microinjection of DNA into pronuclei are particularly valuable for disorders associated with an excess of a protein or an abnormal product. Thus, expression of the mutant transgene will lead to the clinical phenotype.

Embryonic stem cells (ES cells) provide another approach to the transgenic mouse. Since embryonic stem cells are totipotent, they can be

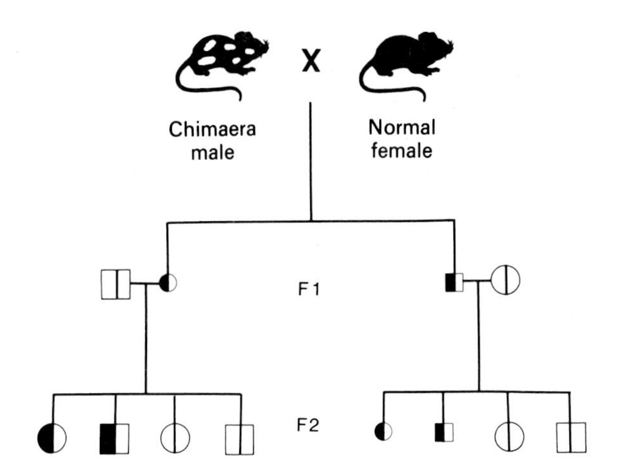

Fig. 10.1 Replacement of wild-type genes by mutant in transgenic mice.
Homologous recombination allows the gene inserted to be targeted to the correct locus. Using embryonic stem cells the gene can be integrated into the germline for future transmission to progeny. A male chimaeric mouse with one mutant insulin-growth factor-2 gene (IGF2) is made by homologous recombination using embryonic stem cells (DeChiara et al 1991). This mouse is crossed with a normal female. Progeny from the union are normal in size or small (indicated by large and small symbols). The *small* progeny result when the only normal IGF2 gene has been maternally transmitted; the *normal sized* progeny when offspring have two normal copies of the IGF2 gene or the only normal gene present is paternal in origin. This can be explained by imprinting, i.e. an active IGF2 gene requires paternal transmission.

genetically manipulated and then reintroduced into the blastocyte of a developing mouse to produce a chimaera. Foreign DNA which has become integrated into the germline of the chimaera will enable the gene to be transmitted to progeny. Appropriate matings will give homozygotes containing the transgene. The embryonic stem cell approach becomes even more powerful when it is used in conjunction with homologous recombination. Genes can now be targeted to their appropriate loci and in this way mutant genes can be used to replace their normal wild-type counterparts (Fig. 10.1).

A mouse model of cystic fibrosis has now been created by targeted disruption of its own CFTR (cystic fibrosis transmembrane regulator) gene. Mice that are homozygous for this defect demonstrate the same electrophysiological defects as seen in the human cystic fibrosis cell, i.e. in both intestinal and respiratory epithelium, the affected mice have an absent cyclic AMP-activated chloride secretory response. This type of mouse will enable a prospective assessment of tissue changes in cystic fibrosis as well as providing a model to test therapeutic substances with more objective parameters, e.g. chloride transport, able to be monitored.

An animal model will also allow a gene with unknown function to be assessed. This can be illustrated by a study of the mouse *hox-1.5* gene (the *hox* genes of mice are closely related to the homeotic genes of *Drosophila*) which was inactivated by homologous recombination. Homozygous mutants for this defect were made through the embryonic stem cell approach. The mice produced had a phenotype similar to the human DiGeorge syndrome, i.e. absent parathyroid and thyroid glands with defects of the heart, major blood vessels and cervical cartilage. In this manner, a systematic study of the various homeobox and related DNA sequences by gene 'knock-out' using homologous recombination and embryonic stem cells will enable their role in development to be determined. A similar approach will allow genetic components of multifactorial disorders to be detected.

At the in vitro level, there is presently a lot of research being undertaken to understand both gene and cell function through manipulations using antisense RNA, antisense oligonucleotides and ribozymes which can be directed to specific gene or cellular activities. In the future, it will be possible to utilise this technology with the relevant in vivo models which have been produced by recombinant DNA means.

Forensic science

As illustrated in Chapter 8, the developments in DNA fingerprinting since 1985 have been rapid. Future progress in this area will require a more reliable method for DNA amplification so that contamination, in particular, does not detract from the utility of this technology in the forensic scenario. Better strategies for fingerprinting will be developed. For example, the internal variability being measured by the digital DNA typing patterns may be more useful than comparisons between individuals within populations. In the long term, molecular medicine in the forensic laboratory will ensure that the scientific components of this discipline continue to develop.

Therapeutics

The present approaches to the production of recombinant DNA drugs and vaccines illustrated in Chapter 7 will appear to be primitive in comparison to future strategies. Knowledge about active sites, methods to improve bioavailability and delay or evade resistance will enable multifunctional drugs to be designed and then produced by recombinant DNA means. The cost and efficiency of vaccines will be improved by genetically engineering vectors which can carry a number of antigens. Infectious agents and malignant cells are able to develop resistance to chemotherapeutic agents in a number of ways. These are being defined and will be overcome in the future with recombinant DNA-derived products which bypass the resistance pathway or even utilise the pathway for a therapeutic effect.

Gene therapy will become a routine form of treatment. The ability to target a gene to its correct locus with no side-effects may reopen the discussion on the role of germline gene therapy in the human. Therapeutics will be an important area to which molecular medicine will make significant contributions. The therapeutic options will only be limited by the imagination of the scientist.

FURTHER READING

Chisaka O, Capecchi M R 1991 Regionally restricted developmental defects resulting from targeted disruption of the mouse homeobox gene *hox 1.5*. Nature 350: 473–479

Clarke L L, Grubb B R, Gabriel S E, Smithies O, Koller B H, Boucher R C 1992 Defective epithelial chloride transport in a gene-targeted mouse model of cystic fibrosis. Science 257: 1125–1128

DeChiara T M, Robertson E J, Efstratiadis A 1991 Parental imprinting of the mouse insulin-like growth factor II gene. Cell 64: 849–859

Erickson R P 1988 Creating animal models of genetic disease. American Journal of Human Genetics 43: 582–586

Hardy J 1992 Framing β-amyloid. Nature Genetics 1: 233–234

Jordan E 1992 The Human Genome Project: where did it come from, where is it going? American Journal of Human Genetics 51: 1–6

McKusick V A 1992 Presidential address, Eighth International Congress of Human Genetics. Human genetics: the last 35 years, the present and the future. American Journal of Human Genetics 50: 663–670

McKusick V A, Amberger J S 1993 The morbid anatomy of the human genome: chromosomal location of mutations causing disease. Journal of Medical Genetics 30: 1–26

Sutherland G R, Richards R I 1992 Anticipation legitimized: unstable DNA to the rescue. American Journal of Human Genetics 51: 7–9

Trends in Biotechnology 1992 vol 10, Nos 1/2 (two issues devoted to many aspects of the Human Genome Project).

US Human Genome Project 1990 Understanding our Genetic Inheritance. NIH Publication No. 90-1590, National Technical Information Service, US Department of Commerce, Springfield, Virginia

Walter M A, Goodfellow P N 1992 Disease and development. Nature 355: 590–591

GLOSSARY AND ABBREVIATIONS

GLOSSARY

Allele Specific Oligonucleotides (ASOs) Oligonucleotides which are constructed with DNA sequences homologous to specific alleles. Two ASOs can be made which differ in sequence at only one nucleotide base thereby distinguishing a mutant allele with a point mutation from its corresponding wild-type allele (see Fig. 2.14).

Alleles Abbreviation for allelomorph and means alternative forms of the same gene.

Allogeneic From one person to another who is genetically dissimilar but of the same species (see autologous, heterologous).

Amino acids The building blocks of proteins. Each amino acid is encoded by a nucleotide triplet (see codon, Fig. 2.1, Table 2.1).

Amniocentesis Aspiration of amniotic fluid during pregnancy.

Amplification Multiple copies of a DNA sequence.

Aneuploid Any chromosome number that is not an exact multiple of the haploid number (23 in humans). Examples of aneuploidy include the presence of an extra copy of a single chromosome, e.g. trisomy 21 (Down syndrome) or the absence of a single chromosome, e.g. monosomy (as found in Turner syndrome 45,X) (see ploidy).

Anneal Formation of double-stranded nucleic acid from single stranded forms.

Antibody More correct term is immunoglobulin. A protein produced by higher vertebrates following exposure to a foreign substance (called an antigen). The Y-shaped antibodies bind to antigens and neutralise them. Antibodies can be polyclonal or monoclonal in origin (the latter is derived from a single cell and so each antibody is identical)(see Fig. 6.12).

Anticipation Increasing severity or earlier age at onset of a genetic disease in successive generations.

Antisense Antisense DNA is the non-coding strand of DNA. The latter functions as the template for mRNA production which then contains the sequence present on the sense strand. Antisense RNA or antisense oligonucleotides have sequences which are complementary to mRNA and so interfere with the latter's function (see Figs 2.1, 7.10).

Assortive mating Sexual reproduction in which the pairing of mates is not random, i.e. members of a particular group which are more (less) likely to mate with other members of that group produce positive (negative) assortive mating.

Attenuated virus A virus which has become less pathogenic following passage outside its natural host.

Autologous From the same person (see allogeneic, heterologous).

Autosomal disease Disease which is the result of an abnormality affecting the 22 pairs of autosomes (non-sex chromosomes).

Bacteriophage, 'phage' A virus which infects bacteria (see clone, vector).

Base pair A measurement of length for DNA. Includes a nucleotide base with its complementary base, i.e. adenine (A) would bind to thymine (T) or cytosine (C) to guanine (G) (see complementary, Fig. 2.1).

Candidate gene A gene which is a likely marker to initiate a search for the genetic basis of an inherited disorder of unknown origin, e.g. the myosin genes in muscle disorders.

Cap Post-transcriptional change to the 5′ end of the growing mRNA molecule in which a modified nucleotide (4-methylguanosine) is added. Has a functional role since recognised by ribosomes as the initiation signal for protein synthesis.

Carcinogen Physical or chemical agent which induces cancer.

Carrier An individual who is heterozygous for a mutant allele which causes a genetic disorder in the homozygous or hemizygous states.

Cell cycle The timed sequence of events occurring in a eukaryotic cell between mitotic divisions. Divided into M (mitotic), S (DNA synthetic), G_1 and G_2 (gap or pause phases) and G_0 (resting phase). The times for each component differ between cell lines (Box 6.1).

Centi*Morgan* (cM) Distance between DNA loci as determined on a genetic map. 1 cM distance indicates that two markers are inherited separately 1% of the time. In terms of the physical map, 1 cM is very approximately equal to 1 Mb (Mb; see megabase). Name is derived from T H Morgan.

Centromere The heterochromatic constricted portion of a chromosome where the chromatids are joined (see heterochromatin, telomeres, Fig. 2.17).

Chimaera An individual composed of a mixture of genetically different cells. A chimaera is distinguished from a mosaic on the basis that the cells in the former are derived from different zygotes, e.g. transgenic mouse formed by the embryonic stem cell approach (see mosaicism, transgenic, Fig. 2.19).

Chorion *villus* sampling (CVS) Biopsy of the chorion frondosum during pregnancy to obtain a source of fetal tissue for prenatal diagnosis (see Figs 4.1, 4.2).

Chromosome walking Directional movement along a chromosome. Used in positional cloning to reach genes. Overlapping clones (contigs) are important for chromosome walking. Genetic distances will allow an assessment of the distance from target DNA. Physical maps e.g. pulsed field gel electrophoresis, will provide more accurate distances for the walk. A faster walk is possible by constructing jumping libraries (see Figs 3.6, 3.7).

Clone Refers to identical cells or molecules with a single ancestral origin. To **clone a gene** means to take a single gene or part of a gene and isolate it from the remainder of genomic DNA. The cloned gene can then be produced in unlimited amounts. (see functional and positional cloning; Figs 2.7, 2.8).

Codon Three adjacent nucleotide bases in DNA/RNA that encode for an amino acid (see Table 2.1).

Complementary The specific binding between the purine-pyrimidine base pairs of double-stranded nucleic acid. Thus, adenine (purine) will covalently bind to thymine (pyrimidine) and guanine (purine) to cytosine (pyrimidine) in a 1 to 1 ratio (see base pair, Fig. 2.1).

Complementary *DNA* (cDNA) DNA which is synthesised from a mRNA template. The enzyme required for this is reverse transcriptase (see Figs 1.1, 6.2).

Compound (heterozygote) An individual with two different mutant alleles at a locus.

Concordance Both members of a twin pair demonstrate the same phenotype or trait (see discordance).

Congenital Present at birth.

Conservation (DNA) The finding that a DNA sequence is present in a wide range of phylogenetically distant organisms suggests functional significance since it is unlikely that during evolution a region of DNA would have remained unaltered unless it had a specific and important function, e.g. it is a gene. The *ras* proto-oncogene illustrates this since it is conserved in organisms as divergent as humans and yeast.

Constitutional (cells) Cells which would be representative of an organism, e.g. in DNA testing for loss of heterozygosity, examples of constitutional cells which would provide a base-line for the DNA polymorphisms would be lymphocytes (if the cancer is non-haematological) or fibroblasts which could be obtained from a skin biopsy.

Constitutive (genes) Genes which are expressed following interaction between a promoter and RNA polymerase without additional regulation. Also called 'housekeeping' genes since often expressed in all cells at low levels.

Consultand The person seeking or referred for genetic counselling (see proband, Box 4.1).

Contigs Overlapping clone sets which represent a continuous region of DNA.

Contiguous gene syndromes A group of disorders which have malformation patterns often in association with mental retardation and growth abnormalities. The clinical heterogeneity found in these disorders may reflect the involvement of a number of physically related but otherwise distinct genes. Examples are given in Table 3.9.

Cosmid Derived from plasmids but contains cos sites from phage lambda. Used as a vector for cloning DNA.

CpG islands Regions of 1–2 kb containing a high density of hypomethylated cytosine residuals associated with guanine. CpG islands are frequently found at the 5′ end of genes (see methylation).

Decoding Identifying the function of a gene from its DNA sequence.

Deletion Loss of a segment of DNA or chromosome (see interstitial deletion, microdeletion, Figs 2.3, 2.4).

Diploid The chromosome number found in somatic cells. In humans this will be 46, i.e. twice the number present in the germ cells (see haploid, Fig. 2.16).

Discordance Members of a twin pair do not demonstrate the same phenotype or trait (see concordance).

Dizygotic twins Twins (fraternal) produced from two separate ova fertilised by different sperms (see monozygotic twins).

Dominant A genetic disorder is said to have dominant inheritance if the mutant phenotype is produced when only one of the two normal (wild-type) alleles at a particular locus is mutated (see recessive, Fig. 3.9).

Dominant-negative effect Inactivation of one of the two tumour suppressor gene loci can produce what appears to be a dominant effect if the mutant protein inhibits the normal product from the remaining normal allele.

e antigen Hepatitis B virus e antigen (HBeAg) – a part of the core antigen of the hepatitis B virus (HBcAg) which is secreted into the serum through cellular secretion pathways. HBeAg correlates strongly with infectivity.

Electroporation The use of a pulsed electric field to introduce DNA into cells in culture.

Embryonic stem cells (ES cells) Totipotential cells that can be cultured from the early embryo. ES cells can be induced to remain undifferentiated in culture. Foreign DNA is transfected into these cells which are then microinjected into blastocysts of developing embryos. A chimaera is produced. If chimaerism also involves the germ cells it will be possible to breed mice which are heterozygous or homozygous for the foreign gene (see Fig. 2.19).

Enhancer DNA sequences which have the following properties: (1) they increase transcriptional activity, (2) they are effective even if inverted in position and (3) they operate over long distances (see Box 1.3).

Env gene Encodes envelope protein of a retrovirus (see Fig. 5.2).

Epigenetic Changes which influence the phenotype but have not arisen from alterations in gene structure, e.g. imprinting.

Episome Plasmid or plasmid-like extrachromosomal DNA which has the ability to integrate into the host's chromosome.

Euchromatin Light appearing bands following staining to produce G (Giemsa) banding of chromosomes. More likely to contain transcriptionally active DNA (see heterochromatin, Fig. 2.16).

Eukaryotes Organisms ranging from yeast to humans which have nucleated cells.

Exon That segment in a gene which codes for a polypeptide and is represented in the mRNA.

Expressivity The severity of a phenotype. Variable expressivity is a feature of autosomal dominant disorders.

Familial A condition which is more common in relatives of an affected individual than in the general population (see breast cancer).

Fingerprints Dermatoglyphic fingerprints: derived from the ridged skin patterns of the fingers. DNA fingerprints: obtained from multilocus minisatellite DNA polymorphisms (see minisatellites, satellites, Figs 8.4, 8.5).

Five prime (5′) The 5′ position of one pentose ring in DNA is connected to the 3′ position of the next pentose via a phosphate group. The phosphodiester-sugar backbone of DNA consists of 5′-3′ linkages and this is the direction in which the nucleotide bases are transcribed (see Fig. 2.1).

Flanking (markers, DNA) DNA markers on either side of a locus: DNA sequences on either side of a gene.

***F*luorescence *in s*itu *h*ybridisation (FISH)** Non-isotopic method to label DNA probes for in situ hybridisation. The ability to utilise multiple fluorochromes in the same reaction increases the utility of this procedure. The resolving power of FISH is further enhanced if interphase chromosomes are studied.

Footprinting Technique which identifies sites where there is protein bound to DNA. This complex then becomes resistant to degradation by nucleases.

Frameshift mutation A mutation in DNA such as a deletion or insertion which interferes with the normal codon (triplet base) reading frame. All codons 3′ to the mutation will have no meaning. For example, the triplets GGT–TCT–GTT code for amino acids glycine, serine and valine respectively. A deletion of one nucleotide, e.g. a G of the GGT, would disrupt the reading frame to give GTT–CTG–TT, etc. The protein product will terminate when a new stop codon is reached.

Functional cloning Cloning strategy in which knowledge of a gene's product (function) is used to clone the gene (see clone, positional cloning, Fig. 2.9).

Gag gene **G**roup specific **a**nti**g**en: encodes core protein for a retrovirus (see Fig. 5.2).

G-banding G (for Giemsa) banding is a commonly used procedure to identify chromosomal bands in a karyotype. Spreads of cells in metaphase are treated with trypsin and then stained with Giemsa (see Fig. 2.16).

Gene A sequence of DNA nucleotide bases which code for a polypeptide.

Genetic engineering Colloquial term for recombinant DNA technology: the experimental or industrial applications of technologies which can alter the genome of a living cell.

Genetic map Constructed by determining how frequently two markers (DNA polymorphisms, physical traits or syndromes) are inherited together. Distances in genetic maps are measured in terms of centiMorgans.

Genome The total genetic material of an organism, i.e. an organism's complete DNA sequence.

Genotype The genetic constitution of an organism. In terms of DNA markers it refers to the genetic constitution of alleles at a specific locus. e.g. the two haplotypes (see haplotype, Fig. 3.10).

Germ cells Cells which differentiate early in embryogenesis to form ova and sperm.

G proteins Abbreviation for guanine-binding proteins. Play important role in relaying messages from the cell surface to the nucleus. Act by binding GTP (guanosine triphosphate) which leads to activation of a second messenger system such as adenylyl cyclase. There are many G proteins including the product of the *ras* proto-oncogenes. G proteins are self-regulating since the GTP-G protein complex is hydrolysed to inactive GDP-G protein by GTPase activity of the G protein. Over 100 receptors convey messages through G proteins (see *ras*, signal transduction, Fig. 6.4).

Guthrie spot Used (incorrectly) to describe the blood spot taken from newborns by heal prick.

The blood spot is then used for newborn screening of genetic and metabolic disorders. The name is derived from the newborn screen for phenylketonuria which uses a test called the Guthrie bacterial inhibition assay.

Haematopoietic Related to the blood; blood forming.

Haemoglobinopathies Genetic disorders involving globin, the protein component of haemoglobin. Divided into the thalassaemia syndromes, e.g. α or β thalassaemia, and the variant haemoglobins, e.g. sickle cell anaemia (HbS).

Haploid The chromosome number found in gametes. In humans this will be 23, i.e. one member of each chromosome pair (see diploid).

Haplotype A set of closely linked DNA markers at one locus which is inherited as a unit (see Box 3.6, Fig. 3.10).

Hemizygous Having only one copy of a given genetic locus, e.g a male is hemizygous for DNA markers on the X-chromosome (see Fig. 3.13).

Heterochromatin Dark-appearing bands following G (Giemsa) banding of chromosomes. Contains predominantly repetitive DNA (see euchromatin, centromere, Fig. 2.16).

Heteroduplex Hybrid DNA involving two strands which are different, e.g. there may be a base-mismatch (see homoduplex, Fig. 2.15).

Heterologous Belonging to another species, e.g. the use of salmon sperm DNA to block non-specific hybridisation by human DNA.

Heteroplasmy The presence of more than one type of mitochondrial DNA in a cell. There are thousands of molecules of mitochondrial DNA per cell. If there is mutant mitochondrial DNA it can be present in varying amounts. Some cells might have predominantly wild-type DNA; others predominantly mutant DNA (called homoplasmy) and others are said to be heteroplasmic because there is a mixture of both. Thus, phenotypic variation between cells is possible.

Heterozygote (ous) An individual with two different alleles (e.g. gene, polymorphic marker) at a single locus (see homozygote, Figs 3.2, 3.4).

HLA Abbreviation for **h**uman **l**eukocyte **a**ntigen. HLA is encoded for by a multigene complex occupying approximately 3000 kb of DNA on the short arm of chromosome 6. Antigens belonging to the HLA system are found on the surface of all cells except the red blood cells. HLA is concerned with normal immunological responses. HLA plays a vital role in graft rejection or acceptance following transplantation. Also known as MHC (**ma**jor **h**istocompatibility **c**omplex).

Homeotic genes Genes which determine the shape of the body along the antero-posterior axis of the embryo. Mutations in homeotic genes cause a part of the body to be replaced by a structure normally found elsewhere. Conserved DNA sequences within homeotic genes are called **homeoboxes**. All vertebrates including humans have four homeotic gene complexes located on different chromosomes. In mammals homeotic genes are called *hox* genes. Another gene family involved in development is the *pax* genes, the conserved sequence for which is called the **pa**ired bo**x** (see *pax* genes).

Homoduplex Hybrid DNA involving two strands which are identical (see heteroduplex).

Homologous recombination A form of gene targeting on the basis of recombination between DNA sequences in the chromosome and newly introduced identical DNA sequences (see homology, Fig. 7.8).

Homology Fundamental similarity, matched, e.g. homologous (the same) chromosomes pair at meiosis; homology between DNA sequences means close similarity.

Homozygote (ous) An individual with two identical alleles (e.g. gene, polymorphism) at a single locus (see heterozygote, Figs 3.2, 3.4).

Hot spots Regions in genes or DNA where mutations occur with unusually high frequency.

Housekeeping (genes) Genes that are expressed

in virtually all cells since they are fundamental to the cell's functions.

Human Genome Project Multicentred, multinational, multibillion dollar project launched in 1988 and estimated to be completed in 2005. The Project aims include construction of a genetic and physical map of the human genome and a number of model organisms. Ultimately there will be complete DNA sequence for the entire human genome. At the same time technology development and training in human gene mapping will be undertaken.

Hybridisation The pairing, through complementary nucleotide bases (A with T and G with C), of RNA/DNA strands to produce an RNA/RNA or RNA/DNA or DNA/DNA hybrid (see Fig. 2.2).

Illegitimate transcription Low transcription of a tissue-specific transcribing gene in non-specific cells, e.g. the detection of mRNA for the β myosin heavy chain gene (a muscle-specific gene) in peripheral blood lymphocytes.

Immunoglobulin See antibody.

Immunophenotyping Typing of cells with immunological markers such as monoclonal antibodies (also called cell marker analysis).

Imprinting Differential effects of maternally and paternally derived DNA (see uniparental disomy, Figs 3.16, 10.1).

Informative (polymorphism) Means a polymorphism is heterozygous and so able to distinguish two alleles. In a parental mating, at least one parent must be heterozygous for a polymorphism to be potentially informative. If both parents are heterozygous, the polymorphism is fully informative (if there is a key individual to help assign which marker co-segregates with disease, etc.) (see Fig. 3.2).

In situ hybridisation Hybridisation of a DNA probe (labelled with ^3H, fluorescein or a chemical such as biotin) to a metaphase chromosome spread or a tissue section on a slide.

Interstitial deletion Loss of DNA or part of a chromosome which does not occupy a terminal position.

Intron Segment of DNA which is transcribed but does not contain coding information for a polypeptide (also called **int**er**v**ening **s**equence or IVS). It is spliced out of the transcript before mature mRNA is formed.

Isozymes (isoenzymes) Different forms of an enzyme.

Karyotype An individual's or a cell's chromosomal constitution. Determined by examination of chromosomes with light microscopy and the use of stains (see Fig. 2.16).

***Kilo*base (kb)** One thousand base pairs in a sequence of DNA.

***Kilo*d*alton* (kDa)** A unit which measures the molecular weight of proteins (=1000 daltons). 1 dalton approximates to the molecular weight of a hydrogen atom. The molecular weight of a protein will be based on the sum of the atomic weights of the elements which comprise it.

Library A large number of recombinant DNA clones which have been inserted into a vector for the purpose of cloning a segment of DNA (see Fig. 2.7).

Linkage The tendency to inherit together two or more non-allelic genes or DNA markers more often than is to be expected by independent assortment. Genes/DNA markers are linked because they are sufficiently close to each other on the same chromosome (see Fig. 3.10).

Linkage disequilibrium Preferential association of linked genes/DNA markers in a population, i.e. the tendency for some alleles at a locus to be found with certain alleles at another locus on the same chromosome with frequencies greater than would be expected by chance alone.

Liposomes Synthetic spherical vesicles with a lipid bilayer. Function as artificial membrane systems to deliver DNA, etc. into cells.

Lod score Statistical test to determine whether a set of linkage data are linked or unlinked. Lod is an abbreviation of the '**l**og$_{10}$ of the **od**ds' favouring linkage. For genetic disorders which are not X-linked, a lod score of +3 (1000:1 odds of linkage) indicates linkage whilst a score of −2 is odds of 100:1 against linkage.

Lymphoproliferative disorders Lymphomas and leukaemias of lymphocyte origin.

***Megabase* (Mb)** One million base pairs in a sequence of DNA.

Meiosis Process in which diploid germ cells undergo division to form the haploid chromosome number (see mitosis).

Messenger RNA (mRNA) Transfers the genetic information from DNA to the ribosomes. Contains the template for polypeptide production.

Metastasis A secondary tumour arising from cells carried from the primary tumour to a distant locus.

Methylation (of DNA) Vertebrate DNA contains a small proportion of 5-methylcytosine which arises from methylation of cytosine bases where they occur in the sequence CpG. The methylation status of DNA correlates with its functional activity: inactive genes are more heavily methylated and vice versa (see CpG).

Microdeletion DNA/chromosomal deletion which is not detectable by conventional techniques such as microscopy (cytogenetics) or Southern blotting (DNA mapping).

Microsatellites As for minisatellites except that the polymorphism allele size is smaller, e.g. <1 kb and the basic core repeat unit involves a two to four nucleotide base pair repeat motif. One example is repeats of the motif CA, e.g. CACACACACA etc. which is also written as AC (ACACACACAC etc.). There is confusion with terminology since the above are identical. To avoid this problem it has been recommended that the microsatellites are written in alphabetical order (the above would be $(AC)_n$) (see minisatellites, satellites, Box 3.10, Fig. 8.1).

Minisatellites Repeat DNA segments which comprise short head-to-tail tandem repeats giving the **v**ariable **n**umber of **t**andem **r**epeat (VNTR) type polymorphisms with approximate size between 1–30 kb. VNTRs can be of two types: single locus or multilocus. The latter are utilised in constructing a DNA 'fingerprint' of an individual (see microsatellites, satellites, Figs 8.2, 8.3, 8.4).

Missense mutation A single DNA base change which leads to a codon specifying a different amino acid, e.g. the base change of GGT (glycine) to GTT (valine).

Mitosis Somatic cell division; process in which chromosomes duplicate and segregate during cell division (see meiosis).

Monoclonal Derived from a single clone, e.g. monoclonal antibody, monoclonal lymphocyte population (see polyclonal, Figs 7.5, 6.14).

Monozygotic twins Genetically identical twins formed by the division into two at an early stage in development of an embryo derived from a single fertilised egg (see dizygotic twins).

Mosaicism A condition in which an individual has two or more cell lines of different genetic or chromosomal constitution. In contrast to a chimaera, both cell lines in a mosaic are derived from the same zygote.

Multifactorial disorders Diseases which result from an interaction of environmental factors with multiple genes at different loci (see polygenic, which is sometimes used in the same sense as multifactorial).

Murine Of the mouse (Latin, mus).

Mutation An alteration in genetic material. This could be a single base change (point mutation) to more extensive losses of DNA (deletions) (see missense mutation, nonsense mutation, Figs 2.14, 2.4).

Nonsense mutation A single DNA base change resulting in a premature stop codon (TAA, TGA, TAG), e.g. TCG (serine) to TAG (stop).

Northern blotting Procedure to transfer RNA from an agarose gel to a nylon membrane (see Southern blotting, Western blotting).

Nosocomial Hospital acquired.

Nucleases Enzymes which break down nucleic acid. There are DNAase (DNase) and RNAase (RNase) enzymes. RNA in particular is susceptible to RNAases so that preparation of RNA requires a lot more care compared to the more robust DNA.

Nucleotide The monomeric component of DNA/RNA comprising a base (A, adenine; T, thymine; U, uracil; G, guanine or C, cytosine), a pentose sugar (deoxyribose or ribose) and a phosphate group (see Fig. 2.1).

Oligonucleotides Small single-stranded segments of DNA typically 20–30 nucleotide bases in size which are synthesised in vitro. Uses include: DNA sequencing, DNA amplification and DNA probes. (see primer, allele-specific oligonucleotides)

Oncogenes Genes associated with neoplastic proliferation following a mutation or perturbation in their expression (see proto-oncogenes, *ras*).

p53 A tumour suppressor gene, mutations of which are frequently found in human cancers (see tumour suppressor gene).

Palindrome Sequence of DNA which is identical in either direction:

e.g. 5′-GTCGAC-3′
3′-CAGCTG-5′.

This is the recognition sequence for the restriction enzyme *Sal*I. Further examples of palindromic sequences are given in Table 2.2. Palindromes involving small to large segments of DNA are found throughout the genome and need not necessarily be sites recognised by restriction enzymes.

Parthenogenesis The development of an egg that has been activated in the absence of sperm.

Pathogenesis The steps involved in development of a disease.

Pax genes Abbreviation from **pa**ired bo**x**. The paired box is a conserved DNA sequence which plays a role in development of the neural crest (see homeotic).

Penetrance All or nothing phenomenon relating to the expression of a gene.

P-glycoprotein Glycoprotein associated with drug resistance. A transmembrane ATP-dependent active transporter which pumps hydrophobic compounds out of cells. Drug-resistant cells demonstrate increased expression of the gene as one mechanism for their resistance.

Phase A term to describe the combination in which polymorphic markers have been inherited within the context of a family study (Box 3.6).

Phenotype The observed appearance of a gene or an organism which is determined by the genotype and its interaction with the environment.

Physical map Can be constructed in different ways but in contrast to genetic maps it represents measurements of physical length (bp, kb, Mb). Types of physical maps include: cytogenetic, pulsed field gel electrophoresis, fluorescent in situ hybridisation, contigs, e.g. cosmid or YAC.

Plasmid Cytoplasmic, autonomously replicating extrachromosomal circular DNA molecule. Used as vectors for cloning. In vivo, plasmids are found in bacteria where they can code for antibiotic resistance factors (see episome, vector).

Pleiotropy Different effects of a gene on apparently unrelated characteristics such as the phenotype, organ systems or functions.

Ploidy The number of chromosomes in a cell. Euploid, the correct number; aneuploid, an abnormally high or low number; polyploid, a multiple of the euploid number.

Pol gene Encodes reverse transcriptase enzyme of a retrovirus (see Fig. 5.2).

Polyclonal Derived from more than one cell (see monoclonal).

Polygenic inheritance Trait which results from an interaction of multiple genes at different loci (see multifactorial disorders).

Polymerase RNA polymerases are enzymes which catalyse the formation of RNA using DNA as a template. DNA polymerases are enzymes which can synthesise DNA from four nucleotide precursors (dATP, dTTP, dCTP and dGTP) provided a template or primer is available to start off the process. Functions of the DNA polymerases include DNA repair and DNA replication. Reverse transcriptase is a DNA polymerase (see Figs 7.10, 2.10).

Polymerase chain reaction (PCR) DNA method which allows amplification of a targeted DNA sequence (see Fig. 2.11).

Polymorphisms (DNA) A part of the DNA sequence that can occur in two forms which can be detected on the basis of variations in the sizes of DNA fragments produced following digestion with restriction enzymes. Polymorphic variations result from point mutations (see RFLP) or insertions of repetitive DNA sequences (see VNTR). In terms of human genetics, polymorphisms are inherited along Mendelian lines in a family and by definition should occur at a 1% or greater frequency within a population (see RFLP, VNTR, Figs 2.5, 3.2, 8.2).

Positional cloning Cloning of a gene on the basis of its chromosomal position rather than its functional properties. Also called 'reverse genetics' (see cloning, functional cloning, Fig. 2.9).

Primer A short oligonucleotide segment which pairs with a complementary single-stranded DNA sequence. The double-stranded segment formed has a free 3′ terminus which provides the template for extension into a second strand (see oligonucleotide, Fig. 2.10).

Proband (or propositus or index case) The affected individual from which a pedigree is constructed (see consultand, Figure 3.13).

Probe A single stranded segment of DNA/RNA which is labelled with a radioactive substance or chemical. The probe will bind to its complementary single-stranded target sequence. Hybrids formed are detectable by autoradiography or by chemical changes. There are a number of different probes: genomic, cDNA, RNA, oligonucleotide. The naming of probes has led to confusion. Therefore, an attempt to induce uniformity has been made by naming loci to which probes will hybridise, e.g. D15S10 indicates human chromosome 15 locus 10. A number of DNA probes could hybridise to this locus (see Fig. 2.2).

Prokaryotes Bacteria and certain algae with cells that are not nucleated.

Promotor DNA sequence located 5′ to a gene which indicates the site for transcription initiation. May influence the amount of mRNA produced and the tissue specificity. Examples of promotors are the TATA, CCAAT boxes (see Cap, Fig. 3.3).

Proto-oncogenes Normal genes involved in cellular proliferation and differentiation. Altered forms of the proto-oncogenes are called oncogenes.

Provirus Virus that is integrated into the chromosome of its host cell and can be transmitted from one generation to another without causing lysis of the host (see retrovirus, reverse transcriptase).

Pulsed field gel electrophoresis (PFGE) A type of gel electrophoresis in which large fragments of DNA can be separated by altering the angle at which the electric current is applied (see Fig. 2.6).

Ras A family of proto-oncogenes (H-*ras*-1, K-*ras*-2 and N-*ras*) which encode for a protein called p21. p21 binds to GTP/GDP and has GTPase activity. *Ras*-derived proteins play a physiological role in regulation of cellular proliferation. Mutations in *ras* are found in a number of cancers (see G proteins, Fig. 6.4).

Recessive The products of both normal (wild-type) alleles at a particular locus are non-functional in a recessive disorder (see dominant, Fig. 3.4).

Recombination Crossing-over (breakage and rejoining) between two loci which results in new combinations of genetic markers/traits at those loci, e.g. one locus has four genetic markers linearly arranged: a-b-c-d and the second locus has: b-b-c-a. Recombination involving these two regions between the b-c markers would give new genetic combinations, i.e. a-b-c-a and b-b-c-d (see Fig. 3.10).

Restriction endonucleases (enzymes) Enzymes which recognise specific short DNA sequences and cleave DNA at these sites (see Table 2.2).

***R*estriction *f*ragment *l*ength *p*olymorphism (RFLP)** Biallelic DNA polymorphism which results from the presence or absence of a restriction endonuclease site (see polymorphisms, Figs 2.5, 3.2, 4.7).

Restriction map A series of restriction endonuclease recognition sites associated with a DNA locus or gene (see Fig 2.3).

Retrovirus RNA virus that utilises reverse transcriptase during its life cycle. After infecting the host cell, the retroviral (RNA) genome is transcribed into DNA which is then integrated into host DNA. In this way the retrovirus can replicate (see provirus, reverse transcriptase, Figs 6.1, 6.2).

Reverse genetics A name for the recombinant DNA strategy which attempts to clone a gene on the basis of its position on the chromosome rather than its functional properties. The name is now being replaced with the more correct description of positional cloning (see cloning, functional cloning, Fig. 2.9).

Reverse transcriptase Enzyme which enables synthesis of single-stranded DNA (called cDNA) from an RNA template (see polymerase, Fig. 1.1).

Ribosomal RNA (rRNA) The nucleic acid content of ribosomes. The latter are small cellular particles which are the site of protein synthesis in the cytoplasm.

Satellite DNA Short head-to-tail tandem repeats which incorporate specific DNA motifs (see microsatellites, minisatellites, Figs 8.1, 8.2, 8.4).

Screening (genetic) Testing individuals on a population basis to identify those who would be at risk for disease or transmission of a genetic disorder.

***S*equence *t*agged *s*ites (STSs)** A way to provide unambiguous identification of DNA markers generated by the Human Genome Project. STSs comprise short, single-copy DNA sequences that characterise mapping landmarks on the genome.

Sequencing (DNA) Establishing the identity and order of nucleotides in a segment of DNA. The 'gold standard' in characterising a mutation (see Figs 2.10, 2.13).

Sibship A group comprising the brothers and sisters (siblings) in a family.

Signal transduction Transfer of signals from extracellular factors and their surface receptors by cytoplasmic messengers to modulate events in the nucleus (see G proteins, Fig. 6.4).

Somatic cells Any cell in an organism which is not a germ cell, i.e. sperm or egg.

Somatic cell genetic disorders One of the four categories of genetic disorders. Defects in DNA are found in specific somatic cells. An example of this type of disorder is cancer. By comparison the three other categories (single gene; multifactorial and chromosomal disorders) all have the genetic abnormality present in all cells including the germ cells.

Somatic cell hybrid A hybrid formed from the fusion together of different cells. These usually come from different species e.g. human and rodent hybrids are frequently used for human gene mapping.

Somatic mutation A mutation which occurs in any cell that will not become a germ cell.

Southern blotting Named after E Southern. Describes the procedure for transferring denatured (i.e. single-stranded) DNA from an agarose gel to a solid support membrane such as nylon (see Northern blotting, Western blotting, Fig. 2.4).

Splicing The removal of introns to produce mature mRNA.

Sporadic No obvious genetic cause.

Start codon Nucleotide codon (ATG) which is positioned at the beginning of a gene sequence in eukaryotes. Prokaryotes do not have such a start codon and so ATG is translated into the amino acid methionine.

Sticky ends Fragments of double-stranded DNA with a few bases not paired, i.e. they anneal with greater efficiency than blunt-ended fragments.

Stop codons Nucleotide codons (TAA, TGA and TAG) are positioned at the 3′ end of a gene sequence and indicate the termination of a polypeptide.

Syntenic genes Genetic loci or genes which lie on the same chromosome.

Tandem repeats Small sections of repetitive DNA in the genome arranged in head to tail formation.

Telomeres The two ends of a chromosome (see centromere, Fig. 2.17)

Transcription Synthesis of a single-stranded RNA molecule from a double-stranded DNA template in the nucleus (see polymerase, translation).

Transduction (gene) Transmission of genetic material from one cell to another by viral infection.

Transduction (signal) See signal transduction.

Transfection Acquisition of new genetic markers by incorporation of added DNA into eukaryotic cells by physical or viral-dependent means (see Fig. 7.2).

Transfer RNA (tRNA) Provides the link between mRNA and rRNA. Each tRNA can combine with a specific amino acid and also bind to the relevant mRNA codon (see codon, mRNA, rRNA, translation).

Transformation (of bacteria) Acquisition of new genetic markers by incorporation of added DNA into bacteria.

Transformation (of cells) Sudden change in a cell's normal growth properties into those found in a tumour cell (see Fig. 6.3).

Transgenic The presence of foreign DNA in the germline. Transgenic animals are produced by experimental insertion of cloned genetic material into the animal's genome. This can be done by microinjection of DNA into the pronucleus of a fertilised egg or through utilisation of embryonic stem cells. A proportion of transgenic animals will express the foreign gene and transmit it to their progeny (see embryonic stem cells).

Transition Change of a purine (i.e. adenine or guanine) to a purine or a pyrimidine (i.e. cytosine or thymine) to a pyrimidine (see transversion).

Translation Cytoplasmic production of a polypeptide from the triplet codon information on mRNA (see transcription).

Translocation The presence of a segment of a chromosome on another chromosome (see Fig. 6.5).

Transversion Change of a purine to a pyrimidine or vice versa (see transition).

Tumour suppressor genes (also called recessive oncogenes, anti-oncogenes, growth suppressor genes). These are normal genes which have the suppression of tumourigenesis as one component of their function (see p53, Figs 6.6, 6.7).

Two-hit model of tumourigenesis A first or predisposing event can be inherited through the germline or arise in somatic cells. A second somatic event must occur to inactivate the remaining functional allele resulting in tumour formation.

Uniparental disomy The inheritance of two copies of a chromosome from the one parent. This can be isodisomy (both chromosomes from the one parent are identical copies) or heterodisomy (the two chromosomes are different copies of the same chromosome). Described with a number of human chromosomes, e.g. 7, 11, 15, 16 (see also imprinting).

V*ariable *n*umber of *t*andem *r*epeats (VNTR) A multiallelic DNA polymorphism which results from insertions or deletions of DNA between two restriction sites (see polymorphisms, Figs 2.5, 8.2).

Vector Cloning vehicle, i.e. plasmid, phage, cosmid or YAC into which DNA to be cloned can be inserted (see Fig. 2.7).

Western blotting A technique used to separate and identify proteins (see Northern blotting, Southern blotting).

Wild-type (gene) The form of the gene normally present in nature.

X-chromosome inactivation Random inactivation of one of the two female X chromosomes during early embryonic development. Thus, cells in a female are mosaic in respect to which of the X chromosomes is functional.

Yeast artificial chromosome (YAC) A cloning vector which allows large segments of DNA, e.g. 300 kb in size, to be cloned.

Zoo blot A way to detect conservation of DNA sequence during evolution. In a zoo blot, a segment of DNA being investigated is used as a probe to hybridise against a series of DNA samples from various species e.g. human, mouse, yeast. If the probe can detect unique DNA sequences in the above, it provides indirect evidence that the DNA sequence has been conserved during evolution, i.e. it has functional significance which might mean it is a gene.

Zoonoses Infections transmitted from animals to humans.

Zygote The diploid cell resulting from union of the haploid male and haploid female gametes, i.e. fertilised ovum.

ABBREVIATIONS

A Adenine nucleotide base

AIDS Acquired immunodeficiency syndrome

ASO Allele specific oligonucleotide

bp Base pair

C Cytosine nucleotide base

CEPH Centre d'Etude du Polymorphisme Humain

CFTR Cystic fibrosis transmembrane regulator, i.e. cystic fibrosis gene

CHO Chinese hamster ovary (cell line)

cM CentiMorgan

CVS Chorion villus sample

cDNA Complementary or copy DNA

DHFR Dihydrofolate reductase

DNA Deoxyribonucleic acid

ES cells Embryonic stem cells

FISH Fluorescence in situ hybridisation

FIV Feline immunodeficiency virus

5′ → 3′ Direction of transcription

G Guanine nucleotide base

GTP Guanosine triphosphate

GDP Guanosine diphosphate

HbF Haemoglobin F (fetal haemoglobin)

HBs Ag Hepatitis B surface antigen

HIV Human immunodeficiency virus

HLA Human leukocyte antigen

HPFH Hereditary persistence of fetal haemoglobin

IgG Immunoglobulin G

IgM Immunoglobulin M

IVS Intervening sequence (=intron)

kb Kilobase

kDa Kilodalton

LDL Low density lipoprotein

LTR Long terminal repeat (of a retrovirus)

Mb Megabase

MHC Major histocompatibility complex

MoAb Monoclonal antibody

mRNA Messenger ribonucleic acid (RNA)

MRSA Methicillin-resistant *Staphylococcus aureus*

^{32}P Radioactive phosphorus

PCR Polymerase chain reaction

PFGE Pulsed field gel electrophoresis

rDNA Recombinant DNA

rRNA Ribosomal RNA

RFLP(s) Restriction fragment length polymorphism(s)

SIV Simian immunodeficiency virus

T Thymine nucleotide base

VNTR(s) Variable number of tandem repeat(s)

YAC(s) Yeast artificial chromosome(s)

Index